T0221023

E-CARGO and Role-Based Collaboration

E-CARGO and Role-Based Collaboration

Modeling and Solving Problems in the Complex World

Haibin Zhu

IEEE Press Series on Systems Science and Engineering
MengChu Zhou, Series Editor

Published by John Wiley & Sons, Inc., Hoboken, New Jersey.
Published simultaneously in Canada.

For general information on our other products and services or for technical support, please contact our Customer Care Department within the United States at (800) 762-2974, outside the United States at (317) 572-3993 or fax (317) 572-4002.

Wiley also publishes its books in a variety of electronic formats. Some content that appears in print may not be available in electronic formats. For more information about Wiley products, visit our web site at www.wiley.com.

Library of Congress Cataloging-in-Publication Data is applied for

Hardback: 9781119693062

Cover Design: Wiley
Cover Image: © metamorworks/iStock/Getty Images, © abadonian/iStock/Getty Images, © sarawuth702/iStock/ Getty Images, © sarawuth702/ iStock/Getty Images

Set in 9.5/12.5pt STIXTwoText by Straive, Pondicherry, India

10 9 8 7 6 5 4 3 2 1

Contents

Author Biography

 **Dr. Haibin Zhu** is a Full Professor and the Chair (July 2019–June 2021) of the Department of *Computer Science and Mathematics*, Founding Director of *Collaborative Systems Laboratory*, member of the *University Budget Plan committee, Arts and Science Executive Committee*, and *the Research Committee, Nipissing University, Canada*. He received a B.S. degree in computer engineering from the Institute of Engineering and Technology, China (1983), and M.S. (1988) and Ph.D. (1997) degrees in computer science from the National University of Defense Technology (NUDT), China. He was a visiting professor and a special lecturer in the College of Computing Sciences, New Jersey Institute of Technology, USA (1999–2002) and a lecturer, an associate professor and a full professor at NUDT (1988–2000). He has published or been accepted over 200 research works including 30 IEEE Transactions articles, five books, five book chapters, three journal issues, and three conference proceedings. He is the most productive author in the category of "Collaboration" https://ieeexplore.ieee.org/search/searchresult.jsp?newsearch =true&queryText=Collaboration.

He is a senior member of ACM, *a full member* of Sigma Xi, and a senior member of IEEE. He is serving as Associate Vice President, Systems Science and Engineering (SSE), and co-chair of the technical committee of *Distributed Intelligent Systems of IEEE Systems, Man and Cybernetics (SMC) Society*, member of *the SSE Technical Activity Committee*, the *Conferences and Meetings Committee*, and the *Electronic Communications Subcommittee of IEEE SMC Society*, Associate Editor (AE) *of IEEE Transactions on SMC: Systems, IEEE Transactions on Computational Social Systems, IEEE SMC Magazine*, and *IEEE Canada Review*. He has been an active organizer for the annual IEEE Int'l Conf. on SMC since 2003, as Special Session Chair, Tutorial Chair, Area Co-Chair, Social Media Co-Chair, Web Co-Chair, Poster Co-Chair, Session Chair, and Special Session Organizer. He is the Publication Chair for the

1st IEEE Int'l Conf. of Human-Machine Systems, Rome, Italy (Online due to COVID-19), 7–9 September 2020 and a Poster Co-Chair for the annual IEEE SMC Conference, Toronto, Canada (Online due to COVID-19), 6–9 October 2020, was the Program Chair (PC) for 16th IEEE Int'l Conf. on Networking, Sensing and Control, Banff, AB, Canada, 8–11 May 2019. He is a PC Chair for 17th IEEE Int'l Conf. on Computer Supported Cooperative Work in Design, Dalian, China, 6–8 May 2020 (postponed due to COVID-19), and was a PC Chair for 17th IEEE Int'l Conf. on Computer Supported Cooperative Work in Design, Whistler, BC, Canada, 27–29 June 2013. He also served as PC members for 120+ academic conferences.

He is the founding researcher of Role-Based Collaboration and Adaptive Collaboration. He has offered over 70 invited talks including keynote and plenary talks on collaboration topics internationally, e.g. Canada, USA, China, UK, Germany, Turkey, Hong Kong, Macau, and Singapore. His research has been being sponsored by NSERC, IBM, DRDC, and OPIC.

He is the receipt of the meritorious service award from IEEE SMC Society (2018), the chancellor's award for excellence in research (2011) and two research achievement awards from Nipissing University (2006, 2012), the IBM Eclipse Innovation Grant Awards (2004, 2005), the Best Paper Award from the 11th ISPE Int'l Conf. on Concurrent Engineering (ISPE/CE2004), the Educator's Fellowship of OOPSLA'03, a 2nd class National Award for Education Achievement (1997), and three 1st Class Ministerial Research Achievement Awards from China (1997, 1994, and 1991).

His research interests include Collaboration Theory, Technologies, Systems, and Applications, Human-Machine Systems, CSCW (Computer-Supported Cooperative Work), Multi-Agent Systems, Software Engineering, and Distributed Intelligent Systems.

Preface

This book was completed during the COVID-19 pandemic, a difficult time when many of us struggled to play our routine *roles* without physical contact. Throughout this period, we saw a large transition toward online collaborative work. Interestingly, these online collaboration tools and platforms were built by the researchers and engineers in the *collaboration* field, or more exactly, the field of Computer-Supported Cooperative Work (CSCW). Initially, CSCW was supposed to be the initial target field for Role-Based Collaboration (RBC). Over the last 15 years, however, RBC has advanced far beyond the field of CSCW. The application of RBC now extends to all the cutting-edge fields in the world, including production, services, management, system engineering, complex systems, cyber-physical systems, big data, and Artificial Intelligence (AI).

During these challenging times, we are experiencing the dynamics and uncertainty of the world. This is natural, after all, the universe is infinite, various, and dynamic. In the human world, people have created many terms and concepts to describe and organize this world. Even for a language specialist, it is almost impossible to grasp all these terms in one's lifetime. In this massive conceptual kingdom, there is a concept that has been used in common vocabulary since the early days of humanity. Despite this, there is no commonly accepted definition of the term even today. This term is *"role."* Everyone understands the concept of a *role*, but it is difficult for anyone to fully define and describe it.

Human society can be abstracted into two types of perceptible units, namely people and things. Roles are the bond between people and people as well as people and things. Roles establish the relationship among people and things. Roles are laws, and the definition and interpretation of a role directly affect the normal operation of society. Confucius reiterated this idea in his famous adages: "名不正，则言不顺；言不顺，则事不成" read as "Míng bùzhèng, zé yán bù shùn; yán bù shùn, zé shì bùchéng," which means "if the name is not right then speech will not be in order, and if speech is not in order then nothing will be accomplished" and "君君臣臣父父子子" read as "Jūn jūnchén chén fù fùzǐ zi," which means "the king is the king, the minister the minister, the father is the father and the son the son."

Human society is inconsistent and, at times, chaotic. The reason for this lies in the lack of proper definitions, understanding, and interpretation of roles. The definition of a *role* should be regulated by mechanisms higher than the human individual (*agent*), but the real world needs people to define roles. Because some individuals are self-interested, they bring unfairness and disharmony to the definitions and explanations of specific roles. Although people try to introduce the concept of organization above individuals, organizations are themselves composed of individuals, and therefore, it is difficult to form a mechanism that is truly higher than individuals. Because of this, the social systems established by different countries around the world are not perfect. Some systems have shown superiority over a period of time, but sometimes they have performed poorly. This is because human society is "Agent-Based." "Role-Based" is an ideal paradigm for the real world. Even though we could not fully implement a role-based world, we may apply role-based structures and management to certain aspects of the world, e.g., a school, a society, an enterprise, a company, an organization, a country, or an international organization.

Another role-related human activity is the economy. In economic activities, a planned economy is composed of a collection of well-organized roles. From the perspective of RBC, a planned economy is inherently more cost-effective than a free economy, because a planned economy is an optimized role design strategy at its ideal state. However, a planned economy does not behave as expected in the real world. For a planned economy to work, the role designers must be absolutely correct and have no direct interest in the system being designed. That is to say, the role designer should not be a role player. However, in reality, the designers of the planned economy often failed to perfectly design the plan, because they design roles and play the roles they design. That is, the role designers are in the state that is similar to a Chinese saying "不识庐山真面目,只缘身在此山中" read as "Bù shí lúshān zhēnmiànmù, zhǐ yuán shēn zài cǐ shānzhōng," which means "the true face of Lushan is lost to my sight, for it is right in this mountain that I reside." This book uses the phenomenon of human society to discuss the design of artificial systems, and it can be taken as a tool to discuss or simulate the management of human society.

I hope this book will assist humankind in clarifying and playing our *roles* better in the universe.

Now, let us look at artificial systems such as computer systems and software systems. For these manufactured systems, because the designer or manufacturer is the master of the system, people can be superior to the agents in the system and manage the agents without joining them. People can design and define roles for the system components, i.e., design or train agents; assign roles to agents, judge the agents' performance while the agents are playing roles; and schedule individual agents while roles are changed in the system. In this way, artificial systems can operate in a coordinated and consistent manner according to the tasks and constraints defined by the roles, thereby achieving the intention of the human manufacturer or designer. In this sense, role-based design is a perfect method for engineering artificial systems, including modeling, analysis, design, construction, assessment, and maintenance.

Collaboration is required when there is a task that cannot be accomplished by an individual. Even people with high intelligence may not know how to handle collaboration well. Learning how to collaborate with others is a life-long task. It requires a systematic and thorough investigation of the challenges in collaboration. Therefore, theories, technologies, practices, and systems to support collaboration should be proposed, tested, verified, validated, and taught.

Collaboration is complex because it is difficult for us to express collaboration activities with formal mathematical expressions. It is impossible to formalize every aspect of collaboration, because each of them, such as intelligence, emotions, or people, is complex enough for researchers to investigate for their whole lives. Therefore, "divide and conquer" is the fundamental method of conducting collaboration research, and we need to provide a pertinent level of abstraction and commit to the aspects that can be handled by an individual researcher or a well-formed team.

RBC is a computational and research methodology that uses roles as a primary underlying mechanism to facilitate collaborative activities. It provides a set of concepts, models, components, and algorithms for establishing collaboration systems. The RBC methodology can be applied to the analysis, design, and development of all kinds of collaboration systems, including societal-technical systems. The models and algorithms can also be applied to industries such as system engineering, production, management, and governance.

RBC provides insights into complex systems through the usage of its kernel Environments, Classes, Agents, Roles, Groups and Objects (E-CARGO) model. The E-CARGO model provides the fundamental components, principles, relationships, and structures for specifying the states, processes, and evolution of complex systems. This book investigates the nature of collaboration, proposes an easy-to-follow process for collaboration, and designs a model to formalize complex problems in collaboration and complex systems.

"Collaboration made easy" is the goal and theme of this book and is supported by RBC and E-CARGO. Please note that while this book has started the research toward this goal, it is still far from reaching it. Readers are welcome to extend, apply, verify, validate, or criticize or even disqualify the proposed models, methods, formalizations, and solutions.

Haibin Zhu, PhD
North Bay, Ontario, Canada
Aug. 1, 2020

A Guide to Reading This Book

This book is written for senior students, graduate students, researchers, and practitioners in relevant engineering fields. This book can be a reference book for senior students, graduate students studying computing, systems engineering, industrial engineering, computational social systems, and any fields that deal with complex systems, who wish to apply computational methodology to their problem solving. The readers are supposed to have a background of object-oriented programming, and discrete mathematics including set theory and predicate logic.

If you have knowledge of formalization and optimization, it should be easy for you to understand the algorithms and problem formalizations. However, you do not have to understand formalizations and optimizations before reading this book. You may choose to learn formalizations and optimization modeling through reading the book because all the symbols are specifically designed for this book. The optimization modeling is straightforward, and you do not have to know Linear Algebra at all. You may also choose to skip the formalization and optimization components, and focus on learning the concepts of role-based collaboration and the contents that are useful for your research and practice. The following chart (Figure 0.1) can help you better obtain the information you need from this book. All the related Java codes are posted in github.com. Readers can access them by https://github.com/haibinnipissing/E-CARGO-Codes.

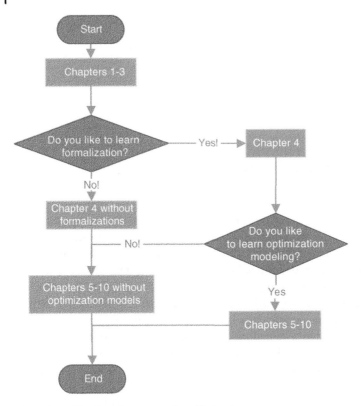

Figure 0.1   The flowchart to reading this book.

Acknowledgments

Writing a book is harder than I thought and more rewarding than I could have ever imagined.

Thanks go to the IBM Eclipse Innovation Awards which made my research on Role-Based Collaboration (RBC) and the Environments - Classes, Agents, Roles, Groups and Objects (E-CARGO) model possible during my challenging start-up research career.

Thanks to Natural Sciences and Engineering Research Council, Canada (NSERC) for its continuous support of my research.

I am eternally grateful to Emerita Distinguished Professor Murray Turoff of New Jersey Institute of Technology, USA, named as the father of computer conferencing. He invited me to New Jersey Institute of Technology as a visiting scholar in 1999 and then recommended me for a full-time position, i.e. Special Lecturer at the same university in 2000. Working at NJIT started my academic career in the western world. The participation in the routine seminars of his group encouraged me to delve into the investigations of the exciting field of collaboration systems. His ideas in designing the EIES (Electronic Information Exchange System) provided me the inspirations for RBC and E-CARGO.

This book would have been impossible without the strong support from my former collaborator, Distinguished Professor MengChu Zhou of New Jersey Institute of Technology, USA, who was ranked as World's top 227th Computer Scientist in 2020. He stood by me during my every struggle and all my successes in my research. Without his collaboration and continuous support, I would not have been able to publish the initial stepping-stone paper on Role-Based Collaboration and E-CARGO in IEEE Transactions on Systems, Man, and Cybernetics, Part C: Applications and Reviews in 2006.

Thanks go to Professor Xianzhong Zhou, Nanjing University, China, and Professor Shaohua Teng, Guangdong University of Technology, China for their cooperation and support.

My visiting PhD students from Nanjing University, China helped me advance this research. They are Dr. Yin Sheng and Dr. Xianjun Zhu. I am grateful to be their co-supervisor for their PhD Dissertations.

To my collaborators in the research of RBC and E-CARGO, Professor Dongning Liu of Guangdong University of Technology, China, and Dr. Hua Ma of Hunan Normal University, China. They helped me extend the applications of RBC and E-CARGO.

To my graduate students, Dr. Linyuan Liu, Luming Feng, Bo Lei, Pinzhi Wang, and Zhe Yu. Their questions make me think deeply and broadly.

Thanks go to the adjunct researcher Dr. Luca Ferrari, visiting researcher Dr. Hongyu Zhang and visiting graduate students Siqin Zhang, Mingjun Lu, and Baoying Huang in the Collaborative Systems Laboratory, Nipissing University, Canada. They help keep my research on RBC and E-CARGO active and productive.

Thanks go to my former colleague, Mike Brewes, retired instructor of Nipissing University, Canada for his long-term support in proofreading and editing my writings in the related research.

Thanks go to my research assistants, Eric Brownlee, and Zikai Wang at Nipissing University, Canada. Their hard work on proofreading the manuscript of this book removed many typos and made the presentation of this book more fluent and understandable. I would like to express my special thanks to the editorial team of Wiley including Mary Hatcher, Victoria Bradshaw, Teresa Netzler, and Gayathree Sekar. Without their hard work, this book could not be presented to the readers in such a graceful style. I should also thank Dr. Senyue Zhang, Xiaohui Li, Lisa Zhao, and Devin Li for their final readings which help polish the writing.

Thanks go to my wife, Jing Zhang, for her continuous patience and encouragement. She has done a lot of the yard work that should have been done by me for the past years. To my son Davis Yu Zhu, he lost many opportunities to play with me under the sunshine during his childhood due to my continuous research effort. He is also a reader and collaborator of my research.

Symbols and Notations (Nomenclature)

Topics	Symbol	Meaning
Acronyms	AC	Adaptive Collaboration.
	AE	Agent Evaluation.
	DGRA	Dynamic Group Role Assignment.
	E-CARGO	Environments - Classes, Agents, Roles, Groups, and Objects.
	GA	General Assignment.
	GAP	General Assignment Problem.
	GBA	Group Budget Assignment.
	GRA	Group Role Assignment.
	GRA^+	Group Role Assignment with Constraints.
	GRA^{++}	Group Role Assignment with Multiple Objectives.
	GRABC	Group Role Assignment with Budget Constraints.
	GRABC-WS	Group Role Assignment with Budget Constraints in the Weighted Sum form.
	GRABC-Syn	Group Role Assignment with Budget Constraints in the Synthesized form.
	$GRABC\text{-}P_{\#}$	Group Role Assignment with Budget Constraints in Form # of Performance, where # = 1, 2, or 3.
	$GRABC\text{-}B_{\#}$	Group Role Assignment with Budget Constraints in Form # of Budget, where # = 1, 2, or 3.
	GRAABD	Group Role Assignment with Agents' Busyness Degree

(*Continued*)

Topics	Symbol	Meaning
	GRAABD-WS	Group Role Assignment Integrated with Agents' Busyness Problem in the Weighted Sum Form
	GRAABD-Syn	Group Role Assignment Integrated with Agents' Busyness Problem in the Synthesized Form
	GRACAR	Group Role Assignment with Conflict Agents on Roles.
	GRACAG	Group Role Assignment with Conflict Agents in a Group.
	GRAP	GRA Problem.
	GMEO	Good at Many things and Expert in One.
	GMEO-1	The GMEO problem with limited training units.
	GMEO-S_1	The simplified GMEO problem.
	GMEO-S_1-CPLEX	The CPLEX based algorithm for the GMEO-S_1 problem.
	GMEO-S_1-S	The efficient algorithm for the GMEO-S problem.
	GMEO-CPLEX	The CPLEX based algorithm for GMEO-1.
	GMEO-GRA	The GRA based algorithm for GMEO-1.
	GMRA	Group Multi-Role Assignment.
	GMRA$^+$	Group Multi-Role Assignment Extended.
	GMRACR	Group Role Assignment with Coupling Roles.
	GMRACR-WS	The Weighted Sum form of GMRACR.
	GMRACR-GP	The Goal Programming form of GMRACR.
	MERA	Most Economical Redundant Assignment.
	MERA$_{GP}$	Most Economical Assignment Problem with Goal Programming.
	MERA$_{GPP}$	The MERA problem with Goal Programming on the required group Performance.
	MERA$_{GPB}$	The MERA problem with Goal Programming on the limited Budget $\overline{C}$.
	MERA$_{GRA}$	The GRA based algorithm for the MERA Problem.

Topics	Symbol	Meaning
	MERA_{NEW}	The new form of the MERA problem.
	$\text{MERA}_{\text{NEWGPB}}$	The new form of the MERA problem with Goal Programming on the limited budget $\overline{C}$.
	RA	Role Assignment.
	RBC	Role-Based Collaboration.
	RGRAP	Rated GRA Problem, i.e., GRAP.
	RT	Role Transfer.
	SGRA	Static group role assignment.
	SimGRAP	Simple GRA Problem
	SRTP	Temporal Role Transfer Problem with Strong restirctions
	WGRAP	Weighted GRA Problem
	WRTP	Temporal Role Transfer Problem with Weak Restrictions
Problem solving	$\overline{o}$	An object.
	$\overline{O}$	The set of all objects in consideration.
	$\overline{p}$	A problem.
	$\overline{P}$	The set of all problems in consideration.
	$\overline{v}$	An operation object
	$\overline{a}$	An argument object.
	$\overline{s}$	A solution.
	$\overline{S}$	The set of all solutions in consideration.
	$\mathcal{N}$	The set of non-negative integers, i.e., $\{0, 1, 2, ...\}$.
Fundamental E-CARGO	$\overline{d_o}, \overline{d_c}, \overline{d_r},$ $\overline{d_a}, \overline{d_e}, \overline{d_m},$ $\overline{d_g}, \overline{d_h},$ and $\overline{d}$	The identification of an object, class, role, agent, environment, message, group, human, and others.
	$\overline{D}$	The set of all identifications in consideration.
	s	The state of an object.
	$\mathcal{D}$	A set of all data structures.
	$\mathcal{C}$	The set of all classes in consideration.
	e	A class.
	$\mathcal{D}_e$	The set of data structures of a class.
	$\mathcal{F}_e$	The set of services of a class.
	$\mathcal{X}_e$	The set of interfaces of a class.
	O	The set of all objects in consideration.
	$\mathcal{O}_r$	The set of objects accessed by a role.
	o	An object in a collaboration system.
	$\mathcal{A}$	The set of all agents in consideration.

(*Continued*)

Topics	Symbol	Meaning
	$\mathcal{A}_p$	The set of agents who can potentially play a role.
	$\mathcal{A}_o$	The set of agents who played a role before.
	$\mathcal{A}_c$	The set of agents who are currently playing a role.
	$\mathcal{A}_g$	The set of agents of group g.
	$a, a_0, a_1, a_2, ...$	Agents.
	$\mathcal{M}$	The set of all messages in consideration.
	♪	A message.
	♪	The message pattern.
	$\mathcal{P}_m$	The object set of parameters matching pattern ♪.
	$\ell_m \in\{$any, some, all$\}$	The label of a message.
	α	The space limit.
	β	The time limit.
	$\mathcal{R}$	The set of roles in discussion.
	$\mathcal{R}_c$	The set of current roles of an agent.
	r_c	The current role of an agent if only one current role is allowed.
	$\mathcal{R}_p$	The set of potential roles of an agent.
	$\mathcal{R}_o$	The set of roles previously played by an agent.
	$\mathcal{R}_x$	The set of current and potential roles.
	$r, r_0, r_1, r_2, ...$	Roles.
	$\bar{r}$	An interface role.
	$\mathcal{M}_{\text{in}}$	The set of incoming messages specified by a role.
	$\mathcal{M}_{\text{out}}$	The set of outgoing messages specified by a role.
	$\mathcal{E}$	A set of all environments in consideration.
	e	An environment.
	q	The role range expressed by $<\ell, u>$.
	ℓ, u	The lower bound and upper bound numbers for the role range q.
	$w \in [0, 1]$	The weight of role.
	$\mathcal{R}_e$	The set of roles of an environment.
	$\mathcal{B}$	The set of tuples $<\overline{d_r}, q, \mathcal{D}_o >$, i.e., role requirement.
	$\mathcal{D}_o$	The set of objects accessible to the corresponding role in $\mathcal{B}$.

Topics	Symbol	Meaning
	$\mathcal{D}_e$	The set of shared objects by all the roles in an environment.
	$\mathcal{D}_g$	The set of groups for an agent to belong to.
	$\mathcal{G}$	A set of groups.
	g	A group.
	$\mathcal{J}$	The set of assignment tuples $<a, \imath>$ of a group.
	s_0	The initial state of a system.
	$\mathcal{H}$	A set of human users.
	$\hbar$	A human being.
	$m = |\mathcal{A}|$	The size of the agent set $\mathcal{A}$.
	$n = |\mathcal{R}|$	The size of the role set.
	$0 \leq i, i_0, i_1, i_2, ... < m$	The indices of agents.
	$0 \leq j, j_0, j_1, j_2, ... < n$	The indices of roles.
	$\sum$	A system.
	V_p^r	The set of supplies of a role
	V_r^r	The set of demands of a role.
	V_p^a	The set of provisions of an agent.
V_r^a	The set of requirements of an agent.	
Role relations	Ω	An inheritance relation.
	Λ	A promotion relation.
	Δ	A report-to relation.
	Θ	A request relation.
	$\in$	A competition relation.
	$\Diamond$	A peer relation.
	Ξ	A conflict relation.
	Γ	$::= \Lambda \cup \Delta \cup \Theta \cup \in \cup \Diamond \cup \Xi$.
GRA	L	The role range vector whose dimension is n.
	$L[j] \in \mathcal{N}$	The lower range of role j.
	U	The upper role range vector whose dimension is n.
	$U[j] \in \mathcal{N}$	The upper range of role j.
	W	The role weight vector whose dimension is n.
	$W[j] \in [0,1]$	The weight of role j.
	Q	The qualification matrix whose dimensions are $m \times n$.

(Continued)

Topics	Symbol	Meaning
	$Q[i, j] \in [0,1]$	The qualification value of agent i on role j.
	T	The role assignment matrix whose dimensions are $m \times n$. It is also used to express the current role matrix in role transfer (RT).
	$T[i, j] \in \{0,1\}$	To express if agent i is assigned to role j, where 1 means yes and 0 no.
	T^*	The optimal assignment matrix for the considered problem, i.e., GRA, GMRA, GRACAR, ...
	τ	The qualification threshold for an agent to play a role.
	σ	$= \sum\limits_{i=0}^{m-1} \sum\limits_{j=0}^{n-1} Q[i,j] \times T[i,j]$, i.e., a group performance.
	σ_w	$= \sum\limits_{j=0}^{n-1} W[j] \times \sum\limits_{i=0}^{m-1} Q[i,j] \times T[i,j]$, i.e., the weighted group performance.
	σ_0	The result of GRA.
	σ_1	The result of GMRA.
	σ_1'	The result of GMRA$^+$.
RT	T	The current role matrix, i.e., the assignment matrix.
	Q_p	The potential role matrix.
	Q	The combination of T and Q_p, i.e., the simple qualification matrix, $Q[i, j] = T[i, j] \vee Q_p[i, j]$ $(0 \leq i < m, 0 \leq j < n)$.
	s	The number of time segments in a period in consideration of temporal role transfer.
	δ	The period in consideration of temporal role transfer, e.g., a day, a week, a month, ...
	$Q_p[i, j] \in \{0,1\}$	To express if agent i can potentially play role j, where 1 means yes and 0 no.
GRACAR/G	A^c	The conflicting agent matrix of a group whose dimensions are $m \times m$.
	$A^c[i_1, i_2] \in \{0,1\}$	To express if agent i_1 is in conflict with agent i_2, where 1 means yes and 0 no.
	n_c	$= \sum\limits_{i_1=0}^{m-1} \sum\limits_{i_2=i_1+1}^{m-1} A^c[i_1, i_2]$, i.e., the number of conflicts.

Topics	Symbol	Meaning
	n_a	$= \sum_{j=0}^{n-1} L[j]$, i.e., the number of required agents.
	p_c	$= n_c/[m \times (m-1)/2]$, i.e., the conflict rate.
	n_{ac}	$= \sum_{i_1, i_2 = 0}^{m-1} \sum_{j=0}^{n-1} T^*[i_1, j] \times T^*[i_2, j] \times A^c[i_1, i_2]$, i.e., the number of assigned conflicts in the T corresponding to GRA.
	λ	The benefit obtained by comparing GRACAR/G with GRA.
GMEO/ MERA	$\overline{Q}$	The capability matrix, or the training plan matrix.
	$\overline{T}$	$= T[i,j] \times \left(1 - \overline{Q}[i,j]\right)$, i.e. the required training matrix.
	$\overline{Q_{-i}}$	The $\overline{Q}$ matrix after removing agent i.
	$\overline{T_{-i}}$	The T matrix after removing agent i.
	V^c	A role training cost vector.
	n_t	The number of the training units, i.e. $\sum_{i=0}^{m-1} \sum_{j=0}^{n-1} \overline{Q}[i,j]$.
	n_d	The number of the trained agents, i.e. $\sum_{i=0}^{m-1} \left\lceil \sum_{j=0}^{n-1} \overline{Q}[i,j]/n \right\rceil$.
	σ_{g1}	The group performance result of GMEO-1.
	σ_{g2}	The result group performance of GMEO-GRA.
	σ^*	The group performance result of MERA$_\#$, i.e. $\sum_{i=0}^{m-1} \sum_{j=0}^{n-1} Q[i,j] \times T^*[i,j]$, where T^* is the T obtained by MERA$_\#$, where # can be GPP, GPB, NEWGPP, NEWGPB, and GRA.
	C^*	The cost result of MERA$_\#$, i.e. $\sum_{i=0}^{m-1} \sum_{j=0}^{n-1} V^c[i] \times \overline{T}^*[i,j]$, where $\overline{T}^*$ is the result $\overline{T}$ obtained by MERA$_\#$, where # can be GPP, GPB, NEWGPP, NEWGPB, and GRA.

(Continued)

Topics	Symbol	Meaning
GRABC	Q'	A budget request matrix.
	Q''	The normalized budget request matrix from Q'.
	ω	The acceptable highest budget limit.
	ω_p	The acceptable lowest group performance.
	P^r	The role performance requirement vector.
	$P^{r'}$	The assigned role performance vector.
	p	The acceptable losing factor for performance.
	p_b	The acceptable losing factor for budget.
	B	The role budget limit vector.
	B'	The assigned role budget vector.
	σ_b	The result of GBA.
	w_1	The weight of the first objective
	w_2	The weight of the second objective
	$\sigma_{(w1,w2)}$	The result of GRABC-WS with $w1$ and $w2$.
	$\sigma_\#$	The result of GRABC-#, where # = Weighted-Sum (WS), Synthesis (Syn), Performances (P_1, P_2, P_3), or Budgets (B_1, B_2 and B_3)
GRAABD	V^b	The busyness vector
	σ_{bw}	The group performance result of GRAIAB-WS
	T^{bw}	The T that makes σ_{bw}
	$Q^{b'}$	$Q^{b'}[i,j] = = w \times Q[i,j] - (1-w) \times V^b[i]$
	Q^r	$Q^r[i,j] = \begin{cases} 3 \times (1-V^b[i]) \times Q[i,j] & (Q[i,j] \geq 0.8) \\ 2 \times (1-V^b[i]) \times Q[i,j] & (0.5 \leq Q[i,j] < 0.8) \\ (1-V^b[i]) \times Q[i,j] & (Q[i,j] < 0.5) \end{cases}$ $(0 \leq i < m, 0 \leq j < n)$, which is supposed to be the actual qualification value for agent i on role j after considering the busyness degrees.
	Q^{nr}	The normalized Q^r, i.e., $Q^{nr}[i,j] = \frac{Q^r[i,j] - min\{Q^r[i,j]\}}{max\{Q^r[i,j] - min\{Q^r[i,j]\}\}}$ $(0 \leq i < m, 0 \leq j < n)$.

Topics	Symbol	Meaning
	σ_{b1}	$= \sum\limits_{i=0}^{m-1} \sum\limits_{j=0}^{n-1} Q^{nr}[i,j] \times T^*[i,j]$, i.e. the actual group performance with GRA.
	σ_{b2}	$= \sum\limits_{i=0}^{m-1} \sum\limits_{j=0}^{n-1} Q^{nr}[i,j] \times T^{bw}[i,j]$, i.e. the actual group performance considering busyness.
	σ_{syn}	$= \max \left\{ \sum\limits_{i=0}^{m-1} \sum\limits_{j=0}^{n-1} Q^{nr}[i,j] \times T[i,j] \right\}$, i.e. the result of GRAABD-Syn.
GMRACR	C^R	The role coupling matrix.
	$C^{R'}$	The normalized role coupling matrix, i.e. $C^{R'}[j_1,j_2] = C^R[j_1,j_2]/max\{C^R[j_1,j_2](0 \le j_1, j_2 < n)\}$.
	T^*_{ws}	The T obtained by GMRACR-WS.
	γ^c	The coupling result of GMRA, i.e. $\sum\limits_{i_1=0}^{m-2} \sum\limits_{i_2=i_1+1}^{m-1} \sum\limits_{j_1=0}^{n-1} \sum\limits_{j_2=0}^{n-1} C^R [j_1,j_2] \times T^*[i_1,j_1] \times T^*[i_2,j_2]$, where T^* is the T obtained by GMRA.
	σ^c_{ws}	The group performance result of GMRACR-WS, i.e. $\sum\limits_{i=0}^{m-1} \sum\limits_{j=0}^{n-1} Q[i,j] \times T^*_{ws}[i,j]$.
	γ^c_{ws}	The coupling result of GMRACR-WS, i.e. $\sum\limits_{i_1=0}^{m-2} \sum\limits_{i_2=i_1+1}^{m-1} \sum\limits_{j_1=0}^{n-1} \sum\limits_{j_2=0}^{n-1} C^R [j_1,j_2] \times T^*_{ws}[i_1,j_1] \times T^*_{ws}[i_2,j_2]$.
	T^*_{GP}	The T obtained by GMRACR-GP.
	σ^c_{GP}	The group performance result of GMRACR-GP, i.e. $\sum\limits_{i=0}^{m-1} \sum\limits_{j=0}^{n-1} Q[i,j] \times T^*_{GP}[i,j]$.
	γ^c_{GP}	The coupling result of GMRACR-GP, $\sum\limits_{i_1=0}^{m-2} \sum\limits_{i_2=i_1+1}^{m-1} \sum\limits_{j_1=0}^{n-1} \sum\limits_{j_2=0}^{n-1} C^R [j_1,j_2] \times T^*_{GP}[i_1,j_1] \times T^*_{GP}[i_2,j_2]$.
GRACCF	$\overline{T}$	$\overline{T}[i,j,i',j'] = T[i,j] \times T[i',j'] = 0$, where

(Continued)

Topics	Symbol	Meaning
		$\overline{T}[i,j,i',j'] \in \{0,1\}$ $(0 \le i, i' < m, 0 \le j, j' < n)$ and $2\overline{T}[i,j,i',j'] \le T[i,j] + T[i',j'] \le \overline{T}[i,j,i',j'] + 1$ $(0 \le i, i' < m, 0 \le j, j' < n)$.
C^f		A *cooperation and conflict Factor (CCF) matrix* C^f is an $(m \times n) \times (m \times n)$ matrix: $(A \times R) \times (A \times R) \to [-1, +1]$, where $C^f[i,j, i',j'] \in [-1, +1]$ expresses the changing degree of agent i's qualification when agent i plays role j and agent i' plays role j' $(0 \le i, i' \le m-1, 0 \le j, j' \le n-1)$, and $C^f[i,j,i',j'] > 0$ (<0) means cooperation (conflict), i.e., an increase (decrease) of the group qualification.
n_{cf}		A *significance number*, i.e., the total number of non-zero elements *in* C^f, *i.e.* $$n_{cf} = \sum_{i=0}^{m-1}\sum_{j=0}^{n-1}\sum_{i'=0}^{m-1}\sum_{j'=0}^{n-1} \lceil |C^f[i,j,i',j']| \rceil).$$
C^{cf}		A *Compact Cooperation & Conflict Factor matrix* C^{cf} is an $n_{cf} \times 5$ matrix, where $C^{cf}[k,4] \in [-1, 0) \cup (0, 0]$ $(0 \le k < n_{cf})$ expresses that the changing degree of agent $C^{cf}[k, 0]$'s qualification on role $C^{cf}[k, 1]$ while agent $C^{cf}[k, 2]$ plays role $C^{cf}[k, 3]$.

Notations

In this book, besides the nomenclature described above, we need to clarify the notations of formalizations. If a and b are objects, $a.b$ denotes b of a or a's b. $\{a, b, \dots\}$ denotes a set of enumerated elements of a, b, and others. If a and b are variables, $a := b$ denotes assign the value of b to a, *and* $a = b$ is to check if the value of a equals to that of b. If a and b are integers, $[a, b]$ denotes the set of all the real numbers between a and b including a and b, and $(a, b]$ is the same, excluding a. If Y is a set, $|Y|$ is the number of elements in set Y. If Y is a vector, $|Y|$ is the size of Y, i.e., how many elements are in the vector, and $Y[i]$ denotes the element at the ith position of Y. If Y is a matrix, $Y[i, j]$ denotes the element at the intersection of row i and column j in Y.

Note

When using symbols, we tried to use different symbols to denote different concepts or terms. However, due to the varieties of real-world applications presented in this book, there might be a few overloading local symbols. We may use local symbols to express corresponding local terms when discussing corresponding problems, especially in Chapters 7 and 8. Please follow the meanings in the specific chapters and sections.

Part I

Backgrounds

1

Introduction

1.1 Collaboration and Collaboration Systems

It is a complex world. In a complex world, there are many problems for people to solve. To solve a problem, we need to establish the concepts, conduct abstraction to grasp the key points of the problem, design a model for the problem, provide a solution, and, finally, apply the solutions in the real world to verify whether the solution is acceptable. Albeit there are many simple problems that can be solved without much effort, researchers are only interested in complex problems. We should first clarify what a complex problem is.

To understand whether a problem is complex or not, we can simply check whether the problem can be formalized using symbols, formulas, equations, or other formal languages, where symbols have unique meanings. We can be assured that if a problem has not yet been formalized, it is a complex problem. On the other hand, if a problem can be formalized but the process to solve it has exponential complexity, it is also a complex problem. Under these criteria, collaboration is a complex problem.

We must admit that the research on collaboration is still in the stage of "infancy" (Miller-Stevens et al. 2016; Morris and Miller-Stevens 2016). The evidence to assert this claim derives from the following facts: (i) there is not a well-accepted definition to the term "collaboration"; (ii) there is not a well-accepted model to completely specify a collaboration system; and (iii) there is even no well-established methodology that can be applied to conduct collaboration in specific fields, e.g. health care or manufacturing.

This book is a first attempt from the systems engineering's viewpoint to kick off the investigation of collaboration and collaboration systems. This book aims to study collaboration, collaboration systems, and complex systems with the support of the model Environments – Classes, Agents, Roles, Groups, and Objects (E-CARGO) and the Role-Based Collaboration (RBC) (Zhu and Zhou 2006b) methodology.

E-CARGO and Role-Based Collaboration: Modeling and Solving Problems in the Complex World, First Edition. Haibin Zhu.
© 2022 by The Institute of Electrical and Electronics Engineers, Inc.
Published 2022 by John Wiley & Sons, Inc.

1.1.1 Collaboration

Collaboration is a polymorphic word. It has many different meanings depending on different viewpoints and contexts. In the 1980s, Gray attempted to define "collaboration" in public administration as "the pool of appreciations and/or tangible resources, e.g. information, money, labor, etc., by two or more stakeholders, to solve a set of problems which neither can solve individually" (Gray 1985, p. 912). Many researchers have defined collaboration very differently since then (Gray 1989; Thomson 2001; Thomson and Perry 2006). Thomson states that "collaboration is a process in which autonomous actors interact through formal and informal negotiation, jointly creating rules and structures governing their relationships and ways to act or decide on the issues that brought them together; it is a process involving shared norms and mutually beneficial interactions" (Thomson 2001). In fact, even Gray's definition has changed over time. In Gray's 1989 paper, Gray redefined collaboration as something that "transforms adversarial interaction into a mutual search for information and for solutions that allow all those participating to ensure that their interests are represented" (Gray 1989, p. 7).

Collaboration is started when all the participants bring something to the table (expertise, money, ability to grant permission), and put their belongings on the table (Thomson and Perry 2006). Collaboration ends when all the parties agree that the result is satisfied (success) or the collaboration cannot continue anymore (failure).

We believe that collaboration is a joint effort of many (>1) people to accomplish a designated task. It is also a problem-solving process that involves many (>1) different problem solvers (Hsieh and O'Neil Jr. 2002).

Definition 1.1 A *team* is a well-organized group of participants that work together to achieve common goals.

Definition 1.2 *Collaboration* is a joint effort to form a team.

To understand these definitions, we need to first realize that collaboration is a joint effort from a group of people, not a job for one person. That is, collaboration requires participants to share, communicate, and interact amongst themselves. Secondly, collaboration needs to set a goal when collaboration starts, i.e., collaboration must produce an artifact, accomplish a task or provide a result. Thirdly, collaboration includes more than one participant, which can be a device, a computer system, a person, an organization, or even a country. Therefore, we can view collaboration as a complex system (Section 1.7).

Collaboration, as a research field, has all the properties of a field that demands investigation:

1) Collaboration is necessary. From Figure 1.1 (UOPX news 2013), it is reported that seven in ten people have worked in a dysfunctional team during their

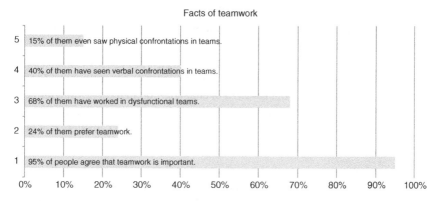

Figure 1.1 How important is teamwork? (ClassesandCareers.com).

lifetime (Item 3). Furthermore, collaboration was necessary for human survival in early human history. Early primitives had to collaborate to build shelters to protect themselves from natural disasters. They had to collaborate to defend themselves from attacks by ferocious animals, such as wolves, tigers, and lions. They collaborated to kill animals to obtain enough food. Therefore, collaboration is a fundamental activity of the human world. No one lives completely alone. People collaborate when a single person cannot accomplish a complex or difficult task, or if a single person can complete a task but it is not economical to do so. For example, to move a piano into a house from the outside is an impossible task for one person to accomplish manually but not very hard for four people to work together. Similarly, people can make tofu dishes in a restaurant, but it is too costly for them to make raw tofu by themselves. By collaborating with a local grocery store to purchase raw tofu, they save considerable time and effort while increasing profit.

2) Collaboration is complex and challenging. It requires many elements and involves a variety of combinatorics. For example, if a team has m members, the number of one-to-one connections between them is $\binom{m}{2} = m!/[2! \times (m-2)!] = m \times (m-1)/2$; the numbers of subgroups of 3 members is $\binom{m}{3}$, ... Therefore, the complexity of collaboration quickly becomes the exponential if we do not carefully organize.

3) Collaboration is interesting. It is fun to investigate the properties of collaboration. It provides researchers with many opportunities to satisfy their curiosities. The intermediate investigation results may be frustrating, surprising, and even exciting. Many collaboration problems may seem bewildering at the beginning. However, after sufficient effort, thinking, reasoning, and experiments are

applied, exciting results may be unveiled. Such a process is similar to the aesthetic appreciation process, i.e. the result of an investigation rewards a feeling similar to sensing the beauty of art.

4) Collaboration is beneficial. Not only is the result of collaboration beneficial, but also is the collaboration process. Many methodologies, algorithms, and technologies developed from the research of collaboration make a fortune for people in different fields. Continuous efforts in collaboration research contribute to both the research community and industries. The following chapters will provide more concrete cases.

Collaboration requires people in a group to fulfill their obligations and respect the rights of others. To collaborate, people generally participate in a group or organization. The related parties must establish common goals by negotiation, divide the whole task into subtasks, distribute subtasks to related parties, and, finally, integrate all complete subtasks to a unified result.

Examining the components of collaboration and the type of interaction between them, collaboration can be classified into the following categories:

- *Natural Collaboration* occurs among people/organizations who are members of a team (Figure 1.2). Looking at Figure 1.2, we can observe how complex

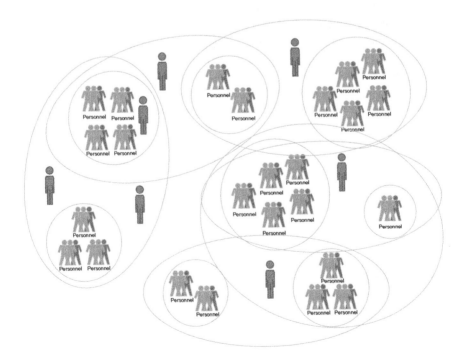

Figure 1.2 Natural collaboration.

collaboration can be. Furthermore, the overlapping and interleaving of teams make collaboration even more complex. Figure 1.2 presents three types of participants: the groups, the small groups, and the manager of a group. We may take the single-large-person icon as a manager, the three-person icon as a small group, and a circle-including-person icons as a group. Please note that, in Figure 1.2, the sharing among the participants is hidden. Most of the problems in natural collaboration arise from the difficulties in sharing resources (Wondolleck and Yaffee 2000) such as knowledge, information, and technologies.

- *Computer-Supported Cooperative Work (CSCW)* (Grudin 1994) is a research topic that mainly focuses on innovations to support collaboration among people through computer systems (Figure 1.3). From Figure 1.3, we observe that people are collaborating through the use of computers. Note that we can include any equipment that facilitates collaborations, e.g. all the devices and equipment presented in Figure 1.4 about Human–Computer/Machine Interaction. Figure 1.3 clearly shows that the CSCW users are sharing the cloud, which is the most important resource in CSCW. If the cloud had unlimited storage capacity and unlimited communication bandwidth, CSCW technology would be easy to implement.

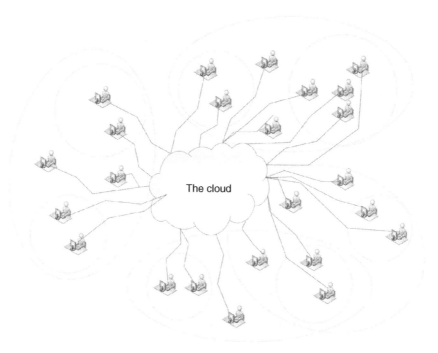

The cloud

Figure 1.3 Computer-supported cooperative work (CSCW).

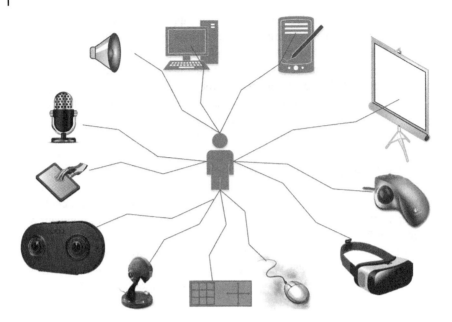

Figure 1.4 Human–computer/machine interaction (HCI/HMI).

- *Human–Computer Interaction (HCI)* (Dix et al. 2003, Preece et al. 1994) *or Human Machine Interaction (HMI)* (Böhmea et al. 2003) is a research field that mainly investigates how to support collaboration between a human and a computer/machine system (Figure 1.4). The aim of HCI/HMI research is to provide more comfortable ways for human users to utilize computers and machines. The challenge for HCI/HMI is understanding the natural interaction ways among human beings and making computers and machines more humanized. Another exciting field in HCI/HMI is augmented interaction (Rekimoto and Nagao 1995), which supports supplementary ways for people to interact with computers and machines other than natural interaction ways. Brain–Machine Interaction (Ramos-Murguialday et al. 2013) is also a hot topic in the HCI/HMI field seeking to help people interact with machines directly through brain neural signals.
- *A distributed system* (Coulouris et al. 2011, Tanenbaum, and van Steen 2016) is a computer system that supports collaboration among computer systems (Figure 1.5), which have their own processors and main memories. The aim of distributed systems' research is to provide more highly available and high-performance computing that acts as though a user is only operating one computer system, i.e. all the components of the distributed system are transparent for the users. Therefore, the cloud metaphor is a very appropriate analogy for

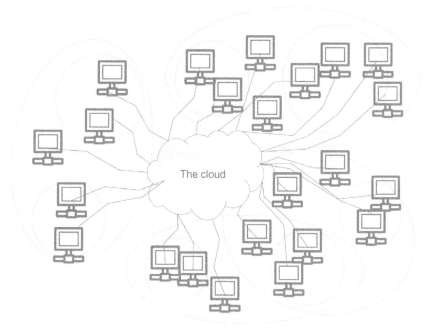

Figure 1.5 Distributed systems.

a distributed system. The challenge of distributed systems lies in providing high processing speed, sufficient communication bandwidth, and a large, safe, stable, and secure storage space for cloud clients.

- *Robot collaboration* is a new topic of research with the development of robot technologies. The goal of this form of collaboration is to form a robot team to accomplish a task that cannot be completed by a single robot. After robots possess human-like abilities, it will be possible to develop a robot team to collaborate to accomplish complex tasks. Figure 1.6 presents the possibility of different robots/agents, groups, and group leaders in collaboration. In robot collaboration, the current major challenge is ensuring that robots possess the necessary abilities for collaboration (Rekleitis et al. 2001), autonomously make decisions, directly sense each other and avoid nearby obstacles. After such robots are built, the methodologies and technologies discussed in this book may be directly applied.

Collaboration can also be classified into such categories: *Mandated collaboration* (Buppert 2010; McNamara 2016) and *voluntary collaboration* (Vella et al. 2016). By mandated collaboration, we mean that the participants are not willing to collaborate but have to do so. On the other hand, voluntary collaboration means the opposite, i.e. the participants are willing to collaborate.

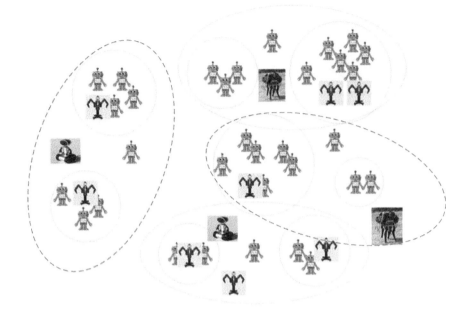

Figure 1.6 Robot collaboration.

For voluntary collaboration, all the participants have the same goal to make the collaboration a success. Therefore, they may put the team's benefits and interests above their individual benefits. However, in mandated collaboration, participants do not hope to collaborate and therefore may be more interested in their individual benefits as opposed to the team's.

An interesting phenomenon is that the difference between mandated and voluntary collaborations is equivocal, that is, we may say that most collaborations are necessary and mandated.

For example, a hospital may wish to investigate a medical technology by itself and requests funds from the government. However, the government requires that at least five hospitals be organized as a team to apply for the government funds. Is this a mandated or voluntary collaboration? Most will answer "mandated collaboration." Another example is a company that hopes to develop a new technology, which is supposed to have a large future market. However, the company lacks key resources, such as technology, financial support, and working spaces. It has to contact several companies to meet the requirements of this project. As a result, it has to yield parts of the future benefits to its collaborators. Is this mandated collaboration or voluntary collaboration? This is also mandated collaboration.

Do we have examples for voluntary collaboration? Yes, we can say that CSCW platform users are conducting voluntary collaboration because they are remotely located and they use such platforms in hope of collaborating. They are intending to collaborate. The discussion in a saloon bar can be taken as a voluntary collaboration because all the participants are voluntary.

"Cooperation" and "coordination" are normally used to describe activities similar to "collaboration." In this book, we will clarify these concepts by formal specifications. We believe that collaboration is composed of coordination and cooperation. The collaboration represents a broader scope than coordination and cooperation. Collaboration involves the initiation, planning, and administration of a team. Cooperation deals with detailed interactions among the members of a team. Coordination deals with the management issues of cooperation among team members in a specific scope other than that of collaboration.

Gray (1989) states that "even though both cooperation and coordination may occur as part of the early process of collaboration, collaboration represents a longer-term-integrated process through which parties who see different aspects of a problem ..., constructively explore their differences ..., and search for solutions that go beyond their own limited visions and implement those solutions jointly."

In her interviews with public agency directors between 1995 and 2000, Thomson (2001) found that, in contrast to the ease with which they define cooperation, agency directors frequently resort to using metaphors to describe collaboration, such as "stepping into other people's shoes," "the combination of hydrogen and oxygen atoms to form water," and "combining yellow and blue circles to form a larger green circle."

Collaboration can be simple or complex. Individual components can accomplish a difficult task by simple collaboration, e.g. one person cannot carry and transport a heavy log manually, but many people together can do it by simply lifting and moving at the same time and pace. One grain of sand can do nothing but tons of sand together can form land and block flooding. This is typical and simple collaboration, i.e. collaboration only by getting together to do something at the same time. Simple collaboration can create considerable results. One participant may not present any special characteristics. However, if there are enough participants, a special property will be produced, i.e. emergence. For example, we do not feel humid from only one molecule of water (H_2O), but we will feel dampness if the humidity reaches 100%, which involves numerous molecules of water. A drop of water can accomplish nothing in most cases, but a bottle of water can help a lot when a person is thirsty and a large quantity of water may form flooding and even destroy lives and buildings. Simple collaboration explains the phenomenon of emergence, which strongly supports the philosophic law, "quantity change may lead to quality change."

A more common type of collaboration is complex collaboration, i.e. complex components get together to accomplish a complex task. Complex collaboration is the major concern of this book. However, the developed models, methodologies, and algorithms discussed in this book can also be used for investigating simple collaboration.

1.1.2 Collaboration Systems

After we discuss collaboration, we need to clarify the terms of "collaboration systems" and "collaborative systems." Conventionally, collaborative systems and collaboration systems have trivial differences. They are often applied interchangeably in different contexts. From our long-term research, we gradually replace the term "collaborative systems" with "collaboration systems." The reason is subtle and seems arbitrary. However, such a choice implies a deep understanding of the nature of collaboration theory and technologies. Therefore, in this book, we differentiate between "collaboration systems" and "collaborative systems."

Definition 1.3 *Collaborative systems* are computer-based systems that assist human users in collaborating remotely with the assumption that the components of the system are collaborative. That is to say, the components of the system, or the participants of collaboration, intend to collaborate, but not the opposite.

For example, we can take Zoom, Google Meet, Cisco Webex, and Microsoft Teams as collaborative systems.

Definition 1.4 *Collaboration systems* are computer-based tools or platforms that support human users to form teams locally or remotely. The users of such systems assume that they may not only collaborate but also compete. Such a system does not require that participants have to be at a distance.

Collaboration systems are the ideal goal of researchers and practitioners of collaboration. That is why we cannot present such examples now. However, we do have tools or platforms that partially meet the requirement of collaboration systems. For example, all the above-mentioned collaborative systems provide partial functions of a collaboration system.

Collaboration tools or platforms enable individuals to share their ideas and talents with collaborators so that a task can be finished efficiently and effectively. This means that the system may include components that do not intend to collaborate or are pushed to collaborate.

We may simply say that collaborative systems are for voluntary collaboration and collaboration systems are for both mandated and voluntary collaboration. Technically, collaboration systems are more complex and difficult than collaborative systems because the former needs to support mandated collaboration and the latter mainly pertain to voluntary collaboration.

Note that a collaboration system supports all the participants to collaborate. It also sets policies to restrain the participants' behavior and establishes criteria to evaluate participants' eligibility to remain in the system. Therefore, we need to consider the problems of cooperation, coordination, competition, and conflict.

We also need to consider the culture or style when organizing collaboration, i.e. individualism or collectivism.

Both collaboration and collaborative systems can also be called collaboration support systems (Cabri et al. 2006).

1.2 Collaboration as "Divide and Conquer"

The fundamental method of software engineering in meeting the challenge of the software crisis in the 1970s was "*divide and conquer*," which is also the basic method for complex problem solving. Therefore, "*divide and conquer*" is taken as the fundamental method used in this book for solving problems in collaboration. In complex problem solving, there are many related concepts that we need to clarify because these concepts help establish a well-accepted methodology and a well-defined model.

In daily life, we often try to obtain our own abstract view, or model, of a problem before a solution is found. In dealing with complex situations, we also need to extract the most important elements. These extractions are an activity of abstraction.

Definition 1.5 A *model* is a template for creating instances and a skeleton to reflect the structure of an object.

A model defines an abstract view for a problem that focuses on only the characteristics related to the problem. From object orientation, a model normally defines data and operations on the data. It is also a simplified representation of a complex reality. This representation is usually for the purpose of understanding reality and containing all the features of reality that are necessary for the task or problem (Zhu and Zhou 2006a).

Definition 1.6 *Integration* is a process used to collect all the related entities in the scope of the stakeholders.

For example, collecting data is a process of integration.

Definition 1.7 *Abstraction* is a process used to understand a problem by separating necessary details from unnecessary ones. The process of modeling is a type of abstraction.

Abstraction is extracting, forgetting, discarding, identifying, and structuring. Informally, abstraction means to "drop" certain "nonessential" features from a representation. We need to be careful about this process because it implies "forgetting." However, this "forgetting" is used to help "remember." In other words, if we are now in an abstraction process, we can say, "don't bother me with the details." This statement has led to the scientific term "information hiding."

Definition 1.8 *Information hiding* is a process used to discard the details of the things in consideration and is also a way to make it impossible for anybody who is not "authorized" to access the hidden details.

At first, information hiding is a purposeful omission of details in the development of an abstract representation. In traditional Chinese education, teachers teach students that there are two phases to read a book. The first phase is to read the book from thick to thin and the second one is to read the book from thin to thick. The meaning of "from thick to thin" is actually an abstraction by information hiding. For example, writing a summary of a book occurs in this step. The next phase of "from thin to thick" means extending by adding details based on the abstraction obtained from the first phase. For example, retelling the details of the book.

From the viewpoint of object orientation, abstraction falls into two general categories: data abstraction and control abstraction. Data abstraction concerns the properties of data, such as numbers, character strings, or other complex entities that are the subjects of computation. This often produces data structures. When we are using a data type, we specify what we can do with this type, but we abstract away how to do it with a given programming language.

Control abstraction deals with the properties of a control transfer, i.e. the modification of the execution path of a program based on the situation at hand. Control abstraction usually produces algorithms. When we are writing a function, we can put calls to other auxiliary functions into it. We specify what these auxiliary functions do, but we abstract away how they do it.

If a problem is too complex to solve or if a thing is too complex to understand and remember, we need to make an abstraction of it. We can view abstraction as a tool to know, create, and understand complex things. For example, a world map is an abstraction of the earth and a provincial map of Ontario is an abstraction of the Province of Ontario. A book of Canada is, in fact, an abstraction of Canada from the viewpoint of the book's authors. Therefore, we can conclude that abstraction has different directions based on different aims and uses.

On the other hand, abstraction is a general methodology to know and study the world and knowledge. When we have many concrete objects to control, we generally remember their common properties and specialties. This is an abstraction process. The result of this process is an abstract concept. For example, dog, horse, and people are abstract concepts, and big data needs abstraction to be useful and valuable.

From these two points of view, we can conclude that "abstraction is the purposeful suppression, or hiding, of some details of a process or artifact, in order to bring out more clearly other aspects, details, or structure" (Budd 2001). Through abstraction, we can understand other relevant concepts, such as information hiding. For example, all the textbooks are abstractions of relevant knowledge.

Definition 1.9 *Modeling* is a process of abstraction, i.e. the process of focusing on those features that are essential for the problem at hand and ignoring those that are not.

Definition 1.10 *Classification* is a general methodology that puts things with similar properties together and differentiates things that have different properties.

Classification is applied in many domains of human lives. Some researchers even categorize classification as a form of abstraction (Budd 2001). Generally speaking, that is true. But in programming, classification and abstraction are different because they concentrate on different aspects of problem solving. An object-oriented system takes classification as a key mechanism to arrange elements in programming.

The "divide" phase in "divide and conquer" is accomplished when each component can be assigned to a single participant.

During the "conquer" phase, we need to understand the following concepts:

Definition 1.11 *Communication*, in general, is the action of exchanging information among different people or autonomous entities.

Definition 1.12 *A connection* is a relationship in which a person, thing, or idea is associated with something else. It is the foundation of communication (Zhu 2008).

Definition 1.13 *An interface* is a medium for two entities to accomplish connection, i.e. a common system of symbols, signs, frameworks, patterns, or behaviors.

Definition 1.14 *Interaction* is a mutual and reciprocal action or influence. It is a situation where two or more participants or things communicate or react to each other (Fischer 2001; Zhu and Hou 2011).

Definition 1.15 *Coordination* is the process of organizing participants so that they work together properly and form a harmonious, functioning part of a team (Nwan et al. 1996; Hou et al. 2014, p.58–59).

Definition 1.16 *Cooperation* is a process in which multiple participants work together in parallel while interacting with each other to obtain the required resources or services by coordination (Hou et al. 2014, p.58–59).

From the above definitions, we differentiate cooperation from coordination and collaboration to clarify the problems of collaboration. Figure 1.7 uses the composition relation, i.e. a concept in a larger circle is composed of the concept in a smaller circle, to show that cooperation includes coordination; coordination needs interaction; interaction requires communication; communication is based on

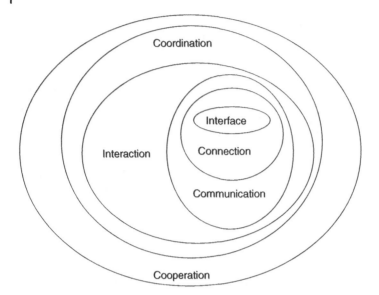

Figure 1.7 The relationships between the related terms.

the support of connections, and connections can be implemented with interfaces, which are foundations. Readers may ask, "where is collaboration?" The answer is that collaboration is composed of cooperation, which is categorized as "role-playing" in RBC (see Chapter 3) and "role-playing" is not the concentration of this book. Or, in other words, this book investigates all the related problems out of Figure 1.7 but aims to simplify the problems in Figure 1.7.

Note that the advocated methodology in this book is *Role-Based Collaboration* and we also propose an application of RBC to interaction, i.e. *Role-Based Interaction* (Zhu and Hou 2011). There may be a question: How do we conduct role-based cooperation, coordination, and communications? We have to admit that we have not yet accomplished solid and sufficient research on these aspects. There are still research potentials in these aspects. E-CARGO and other modeling methods in the subsequent chapters might also be applied in answering this question. Readers who are interested in this question are welcome to delve into related investigations.

1.3 Key Components of Collaboration

Collaboration, as a system, is composed of components and can be divided and conquered. However, collaboration is complex and not easy to divide. Few researchers have provided a well-accepted decomposition of collaboration. The following components of collaboration are proposed by the field of public administration (Mayer and Kenter 2016):

Communication: Each team member must communicate with each other. Such communication increases understanding. Good communication promotes collaboration and poor communication hinders it.

Consensus decision making: The team needs to take action in a consistent way. In collaboration, decision making may be individual, collective, local, and global. All decisions and decision steps should be in consensus. Discordances and conflicts may make a mess.

Roles: For participants to collaborate effectively, there should be different positions, or roles, available for the team members to take. Each member should play specific roles that match the member's interests, abilities, and skills.

Goals: Each collaboration should have a result, i.e. a product, a state, an achievement, or an accomplishment. Such a goal should be shared by all team members.

Leadership: A leader (Day et al. 2014; Goleman 1998) is a special role in a team. A leader needs to foster a sense of trust and loyalty in the team. The leader role requires the role player to possess a strong self-driven personality. The team should accomplish the task with the driving force of the leader.

Sharing: Sharing is essential to collaboration. It reflects the nature of collaboration (Section 1.4). Sharing includes resource sharing and view sharing. In collaboration, resources should be shared and administrated in an effective way so as to avoid conflicts among the team members. In collaboration, all the team members should share a common vision or view related to the goal.

Social capital: Social capital is a complicated factor that refers to the value of collaboration, which includes all the involved elements, the relationships among these elements, and the impact that these relationships have on the whole collaboration.

Social capital (Adler and Kwon 2002) can explain the achievement of collaboration, i.e. the improved performance of teams, the growth of enterprises, winning games, the successful products from a factory, the profit of an organization, or the evolution of communities. Social capital is considered the investment of resources with an expected return (Lin 2011).

Trust: Trust is a particular level of confidence for one person to assess that another person or group of people will perform a particular action before it is done (Dasgupta 1988; Gambetta 1988; Luhmann 2017; Ramchurn et al. 2004). Trust can be interpreted as both a property and a relationship. From the viewpoint of a property, trust is highly related to credibility, quality, reputations, reliability, validity, utility, robustness, and false-alarm rate.

From the relationship's viewpoint, trust can be set up, evolved, and broken up and trust can be divided into long-term and stable trust; transient trust; and the ways of how trust evolves across time. In collaboration, each team member should trust each other. The loss of trust may lead to a failure of collaboration.

In addition to the list above, we need to add one more essential component to collaboration.

Role players: participants or role players are essential components that cannot be ignored in collaboration. Without role players, there would be no collaboration. They are essential and necessary for collaboration. Different participants can both contribute to and create trouble in collaboration.

In this book, we will develop models and methodologies for collaboration. We believe that these models and methodologies will reflect the components of collaboration.

1.4 The Nature of Collaboration

Collaboration is like a "Man of a Thousand Faces." It is easy to tell the properties of it, e.g. Section 1.3, but it is difficult to explore its nature. To really understand collaboration, we need to delve into its nature. Following the idea of "divide and conquer," we will clarify the nature of collaboration.

We can describe natural collaboration as a class (Figure 1.8). When collaboration is started, we first encounter scattered *people*. All these people need to take responsibility and hold rights in order to contribute to the goal of the requested or designated task, i.e. playing *roles*. These two parts are essential attributes of collaboration. During a collaboration, people may bring in their physical and mental resources. All the participants need to *share* these pooled resources.

Such sharing can be done by *integrating* and *distributing* resources. That is, *sharing* is the essential activity of collaboration. Therefore, we believe that the nature of collaboration is that *People* play *Roles* by *Sharing*. In the narrowest sense, sharing refers to the joint or alternating use of inherently finite goods, such as a common pasture or a shared residence. We can restate that the nature of collaboration is a team of *participants* playing different *roles* by *sharing* the common goal and designated task of the team.

In collaboration, *sharing* is a highly abstract function of the collaboration class that includes everything related to collaboration. Sharing includes sharing the input, the process, and the output. Therefore, sharing can be categorized into responsibility sharing, view sharing, data sharing, information sharing, knowledge sharing, resource sharing, intelligence sharing,

Figure 1.8 The nature of collaboration.

Collaboration
People
Roles
Sharing

benefit sharing, and more. Each kind of sharing needs a nontrivial effort to ease collaboration.

For example, in CSCW, view sharing can be supported by "What You See is What I See" (WYSIWIS) technology. Data sharing, information sharing, knowledge sharing, resource sharing, and intelligence sharing can be supported by information technology. Benefit sharing means that collaboration concentrates on mutual benefit. Joining a team is better than being alone, i.e. if one can gain more than what they can gain by themselves, they are likely willing to participate in the collaboration.

Problems and challenges occur in sharing, i.e. synchronization, mutual exclusion, deadlock, information leaking, security, and conflicts. To avoid problems associated with sharing, we need the process of dividing and distributing. Well-done distribution can avoid problems or at least ease the problem solving. That is to say, sharing can be further divided into two major aspects: distribution and execution (Figure 1.9). In Figure 1.9, we use UML (Fowler et al. 2003) symbols, i.e. classes and compositions, to express their relationships. It is generally accepted that collaboration is composed of distribution and execution.

To facilitate distribution, we need the processes of integration, negotiation, specification, evaluation, and assignment. Integration is used to collect all the ideas and resources brought in by the participants. Negotiation means all the participants need to negotiate their common goals, benefits, resources, processes, and tasks. Specification means that all the negotiated entities must be specified by

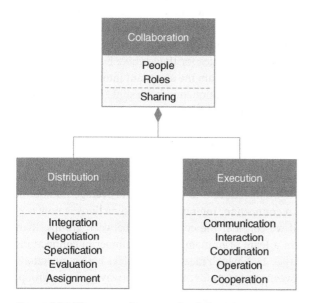

Figure 1.9 The two major parts of collaboration.

formalizations to avoid ambiguity and increase clarity. Evaluation is used to check all the requirements and the provisions and digitalize the matching degrees between the two. Assignment (Burkard et al. 2009) is used to couple the most suitable requirements to the required provisions from the viewpoint of collaboration but not any individual participant's viewpoint.

To conduct execution, we need the processes of communication, interaction, coordination, operation, and cooperation as defined in Section 1.2. We need to point out that the complexity of collaboration mainly comes from communication, interaction, coordination, and cooperation because of the sharing requirement. If the distribution is accomplished well enough, the execution will become less complex by that all the players concentrate on their individual operations. That is why this book's major contents concentrate on the models, algorithms, and methodologies to support distribution.

1.5 The Complexity of Collaboration

It is not very hard to assert that collaboration is complex because collaboration involves various dilemmas. These dilemmas may involve multiple participants and the participants may have shared interests but insufficient resources. They may dispute over the ways to solve a problem. They may have different degrees of skills and experiences. There might be inadequate administrative processes ready to deal with a challenge (Gray 1989; Poutanen 2016). Collaboration is complex because there are many things in the world and there are many connections between those things. In conventional thinking, collaboration is a complex skill that people need to learn, study, and practice for their whole lives.

The complexity of collaboration comes from the abundant interactions between team members. When we look at collaboration, our focus shifts from the individual components to the whole team. The complexity emerges as a result of the patterns of interactions between the components. Furthermore, the complexity also originates from the numerous ways of forming a group.

As we discussed in Section 1.1, we know that the number of undirected connections for a team of m members is $m \times (m - 1)/2$. In one connection, we may conduct different interactions. If we introduce n to represent the average number of ways of interactions for each connection, the number of interactions is now $n \times m \times (m - 1)/2$. Moreover, if we consider the connections between groups, then suppose we have p groups in a team of m members, the number of connections is now $m \times (m - 1)/2 + p \times (p - 1)/2$, which means we consider both connections between individuals and between groups. As for the number of ways to form a group, we have $\sum_{i=1}^{m-1} \binom{m}{i} = 2^{m-1} - 1 \, (m > 1)$ ways for dividing the team into two groups.

The cases may be more complex if we consider the ways to divide the team into 3 groups with k_1, k_2, and k_3 members, $m = k_1 + k_2 + k_3$. The number of group formations is: $m!/(k_1!k_2!k_3!)$ (Brualdi 2008). If we continue, one may find that collaboration can be significantly more complex by introducing more different cases.

To overcome the complexity challenge of collaboration, we use the "divide and conquer" methodology to investigate the nature of collaboration as discussed in Section 1.4:

Collaboration = Distribution + Execution.

Distribution is to assign different tasks to the participants of collaboration, and execution then allows all the participants to concentrate on accomplishing their own tasks and they may need to cooperate with others through sharing, communications, interactions, and coordination.

Distribution has been investigated for decades as a complex problem (Burkard et al. 2009). Related terminologies include "assignment" and "scheduling." Distribution needs integration and division. From software engineering's viewpoint (Pressman 2014), "divide and conquer" is the fundamental approach to producing a high-quality software product. To manage task distribution, we must first divide the whole task of collaboration into subtasks. In many cases, it is very hard to divide a task into subtasks (see Section 1.7). Therefore, distribution is a complex problem.

Execution is complex due to the sharing, concurrency, and parallelism among the collaborating participants. From a computer systems' viewpoint, it is a multi-process and multi-thread execution. That is why we need to deal with the complexity of sharing, communications, interactions, coordination, and cooperation.

If people accept that game theory (Osborne and Rubinstein 1994) deals with complex problems in economies and societies, it is easy for them to accept that collaboration deals with complex problems because collaboration involves scenarios with multiplayer games.

Readers may find that after reading the following chapters, even though collaboration is made easier with the assistance of the proposed formalized methodologies, algorithms, and models, many problems in collaboration are still very complex.

1.6 Collectivism or Individualism

When we talk about natural collaboration, we cannot avoid the consideration of the participants' interests. We have to make a choice: collectivism or individualism.

Individualism emphasizes the individuals' interests and benefits. With individualism, participants act on their own judgments, keep and use the product of their own efforts, and pursue the choices they value. That is to say, the individual is sovereign and the fundamental unit of moral concern. Individualism emphasizes the moral worth of an individual. That is why investigators emphasize agents' individualism in developing multi-agent systems (Alonso 1999; O'Sullivan and Haklay 2000).

Individualism is reasonable (Biddle 2012), we can observe the world and people and immediately see different individuals. The individuals may be in groups (e.g. on a hockey team), but as individual people. People have their own bodies, minds, and lives. Groups are nothing more than individuals who have come together for some purpose.

Collaboration has many individual participants. Intuitive thought is that we need to consider individual participants' benefits when conducting sharing, i.e. we need to be fair in distribution. Here we have an important criterion, i.e. what we mean by *fairness*. We must admit that *fairness* does not simply mean that everybody shares the same amount of resources and tasks.

When we are thinking of collaboration, we need to accomplish the task designated to the team of people involved in the collaboration. Accomplishing the task is the top goal of the team. If the team accomplishes the task, every participant in the team will definitely obtain benefits. If the team fails, each member of the team will lose. If the team fails, fairness has no meaning for participants in the collaboration. Now, we need to shift our concern to collectivism.

Collectivism indicates a phenomenon in which collaboration is powered by the interest or benefit of the whole team. We believe that collectivism does not destroy all individuals' interests and benefits. Collectivism is to maximize the team's interest and benefit. When conducting collaboration, individual interests and benefits are second to the team's interest and benefit. If an individual's interest is in conflict with the team's, the individual should follow the regulation of the team to maximize the team's benefit. Such a choice may seem unfair to the individual in a short period. However, maximizing the team's benefit can make each individual feel fair for a long period due to adaptation, i.e. in the series of activities of collaboration, these individuals can be offered better shares by playing roles. An extreme solution to alleviate such a bad feeling is to allow the individuals to leave the team if the individuals feel that they are unfairly treated.

According to collectivism, one statement is "the group or society is the basic unit of moral concern and the individual is of value only insofar as they serve the group." That is why collectivism is not welcome in the western world in general (Biddle 2012).

Note that, in this book, we are discussing *collaboration*. We should advocate collectivism, at least during the period of collaboration. We definitely emphasize that the team is the only entity we care about during collaboration. Successful collaboration is for the team but not for individual participants as a starting point. However, individuals definitely gain from the successful collaboration by joining the team. For example, in the National Basketball Association (NBA) season, a basketball player will receive a significant remuneration after the team wins the final game, even though the initial goal of the team is to win a set of games without considering the individual players' interests.

Therefore, we encourage collectivism first and then consider individualism second. That is to say, in developing collaboration models, methodologies, algorithms, and structures, we are aiming at advocating collectivism first, and at the same time, also encouraging the investigation of those policies by considering individualism if possible (Zhu 2020).

Collaboration is a result of decision making. To determine whether we conduct collaboration may take different viewpoints:

From the viewpoint of individualism, we can use a simple formula (1.1) (Grasse and Ward 2016) to judge whether participant i chooses to join collaboration:

$$B_i \times P_s > C_i, \tag{1.1}$$

where B_i is the expected benefits in dollars for participant i, P_s is the success probability of collaboration, C_i is the cost in dollars paid by participant i, i.e. $C_i = T_i \times H_i$, and T_i is the time used by participant i and H_i is the hourly rate of participant i.

However, from the viewpoint of collectivism, we need to use formula (1.2) to judge whether to conduct collaboration:

$$B_t \times P_s > \sum_{i=0}^{m-1} C_i \tag{1.2}$$

where B_t is the expected benefit in dollars for the team.

For example, John, a software developer, expects to obtain \$100 000 (B_i) when the software development project successfully delivers and the probability of success (P_s) is 80%. If he needs to pay a total time (T_i) of 1000 hours in the project including meetings, coding, testing, and maintenance, and his current hourly rate (H_i) is \$60, he can easily decide to join the project because $C_i = \$60\,000$, and $B_i \times P_s = \$80\,000$. If he needs to pay 1400 hours, he will not. This is a decision decided by individualism.

On the other hand, Ann, the Chief Executive Officer (CEO) of a consulting company, expects to obtain the benefits of \$1M to sign a new contract. If she needs 10 employees in her company to work 1000 hours each, the average hourly rate of the employees is \$60, and the success probability is 80%. She would like to sign the contract. If she needs 15 employees, she would not. This is a decision decided by collectivism.

We believe that collaboration brings in benefits that will be shared by individual participants in the collaboration. Therefore, successful collaboration is a common goal for all participants. It is true that in some situations, maximizing the benefit of collaboration from the viewpoint of collectivism may be detrimental to the benefit of some individuals. However, we also believe that sharing the benefits obtained by successful collaboration may significantly alleviate an individual's losses caused by collectivism.

1.7 Collaboration and Complex Systems

1.7.1 What Are Complex Systems?

A complex system consists of numerous elements, which can function independently and may interact with each other. It is also defined as a system that cannot be easily decomposed into simpler subsystems. Complex systems have behaviors that are difficult to model due to the dependencies, competitions, relationships, or other types of interactions between their parts or between a given system and its environment. A system is complex if it has distinct properties that arise from the relationships among the components of the system. The properties of a complex system normally include nonlinearity, emergence, spontaneous order, adaptation, and feedback loops.

A complex system involves collaboration, at least at the level of distributed systems, i.e. the collaboration, cooperation, coordination, and interactions among system components.

The complexity of a system is the degree of difficulty in predicting the properties of the system, given the properties of the system's components. To understand the complexity of a system, we need to build an intuition of how components of the system interact and how these interactions affect the overall system behaviors, responses, and outcomes (Czerwinski et al. 2000).

A complex system should possess the following typical characteristics (Bar-Yam 2002; Cilliers 1998):

1) *A large number of components*: Complex systems are composed of a large number of components. When the number is relatively small, the behavior of the elements can often be given a formal description in conventional terms. However, when the number becomes sufficiently large, conventional means (e.g. a system of differential equations) not only become impractical, they also cease to assist in any understanding of the system.
2) *Required connections*: All the components are connected in various ways. A connection is required if the system will never work or cannot work well when we remove the connection. For example, the human body has many interconnected components. If a component of the human body is disconnected, the body will often become injured and will not work well. In order

to constitute a complex system, the elements have to interact and this interaction must be dynamic. A complex system changes with time. The interactions do not have to be physical; they can be nonphysical, like the transference of information. Due to the presence of the required connections, there is a paradox *to divide a complex system into independent components*, i.e. if a system is complex, we cannot divide the system; if we can divide a system into independent components, then the system is not complex. This paradox confirms that complex systems are indeed complex.

3) *Collaboration:* After connecting with each other, components collaborate to provide the required functions of a complex system. Such collaboration can occur in different phases, such as communication, interaction, team formation, coordination, and cooperation. The aim of communication is sharing. Communication is the prerequisite of interaction. The objective of interactions is to influence each other. In a complex system, the communications and interactions are *spontaneous* and have a short range, e.g. email exchange, phone calls, or chat groups. Communications and interactions also have *recurrences* due to requirements. Team formation is required when components need to work together to function. To work as a team, participants must conduct coordination and cooperation.

4) *The locality of components*: Each component is not aware of the whole system behavior and it can only respond to the locally available information. If each element knew what was happening to the system as a whole, all of the complexity would have to be present in that element. This would either entail a physical impossibility, in the sense that a single element does not have the necessary capacity, or constitute a metaphysical move to the viewpoint that the consciousness of the whole is contained in one particular unit.

5) *Dynamics*: Dynamics means openness, development, and changing. Complex systems are usually open systems, which interact with other systems. In other words, it is not easy to define the boundary of a complex system. Complex systems are related to time. They have different states at different times. They evolve over time. The past states may affect the present and future behavior of the system. Changing is the only thing that is determined in complex systems. All the above characteristics will change with the development of the system along its timeline.

Please note that it is hard to divide a complex system into components. Such a situation seems in conflict with the first property in the above list. However, it is actually the two extremes of a complex system. A Chinese idiom saying "物极必反" read as "Wùjíbìfǎn," which means "things will develop in the opposite direction when they become extreme," clearly unveils this fact. Specifically, a complex system may have too many components for us to grasp and master them by classifying them into a limited number of components. Therefore, dealing with such a system as a whole might be an effective way.

1.7.2 Examples of Complex Systems

Now we discuss some examples of complex systems, which are on the cutting edge of research fields and could be modeled and handled by the proposed methodologies and models in this book.

1) The human body (Rettner 2016) is a typical complex system composed of subsystems, i.e. the circulatory, the digestive, the endocrine, the immune, the lymphatic, the nervous, the muscular, the reproductive, the skeletal, the respiratory, the urinary, and the integumentary systems; and organs, i.e. brain, heart, kidneys, liver, and lungs. These systems and organs are formed by the connections and collaboration of cells. The connections between these systems and organs are required. The major methodology of Western medicine is to "divide and conquer" and that of Eastern medicine is to consider the human body as a whole.

2) The Internet of Things (IoT) is defined as a phenomenon/system that every component in the context/system is connected with each other by a high-speed network. Such connections bring new and unforeseen benefits, advantages, and challenges. Everything such as cellphones, coffee makers, washing machines, headphones, lamps, and wearable devices can be connected with each other through an on-and-off switch to the Internet. In IoT, connections are obviously required, either wired or wireless.

3) A Cyber-Physical System (CPS) is defined as an integration of computation, networking, and physical processes. Embedded computers and networks can monitor and control physical processes, with feedback loops where the physical processes affect computations and vice versa (Lee 2008). The challenges of CPSs include: an appropriate level of abstraction, concurrency, interfacing, synchronization, and real-time control, where abstraction in a CPS needs to provide the right models for components and make the CPS easily analyzable, designable, implementable, and maintainable. Concurrency emphasizes that many components are running along their own threads; interfacing means that properly designed interfaces significantly affect the quality of a CPS; synchronization means that components in the CPS need to be synchronized in time to accomplish a designated task; and real-time control offers instant, stable, and anticipated behaviors of corresponding components.

4) A System of Systems (SoS) (Boardman and Sauser 2006; Maier 1998) is a collection of systems that share their resources and capabilities among each other to form a new, more complex system that offers more functionality and performance than simply the sum of the constituent systems. At present, an SoS is a promising research discipline that lacks models, frameworks, formal processes, quantitative analyses, effective tools, and well-accepted design methods.

An SoS presents the characteristics of autonomy, belongings, connectivity, diversity, and emergence in different forms other than those in a system of subsystems. It is different from a system of subsystems in that the components of an SoS are systems themselves but the components of the latter are just subsystems, which cannot exist independently and may lose one or more of the properties of a system. Therefore, a team of people is an SoS but a person is not because the components of a person cannot be an independent system itself, e.g. the head of a person cannot be an independent system. Robot teams are SoSs because a component or robot is an independent system. A software development team is an SoS because it is composed of components that can be independent systems, such as people, projects, products, and processes (Pressman 2014).

5) A social system is composed of numerous people. People interact with various relationships. The behaviors of social systems are dynamic and adaptive. Conventional ways of dealing with social systems are mainly from humanities, i.e. informal ways, because many problems in social systems are hard to formalize. A social system possesses all the properties of a complex system. The complexity of social systems comes mainly from the aspects of collaboration, i.e. sharing, connections, communications, interactions, coordination, and cooperation. Computational social systems (Wang 2014) are promising to develop better ways to deal with the problems of social systems.

We can continue this list, but it may deviate from our main topic on collaboration. The major idea we are pointing out is that complex systems definitely include collaboration. We believe that the models, methodology, and algorithms discussed in this book can be applied to all the areas discussed above, but we will need to use more effort and time to investigate and develop this.

1.8 Collaboration and Problem Solving

As we stated at the beginning of this chapter, collaboration is a complex problem. However, it is not only a problem but also a way of problem solving (Zhu 2009). Following the idea of object orientation, we may define the related concepts in problem solving. To differentiate objects after the concept of class is introduced, we need to define an *ordinary object*.

Definition 1.17 Everything in the world is an *ordinary object*.

We use $\bar{o}$ to express an ordinary object and $\overline{O}$ to express the set of all the ordinary objects in the world. In this chapter, objects are ordinary objects in the observable scope.

Definition 1.18   A problem $\bar{p}$ is defined as a tuple, $\bar{p}:: = <\bar{v}, \bar{a}>$, where $\bar{v}$ is an operation object, and $\bar{a}$ is an object called argument, $\bar{a}$ may be null. It can be also expressed as $\bar{v}(\bar{a})$.

We use $\bar{P}$ to express the set of all the problems. In a natural language, $\bar{v}$ is a verb and $\bar{a}$ is a noun. For example, "do(homework)" is a problem.

Definition 1.19   A *problem* $\bar{p}$ is complex if it has not yet been formalized or if the solution to it has exponential complexity.

Definition 1.20   A solution $\bar{s}$ is a way to solve a problem $\bar{p}$ or a concrete result of the way to solve $\bar{p}$. The symbol $\bar{s}$ denotes a specific solution and $\bar{S}$ denotes the set of all the solutions.

Definition 1.21   *Problem solving* for a problem $\bar{p} \in \bar{P}$ can be defined as an action to find a map $<\bar{p}, \bar{s}>$, where $\bar{s} \in \bar{S}$.

In Definition 1.18, a problem $\bar{p}$ may be very simple or complex. For a specific problem $\bar{p}$, its map $<\bar{p}, \bar{s}>$ may or may not exist.

Definition 1.22   *Collaborative problem solving* (Hsieh and O'Neil 2002) for a problem $\bar{p} \in \bar{P}$ can be defined as the activity of a group of people or intelligent entity to find a map $<\bar{p}, \bar{s}>$, where $\bar{s} \in \bar{S}$.

Collaborative problem solving is one area for this book to contribute to.

Definition 1.23   A *problem solver* for a problem $\bar{p} \in \bar{P}$ can be defined as an active entity that conducts the action specified by $\bar{p}.\bar{v}$.

A problem solver can be a machine, a computer, a robot, a person, a team of people, or an organization.

Definition 1.24   A problem $\bar{p}$ is *solvable* if there is one solution $\bar{s}$ making $<\bar{p}, \bar{s}>$ exist.

The **properties** of problems and problem solving are as follows:

1) One problem may or may not have a solution.
2) One problem may have many (>1) solutions.
3) A solution may be concrete or abstract.
4) The set of solutions $\bar{S}$ is expanding.
5) There are many forms of solutions.
6) A problem may be a compound problem, i.e. problem $\bar{p_1}$ is composed of n sub-problems $(\bar{p_{11}}, \bar{p_{12}}, ..., \bar{p_{1n}})$. Note that we avoid using "complex," because a compound problem may not be a complex problem and an undividable problem may be a complex problem.

A problem may have no solution for the time being but may have a solution at a later time. For example, the problem "landOnByACar (theMoon)" has no current solution. "The homework done" and "learn, understand and do" are solutions to the problem "do(homework)"; "the homework done" is a concrete solution but "learn, understand and answer" is an abstract one; the problem "callOnAPlane (aFriend)" had no solution in the 1960s but has one now; mathematical problems have solutions in different forms, such as numbers, formulas, equations, tables, and graphs; health care problems also have solutions in different forms, such as pills, surgeries, and transplants; and software problems have solutions in the form of software systems.

From the above definitions, it is evident that "*collaboration*" is both a solution and a problem. It is a solution to a problem $\overline{p_s}$ that cannot be accomplished by a single person. It is also a problem $\overline{p_c}$ for the team of people who try to solve the previous problem $\overline{p_s}$. What we will discuss in this book is one specific solution, Role-Based Collaboration (RBC).

How to collaborate, or simply collaboration, is a complex problem, because it has not yet been formalized completely. Even though we divide collaboration into subproblems, these subproblems are still complex due to nonexistent formalization, or the high complexity of the formalized problem.

By reading this book, one may find that collaboration is an effective way to solve complex problems.

1.9 Summary

To understand the concept of collaboration as discussed in this book, we may need to take different standing points.

Collaboration can be a task for an administrator who will manage a group of people to work, i.e. collaborate. From the viewpoint of an administrator, collaboration concentrates on collectivism while considering individualism. The aim of this book is to provide a set of theories, models, methodologies, algorithms, and computer-based tools to support a group of components conducting teamwork.

Collaboration can also be viewed from an individual participating in collaboration, i.e. we may play a role as a participant. The models and formalizations in this book do not oppose this viewpoint, but rather encourage more research and investigations from this viewpoint. As a matter of fact, we investigated "how one could obtain a preferred position in a group from one agent's viewpoint," i.e. individualism (Zhu et al. 2018).

The fundamental goal of this book is to make collaboration easy. After we introduce the model and the methodology, we will be able to clarify the terminologies around collaboration from a more systematic and technical point of view. "Collaboration made easy" is the claim of this book.

Concepts are the first step of scientific research. They provide a basis in the formation of new models. Problem solving involves many fundamental concepts and mechanisms such as induction, deduction, abstraction, classification, and decomposition. Abstraction is the first thinking methodology for people to use when attempting to compose a new concept. It is necessary to define fundamental and abstract concepts and understand the basic properties of problem solving.

A problem can also be defined as a tuple of an operation and a real-world object. Such a definition may provide a guide on how to solve a problem, i.e. taking action to do the operation. Hence, collaboration is an anticipated action for solving a complicated problem.

In the following chapters, we mainly use agents as the participants of collaboration. We may also use people to emphasize natural collaboration or members to emphasize the relationships of parts (members) and the whole (the group).

It should be noted that readers may find that the following chapters try to resolve the issues discussed in this chapter. However, due to the complexity of collaboration, the author admits that there are still many related problems that have not been solved satisfactorily. The author welcomes interested readers to join the collaboration system research community to investigate more deeply and broadly to promote collaboration research.

References

Adler, P.S. and Kwon, S.-W. (2002). Social capital: prospects for a new concept. *The Academy of Management Review* 27 (1): 17–40.

Alonso, E. (1999). An individualistic approach to social action in multi-agent systems. *Journal of Experimental & Theoretical Artificial Intelligence* 11 (4): 519–530.

Bar-Yam, Y. (2002). General features of complex systems, *UNESCO - Encyclopedia of Life Support Systems*. http://www.eolss.net/sample-chapters/c15/E1-29-01-00.pdf (accessed 10 August 2020).

Biddle, C. (2012). Individualism vs. collectivism: our future, our choice, *The Objective Standard*, Spring 2012. https://www.theobjectivestandard.com/issues/2012-spring/individualism-collectivism/ (accessed 10 August 2020).

Boardman, J. and Sauser, B. (2006). System of systems – the meaning of of. *Proceedings of the IEEE/SMC International Conference on System of Systems Engineering*, Los Angeles, CA, USA (April 2006), pp. 118–123.

Böhmea, H.-J., Wilhelma, T., Keya, J. 1 et al. (2003). An approach to multi-modal human–machine interaction for intelligent service robots. *Robotics and Autonomous Systems* 44 (1): 83–96.

Brualdi, R.A. (2008). *Introductory Combinatorics*, 5e. Upper Saddle River, NJ: Prentice Hall.

Budd, T.A. (2001). *An Introduction to Object-Oriented Programming*, 3e. Boston, MA: Addison-Wesley Longman Publishing Co., Inc.

Buppert, C. (2010). The pros and cons of mandated collaboration. *The Journal for Nurse Practitioners* 6 (3): 175–176.

Burkard, R.E., Dell'Amico, M., and Martello, S. (2009). *Assignment Problems, Revised Reprint*. Philadelphia, PA: Siam.

Cabri, G., Zhu, H., and Yang, J.B. (2006). Guest editorial special issue on collaboration support systems. *IEEE Transactions on Systems, Man, and Cybernetics - Part A: Systems and Humans* 36 (6): 1042–1043.

Cilliers, P. (1998). *Complexity and Postmodernism: Understanding Complex Systems*. New York, NY: Routledge.

Coulouris, G.F., Dollimore, J., and Kindberg, T. (2011). *Distributed Systems: Concepts and Design*, 5e. London, UK: Pearson.

Czerwinski M., Cutrell E., and Horvitz E. (2000). Instant messaging and interruption: influence of task type on performance. *Proceedings of OZCHI*, Brisbane, Australia (22–26 November 2010), pp. 356–361.

Dasgupta, P. (1988). Trust as a commodity. In: *Trust: Making and Breaking Cooperative Relations* (ed. D. Gambetta), 49–72. Blackwell.

Day, D.V., Fleenor, J.W., Atwater, L.E. et al. (2014). Advances in leader and leadership development: a review of 25 years of research and theory. *The Leadership Quarterly* 25: 63–82.

Dix, A., Dix, A.J., Finlay, J. et al. (2003). *Human-Computer Interaction*. Upper Saddle River, NJ: Pearson/Prentice-Hall.

Fischer, G. (2001). User modeling in human–computer interaction. *User Modeling and User-Adapted Interaction* 11: 65–86.

Fowler, M., Kobryn, C., and Scott, K. (2003). *UML Distilled: A Brief Guide to the Standard Object Modeling Language*. Hoboken, NJ: Addison-Wesley.

Gambetta, D. (1988). Can we trust trust? In: *Trust: Making and Breaking Cooperative Relations* (ed. D. Gambetta), 213–237. Basil, UK: University of Oxford.

Goleman, D. (1998). What makes a leader? *Harvard Business Review* 76: 93–102.

Grasse, N.J. and Ward, K.D. (2016). Applying cooperative biological theory to nonprofit collaboration. In: *Advancing Collaboration Theory: Models. Typologies, and Evidence* (eds. J.C. Morris and K. Miller-Stevens), 89–115. New York, USA: Routledge.

Gray, B. (1985). Conditions facilitating interorganizational collaboration. *Human Relations* 38 (10): 911–936.

Gray, B. (1989). *Collaborating: Finding Common Ground for Multiparty Problems*. San Francisco, CA: Jossey-Bass.

Grudin, J. (1994). Computer-supported cooperative work: history and focus. *IEEE Computer* 27 (5): 19–26.

Hou, M., Banbury, S., and Burns, C. (2014). *Intelligent Adaptive Systems: An Interaction-Centered Design Perspective*. Boca Raton, FL: CRC Press.

Hsieh, I.-L.G. and O'Neil, H.F. Jr. (2002). Types of feedback in a computer-based collaborative problem-solving group task. *Computers in Human Behavior* 18 (6): 699–715.

Lee, E.A. (2008). Cyber physical systems: design challenges. *Proceedings of the 11th Int'l Symposium on Object-Oriented Real-Time Distributed Computing*, Orlando, FL, USA (5–7 May 2008), pp. 363–369.

Lin, N. (2011). *Social Capital: A Theory of Social Structure and Action*. Cambridge, England: Cambridge University Press.

Luhmann, N. (2017). *Trust and Power*. Konstanz, Germany: Wiley.

Maier, M.W. (1998). Architecting principles for system of systems. *Systems Engineering* 1 (4): 267–284.

Mayer, M. and Kenter, R. (2016). The prevailing elements of public-sector collaboration. In: *Advancing Collaboration Theory: Models. Typologies, and Evidence* (eds. J.C. Morris and K. Miller-Stevens), 43–64. New York, USA: Routledge.

McNamara, M.W. (2016). Unravelling the characteristics of mandated collaboration. In: *Advancing Collaboration Theory: Models. Typologies, and Evidence* (eds. J.C. Morris and K. Miller-Stevens), 65–86. New York, USA: Routledge.

Miller-Stevens, K., Henley, T., and Diaz-Kope, L. (2016). A new model of collaborative federalism from a governance perspective. In: *Advancing Collaboration Theory: Models. Typologies, and Evidence* (eds. J.C. Morris and K. Miller-Stevens), 148–174. New York, USA: Routledge.

Morris, J. and Miller-Stevens, K. (2016). The state of knowledge in collaboration. In: *Advancing Collaboration Theory: Models. Typologies, and Evidence* (eds. J.C. Morris and K. Miller-Stevens), 3–13. New York, USA: Routledge.

Nwan, H.S., Lee, L., and Jennings, N.R. (1996). Coordination in software agent systems. *BT Technology Journal* 14 (4): 79–89.

Osborne, M.J. and Rubinstein, A. (1994). *A Course in Game Theory*. Cambridge, MA: The MIT Press.

O'Sullivan, D. and Haklay, M. (2000). Agent-based models and individualism: is the world agent-based? *Environment and Planning A: Economy and Space* 32 (8): 1409–1425.

Poutanen, P. (2016). Complexity and collaboration in creative group work. Academic Dissertation. Unigrafia, Helsinki.

Preece, J., Rogers, Y., and Sharp, H.C. (1994). *Human-Computer Interaction*. Essex, UK: Addison-Wesley Longman Ltd.

Pressman, R.S. (2014). *Software Engineering: A Practitioner's Approach*, 8e. New York, NY: McGraw-Hill Education.

Ramchurn, S.D., Huynh, D., and Jennings, N.R. (2004). Trust in multi-agent systems. *The Knowledge Engineering Review* 19 (1): 1–25.

Ramos-Murguialday, A., Broetz, D., Rea, M. et al. (2013). Brain–machine interface in chronic stroke rehabilitation: a controlled study. *Annals of Neurology* 74 (1): 100–108.

Rekimoto, J. and Nagao, K. (1995). The world through the computer: computer augmented interaction with real world environments. *Proceedings of the 8th annual ACM symposium on User Interface and Software Technology*, Pittsburgh, Pennsylvania, USA (15 November 1995), pp. 29–36.

Rekleitis, I., Dudek, G., and Milios, E. (2001). Multi-robot collaboration for robust exploration. *Annals of Mathematics and Artificial Intelligence* 31: 7–40.

Rettner, R. (2016). The human body: anatomy, facts and functions, *Live Science*. https:// www.livescience.com/37009-human-body.html (accessed 10 August 2020).

Tanenbaum, A.S. and van Steen, M. (2016). *Distributed Systems: Principles and Paradigms*. Scotts Valley, CA: CreateSpace Independent Publishing Platform.

Thomson, A.M. (2001). Collaboration: meaning and measurement. PhD Dissertation, Indiana University.

Thomson, A.M. and Perry, J. (2006). Collaboration process: inside the black box. *Public Administration Review* 66: 20–32.

UOPX News (2013). University of Phoenix survey reveals nearly seven-in-ten workers have been part of dysfunctional teams. https://www.phoenix.edu/news/releases/ 2013/01/university-of-phoenix-survey-reveals-nearly-seven-in-ten-workers-have- been-part-of-dysfunctional-teams.html (accessed 10 August 2020).

Vella, K., Butler, W.H., Sipe, N. et al. (2016). Voluntary collaboration for adaptive governance: the southeast Florida regional climate change compact. *Journal of Planning Education and Research* 36 (3): 363–376.

Wang, F.Y. (2014). Computational social systems in a new period: a fast transition into the third axial age. *IEEE Transactions on Computational Social Systems* 4 (3): 53–54.

Wondolleck, J.M. and Yaffee, S.L. (2000). *Making Collaboration Work: Lessons from Innovation in Natural Resource Management*. Washington, DC: Island Press.

Zhu, H. (2008). Fundamental issues in the design of a role engine. *Proceedings of The 9th Int'l Symposium on Collaborative Technologies and Systems (CTS 2008)*, Irvine, CA, USA (19–23 May 2008), pp. 399–407.

Zhu, H. (2009). Granular problem solving and its applications in software engineering. *Internation Journal of Granular Computing Rough Sets and Intelligent Systems* 1 (2): 150–163.

Zhu, H. (2020). Computational social simulation with E-CARGO: comparison between collectivism and individualism. *IEEE Transactions on Computational Social Systems* 7 (6): 1345–1357.

Zhu, H. and Hou, M. (2011). Role-based human-computer interaction. *International Journal of Cognitive Informatics and Natural Intelligence* 5 (2): 37–57.

Zhu, H. and Zhou, M.C. (2006a). *Object-Oriented Programming with C++: Aproject-Based Approach*. Beijing, China: Tsinghua University Press.

Zhu, H. and Zhou, M.C. (2006b). Role-based collaboration and its kernel mechanisms. *IEEE Transactions on Systems, Man and Cybernetics, Part C* 36 (4): 578–589.

Zhu, H., Ma, H., and Zhang, H. (2018). Acquire the preferred position in a team. *The IEEE Conference of Computer-Supported Cooperative Work in Design*, Nanjing, China (9–11 May 2018), pp. 116–121.

Exercises

1 Try to differentiate the following terminologies: collaboration, cooperation, coordination, communication, connection, interaction, interface, integration, sharing, and team.

2 Why do we conduct research in collaboration?

3 Try to differentiate collaborative systems from collaboration systems.

4 Explain why "collaboration" is a polymorphic word. Hint: You may consider the term "collaboration" as a problem, a system, a complex system, a problem-solving method, or a "divide and conquer" method.

5 Try to describe the differences and connections between the following terminologies: abstraction, model, information hiding, encapsulation, modeling, and classification.

6 What are the key components of collaboration?

7 What is the nature of collaboration?

8 Do you prefer collectivism or individualism in collaboration? Why?

9 Why is collaboration complex?

10 Give examples of complex systems and their connections to collaboration.

11 Explain why collaboration is both a problem and a solution to a problem.

2

Role Concepts

2.1 Terminology

Roles are important in our social lives. They have been applied in China to help people manage and understand societies since more than 2000 years ago (Zhu and Zhou 2006). As a common word, "role" is easily understood and everybody can give examples to describe what a role is. However, as a concept, there is no generally accepted definition (Partsakoulakis and Vouros 2004; Steimann 2000b). This is also a conclusion we have reached after mining the relevant literature on roles.

Role concepts have been applied widely in management, sociology, and psychology for many years (Ashforth 2001; Biddle and Thomas 1966; Botha and Eloff 2001; Hawkins et al. 1983; Hellriegel et al. 1983; Zigurs and Kozar 1994). Roles are very useful for modeling authority, responsibility, functions, and interactions associated with manager positions within organizations. Roles are given attention in a variety of areas relevant to information systems such as modeling (Albano et al. 1993; Bachman and Daya 1977; Bäumer et al. 2000; Genilloud and Wegmann 2000; Gottlob et al. 1996; Gregg and Walczak 2000; Greenberg 1991; Kendall 1999; Kristensen 1995; Kristensen and Østerbye 1996; Kuncak et al. 2002; Murdoch and McDermid 2000; Patterson 1991; Pernici 1990; Reenskaug et al. 1995; Reimer 1985; Riehle and Gross 1998; Riehle et al. 2000; Role Modellers Ltd 2004; Schrefl and Thalhammer 2004), software design (Dafoulas and Macaulay 2001; Holt et al. 1983; Kendall 1999; Murdoch and McDermid 2000; Reenskaug et al. 1995; Van-Hilst and Notkin 1996; Zhu and Zhou 2006), access control (Ahn and Sandhu 2000; Ahn et al. 2003; Bertino et al. 2001; Covington et al. 2000; Ferraiolo and Kuhn 1992; Ferraiolo et al. 1995; Ferraiolo et al. 2001; Joshi et al. 2005; Nyanchama and Osborn 1999; Park et al. 2001; Sandhu et al. 1996, 1999), system administration (Lupu and Sloman 1997; Oh and Sandhu 2002; Sandhu et al. 1999), agent systems (Becht et al. 1999; Cabri et al. 2004, 2005, 2005a; Depke et al. 2001; Ferber et al. 2004; Odell 2002; Odell et al. 2003, 2005; Partsakoulakis and Vouros 2004; Stone

E-CARGO and Role-Based Collaboration: Modeling and Solving Problems in the Complex World, First Edition. Haibin Zhu.
© 2022 by The Institute of Electrical and Electronics Engineers, Inc.
Published 2022 by John Wiley & Sons, Inc.

and Veloso 1999), database systems (Chigrik 2001; Oracle 2002; Reimer 1985), and computer-supported cooperative work (CSCW) systems (Bandinelli et al. 1996; Edwards 1996; Greenberg 1991; Guzdial et al. 2000; Leland et al. 1988; Turoff and Hiltz 1982; Ye et al. 2001; Zhu 2006; Zhu and Zhou 2006).

Why are roles widely applied in information systems? It is because such systems are tools for helping people perform their tasks better and more efficiently. In daily life, working together in harmony requires that all the people fulfill their obligations and respect everybody else's rights. In other words, people are required to respect the social laws of their community (Ashforth 2001). Building a system that can be trusted and used by participants in organizations requires the same rule. Within organizations, individuals fill specific positions or roles. This position or role represents a specific "seat" that has certain privileges and responsibilities (Biddle and Thomas 1966).

When roles are not clearly specified, the issue of role ambiguity befalls an organization, resulting in conflicting expectations and demands. Role conflicts can occur due to individuals' internal standards, required job (or role) behaviors, capabilities of individuals, and expectations about the roles (1983). Role adjustment is required to help workers cope with ambiguity and conflict in their roles. Based on their experience with a speech-act-based office model, Auramäki et al. (1988) state: "All tasks cannot be predefined and roles evolve. Sometimes the actual role performance depends on a role holder's skills and capabilities. It can be difficult to assign responsibilities to the roles without a detailed understanding of the role players' skills and capacity."

The actual situation of role applications in information systems is definitely in chaos. There are too many papers to cite, which offer no clear statement of what roles are. We encounter many difficulties in classifying the relationships of role concepts and mechanisms in different areas of applications. By "role mechanisms," we emphasize that roles can support the life cycle of a system including its execution.

Many early information system papers using roles do not refer to the research on roles in social psychology, in which role theory has been investigated systematically. Roles seem to be phantoms. They are everywhere and every explanation seems reasonable, but it is difficult to grasp, identify, and completely specify them. Hence, there is still a need to clarify role concepts in order to support the design of better information systems. From experience, we have learned that many good managerial methodologies are well-adapted in developing computer-based systems, and many concepts in people's lives have been successfully introduced into these systems. We believe that roles should be such a concept as well.

All individuals should have clear positions within a group and their roles should be related but not interfere with each other. To form a successful organization, we need to have the members in the organization play roles. With roles, collaboration is made easier and more fruitful. Roles are fundamental tools to support people's activities. A role can also be applied as an abstraction tool when describing concepts at a position between a class and an object, i.e. separation of concerns (Kendall 1999; Pernici 1990).

In collaboration, people interact with other participants and their environment. To collaborate with other people or to be cooperative, they must know what they can do and what they are willing to do. To work efficiently, they must know what objects in their environment they can access. When collaborating, working efficiently and being cooperative involve playing roles. Therefore, in general, a person has two kinds of existence: one as a service provider and the other as a client. When people play roles, they provide specific services and possess certain rights to request services. With this in mind, a role can be considered a way of viewing people in a collaborative environment. Their role in a collaborative environment is a wrapper with a service interface and a request interface shown in Figure 2.1. Therefore, we can separate a role into two parts:

1) The service interface, including incoming messages; and
2) The request interface, including outgoing messages.

In collaboration without role mechanisms, people cooperate with other participants by sending messages to each other and to objects in the system. People may ask the system for services, retrieve the states of objects, and send messages to objects. In such a scenario, the system and its objects are the collaboration media of the participants. In a system with role mechanisms, participants play roles when they interact with the collaborative system. Roles include two major components: rights and responsibilities.

Our motivation to investigate the role concept and its applications is to obtain deeper insights into how and why roles are utilized. We believe that a common understanding of the terminology of the role is necessary, even though different contributors may interpret the term differently. Consequently, we can better apply

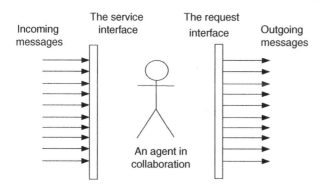

Figure 2.1 A role is a wrapper of an agent.

roles as an underlying mechanism to system analysis, design, and construction. The role concept and mechanism should be understood in a complete and systematic way to avoid the parable of the "blind men and an elephant." In this chapter, we emphasize that roles can be utilized as an underlying mechanism in system design and implementation. We hope to help system designers and developers understand and apply roles to their information systems.

We admit that even though we have cited many references, we must ignore many that mention roles, partially because they use "role" as a common word but not as a specified concept or mechanism. However, we believe that the literature we cite still conveys the common and basic ideas and applications of roles. To clarify different role concepts, we classify them into five categories: *object modeling*, *agent systems*, *Role-Based Access Control (RBAC)*, *CSCW*, and *social psychology and management*.

This chapter uses the following terminologies: *Data* are plain digital bits or bytes stored in a system in different formats or *data structures*. *Objects* are composed of data and *operations* that access the data. *Classes* are used to describe objects' common operations and data structures. *Objects* are also instances of their classes. Objects are passive and normally accessed by active entities. *Agents* are active entities that possess intelligent properties, such as autonomy, social ability, reactivity, and proactiveness (Wooldridge and Jennings 1995). *Users* are *human beings* who use information systems. *A group* is a set of *entities* such as *objects*, *agents*, or *human beings*. *Interfaces* are the methods of interaction between system *components* such as *objects*, *agents*, and *software or hardware components*. *User interfaces* are the methods of interaction between users and systems. A *system* is an assemblage of parts forming a complex or unitary whole that serves a useful purpose (Biddle and Thomas 1966). We emphasize that a system is considered a group of interdependent *entities* that interact regularly to perform a task. Here, *entities* can be expanded to express *objects*, *agents*, *roles*, *software or hardware components*, *software systems*, *hardware devices*, *human beings*, and *groups*.

After studying the contributions of roles in management, sociology, and psychology, we conceive that the idea of roles in social psychology covers what a role should be because an information system is a virtual community that simulates a real one. To understand and compare these different types of roles, we classify them based on the following aspects:

By the *reasons* why roles are introduced, we have:

- *Evolution*: roles express the states of objects at different stages of evolution.
- *Separation of concerns*: roles are entities that express different aspects of a role player in different contexts and at different times.
- *Interaction*: roles facilitate interactions between entities such as people, agents, or objects.
- *Classification*: roles classify entities to simplify their administration.

- *Authorization*: it is a special classification wherein roles are used to assign rights that allow entities to access system resources.

By *role players*, we have:

- *Operation*: the roles are played by a method of an object.
- *Object*: the roles are played by objects.
- *Agent*: the roles are played by agents.
- *Human:* the roles are played by humans.
- *Group:* the roles are played by groups of the above players.

By the *contents* roles contain, we have:

- *Requests*: Roles are entities that facilitate role players to access system resources (files, objects, and devices) or request services. Roles are tags or tickets for role players to access objects or request servers to serve. In this category, the *rights* of RBAC-roles (see Section 2.4) are evidently special forms of requests.
- *Services*: Roles are entities that facilitate role players to provide specific groups of services to the outside world. *Responsibilities* are, in fact, the interface part of services.

By whether roles have detailed *specifications,* we have:

- *Interface*: Roles are abstract entities that express the interfaces between role players. Roles only specify what their services (responsibilities) and requests (rights) are. How they are processed depends on the players. Here, we extend the *interface* to a generalized term that includes a *user interface*.
- *Process*: Roles are detailed descriptions in specifying the behaviors of their players. They specify not only what their services (responsibilities) and requests (rights) are but also how they are processed.

By the meaning of *role assignment*, we have:

- *Instantiation:* to create an instance based on a role, i.e. a role is a template to create a part of an object.
- *Allocation*: to attach a new entity to a player, i.e. a role is a different entity (with behaviors and structures) from what the player currently has.
- *Constraint*: to mandate constraints on a player, i.e. a role is a wrapper for the player to interact with outsiders.

By how roles are *related* to each other (*role relationship*), we have:

- *Conflict*: some roles cannot be played at the same time or in a period of time by one player.
- *Qualification*: to be qualified to play some roles, a player must have played some other roles first or possessed some required abilities.

- *Inheritance*: some roles may inherit from other roles in a class hierarchy.
- *Place*: one role may be located among many other roles to serve some and request the others.

2.2 Modeling-Roles

To model objects in a problem's domain, we require direct mapping and abstraction. Roles are a kind of abstraction in modeling. We call these models "role-based models" and the roles in them "modeling-roles."

A modeling-role is considered an abstraction and decomposition mechanism related to objects. When an object plays a role, it accepts messages and provides services related to its role. A role constitutes a part of an object's behavior that is obtained by considering only the interactions of that role and hiding all others. The following characteristics of roles are commonly acknowledged in the literature on modeling methodologies:

- Roles model a perspective on a phenomenon. Roles are a tool for conceptualization.
- Roles can be acquired and abandoned independently of each other.
- Roles can be organized in hierarchies, and generalized or specialized. An object's various roles may share a common structure and behavior.
- Roles are bound to an existing object and an object may play several roles at a time.
- Roles are used to emphasize how entities interact with each other.
- Roles can be dynamic and involve sequencing, evolution, and role transfer.
- The states and features of an object can be role-specific.

The following characteristics of roles are not commonly accepted:

- An object and its roles have different identities (Dahchour et al. 2004) or share an identity (Albano et al. 1993).
- Roles can (Kristensen 1995; Kristensen and Østerbye 1996) or cannot (Steimann 2000b) play other roles.
- Access to an object is (Demsky and Rinard 2002; Kuncak et al. 2002; Reimer 1985) or is not (Steimann 2000a) restricted by a role.
- A role has meaning only in a role model (Kendall 1999) or has meaning generally (Dafoulas and Macaulay 2001; Kristensen 1995; Kristensen and Østerbye 1996).
- Roles have (Kristensen 1995; Kristensen and Østerbye 1996) or do not have (Steimann 2000a) their own states and behaviors.

The comprehensive studies on roles as fundamental concepts and mechanisms (Brachman and Schmolze 1985; Fortino et al. 2004, 2010; Fortino and Russo 2012;

Gottlob et al. 1996; Guarino 1992; Kendall 1999; King and Sethi 1998; Kristensen 1995; Kristensen and Østerbye 1996; Kühn et al. 2015; Shah et al. 2007; Steimann 2000b) provide useful guidelines for role applications in object modeling.

Some authors take roles as ways of describing *behavior* (Pernici 1990); some as objects that possess both *data* and *operations* (Gottlob et al. 1996; Schrefl and Thalhammer 2004); some as *interfaces* (Steimann 2001); and some as *external properties* of objects (Kristensen 1995; Kristensen and Østerbye 1996). To better understand modeling-roles, we categorize the major reasons why roles are introduced as: (i) evolution of objects; (ii) a fundamental modeling concept; (iii) the interfaces between objects; and (iv) the separation of concerns.

To express the evolution of objects, roles are considered states and tags. The major concern is how to organize the roles in a well-defined structure and associate them with objects (Albano et al. 1993; Bachman and Daya 1977; Reimer 1985; Gottlob et al. 1996; Pernici 1990; Gottlob et al. 1996; Schrefl and Thalhammer 2004).

Albano et al. (1993) implemented an object-oriented programming language, called Fibonacci, with roles. Their role concept is similar to that of Bachman and Daya (1977) but concentrated on the evolution of data objects. Roles are used to express the different states of an object. An object may change its identification to express the different roles it plays. These roles are designed as classes. An object is copied in part to form a new instance with the new role class. Therefore, the new instance and the original instance are combined to form the state of the object. This emphasis that an object has a set of roles is similar to the idea of Pernici (1990). We can call them Object-Oriented Database-roles (OODB-roles).

Roles as interfaces only specify the responsibilities of the role players (or objects) and the details of how to accomplish tasks are entirely up to the role players (or objects). This can protect system analysts and designers from being concerned with the unnecessary details of objects (Genilloud and Wegmann 2000; Steimann 2001).

The collaborative behavior of a role represents the specific context in which it is defined, together with other roles in UML (2000a). All the actions in collaborative behavior belong to one or more of its roles. In object modeling, roles are taken as dependent concepts of objects. In other words, roles cannot exist without objects. Every object may collaborate with other objects and should play a role in collaboration.

Separation of concerns of objects is another aspect for roles to consider. This idea can help "divide and conquer" and provide a new way for analysts and designers to divide and distribute tasks (Pernici 1990; Kendall 1999).

In specification and design, roles are taken as concepts or tools used to specify processes and designs in mapping a problem domain to a solution domain (Bäumer et al. 2000; Holt et al. 1983; Kendall 1999; Kristensen 1995; Kristensen and Østerbye 1996; Murdoch and McDermid 2000; Ould 1995; Reenskaug et al. 1995; Role Modellers Ltd 2004; Steimann 2000a; VanHilst and Notkin 1996; Riehle and Gross 1998; Reenskaug et al. 1995; Zhao and Kendall 2000).

There are two major limitations in a modeling-role. One is that the role modelers stand among objects. When they model roles, they mainly consider the collaboration among objects. They do not consider the aspects of human users within the systems. Few contributions in modeling methodology inherit the good ideas from social-roles that will be discussed later. The other limitation is that roles are taken as a dependent concept of objects, i.e. objects are composed of roles. In fact, roles can be defined independently, similar to how classes can be defined before defining objects. Roles can be specified by system analysts and designers before objects are defined and created. Therefore, based on the traditional view of data modeling, it is difficult to obtain the whole view of a role in collaboration. However, when an object plays a specific role, it may have some special rights to access other objects. When we design a class of objects, we may consider another viewpoint wherein we need to know what services other classes of objects have provided. In modeling, roles should be identified and specified before designing classes. Roles are guidelines for modelers to design classes. It is therefore arguable for roles to have such properties: dependency, i.e. a role cannot exist without an object; identity, i.e. the role and the object may have the same identity, and locality, i.e. a role may only have a meaning in a role model (Kendall 1999).

2.3 Roles in Agent Systems

Different aspects of the role mechanisms are introduced into agent systems. The role concepts are generally used in agent systems to describe the collaboration among cooperative processes or agents. We call these roles agent-roles. Because agent concepts evolve from objects (Odell 2002), agent-roles inherit the properties of modeling-roles.

Becht et al. (1999) pioneer the concept of considering roles as interfaces and developed a multi-agent system ROPE. The cooperative processes are described with roles. An agent that wants to take part in the cooperation must fulfill the service requirement specified by a role. In their system, a role is formally defined as an entity consisting of a set of required permissions, a set of granted permissions, a directed graph of service invocations, and a state visible to the runtime environment but not to other agents.

Taking roles as specific processes is the main idea of Stone and Veloso (1999)'s soccer team formation. In this formation, a role consists of a specification of an agent's internal and external behavior. They use roles to obtain a flexible team agent structure and a method for inter-agent communication. First, the team agent structure allows agents to capture and reason about team agreements. They express collaboration between agents by formations that decompose the task space into a set of roles. In their solution, agents can flexibly switch roles within formations, and agents can change formations dynamically according to prede-fined triggers to be evaluated at run-time. This flexibility increases the perfor-mance of the overall team. Furthermore, their teamwork structure includes pre-planning for frequently occurring situations.

In agent-oriented software engineering, the Gaia methodology (Wooldridge et al. 2000; Zambonelli et al. 2003) is proposed to support system analysis and design by viewing multi-agent systems as organizations. Analysis and design are separated phases, i.e. analysis aims to develop an understanding of the system structures by role and interaction models while the design phase aims to define the actual details of the agents in the system. An important contribution of Gaia is that its roles accommodate rights or permissions by overcoming the weakness of roles in modeling-roles, responsibilities. In Gaia, roles are described with responsibil-ities, permissions, interaction protocols, and activities. The ways of roles being applied in agent-oriented software engineering are further developed to the ideas presented in the work of Zhu (2003), Zhu (2003a), Zhu (2006a), and Zhu and Zhou (2006a).

In Organization-Centered Multi-Agent Systems (OCMAS) (Ferber et al. 2004), roles are emphasized as an important element of organizations. A role is the abstract representation of a functional position of an agent in a group. Roles can help overcome the drawbacks of agent-centered multi-agent systems, where an agent may communicate with any other agents; an agent provides services available to every other agent; there are no constraints for agents to access each other; and agents interact with each other directly. The roles in OCMAS are similar to those in (Wooldridge et al. 2000). Group roles are mentioned in Ferber et al. (2004) to mean the roles in a group, i.e. an agent must play a role in a group. This incurs the argument of which entity should be based on the other: roles or groups (Zhu and Zhou 2006).

Cabri et al. (2004, 2005, 2005a) apply both aspects of agent-roles: processes and interfaces. A role imposes defined behavior to the objects playing it, thereby reflecting the process aspect. It also allows a set of capabilities, which can be used by agents to carry out their tasks, reflecting the interface aspect. This allows a role to serve as an abstract description of the functions an agent is responsible to fulfill in order to reach an assigned goal. From the agent-oriented approaches, roles are a proper method for refining agent-oriented models. Their work (Cabri et al. 2005a) considers rights as part of their agent-roles.

Based on the work of such psychologists as Biddle and Thomas (1966) and Sha-kespeare's role concepts in theaters, Odell (2002) and Odell et al. (2005) proposed a superstructure specification that defines the user-level constructs required to model agents, their roles, and their groups. Their modeling constructs provide the basic foundational elements required for multi-agent systems to foster dynamic group formation and operation. Roles specify normative behavioral repertoires for agents and provide the building blocks for social agents and the requirements by which the agents interact. Individual organizations can thus track and control their behaviors by applying these role concepts to facilitate dynamic, controlled, and task-oriented group formation. The idea that some roles are group roles (Odell et al. 2003) is similar to that of (Ferber et al. 2004), i.e. group roles are only meaningful in a group (Edwards 1996).

Partsakoulakis and Vouros (2004) review role concepts and mechanisms in multi-agent systems and view roles as tools to manage the complexity of tasks and environments. A high degree of interactions, environment dynamics, and distributivity is emphasized to show the importance of roles. They use five characteristics to analyze the past research on agent-roles: specification, dynamics of assignments, dynamics of roles, the cardinality of roles, and life-span of roles. From their analysis, most roles in this area are used both as (i) an intuitive concept in order to analyze multi-agent systems and model interagent social activity; and (ii) a formal structure used to implement coherent and robust teams.

In many contributions, both aspects of roles: interfaces and processes (Cabri et al. 2004, 2005, 2005a; Odell et al. 2003, 2005) have been discussed. *Interface roles* are used in system analysis and design when architecture and structures are more important, while process roles are used in system implementation when concrete jobs should be specified clearly and exactly.

Roles can be abstract as *interfaces* and concrete as *processes*. Roles have many aspects to consider in different usages. That is why roles can be applied to every stage of system development: analysis, design, and implementation.

However, most authors fail to clarify when and how to separate interface roles and process roles. This is an important specification in agent systems because this ambiguity may bring difficulties to system design and maintenance. Gaia (Wooldridge et al. 2000; Zambonelli et al. 2003) makes this clear that role models are composed at the analysis phase, while agent models are composed at the design phase. From the viewpoint of systems engineering, interface roles should be used in the system design to specify the relationships among system components while process roles should be used in system construction to implement components. This idea is also discussed in our previous work (Zhu 2006a, Zhu and Zhou 2006a).

In agent systems, the following consensus is reached (Becht et al. 1999; Cabri et al. 2004, 2005, 2005a; Odell et al. 2003, 2005; Partsakoulakis and Vouros 2004; Stone and Veloso 1999):

- A role can be taken as an interface.
- A role can be taken as a process.
- A role instance is deleted when an agent is destroyed, i.e. its lifetime depends on its agents.
- Roles are used to forming different interfaces for agents in order to restrict the visibility of features and handle permissions to access the internal state and role services of agents.

- Roles have three functions: comprise special behavior, form the behavior of an agent, and allocate a place for an agent in a group (or define the inter-relationships among agents).
- Roles can be used to express the organizational structure of a multi-agent system.
- Roles can be used to specify interactions in a generic way.
- Roles can be used as agent-building blocks in (Unified Modeling Language) class diagrams.

2.4 Role-Based Access Control (RBAC)

Owing to numerous publications on RBAC (Ahn and Sandhu 2000; Ahn et al. 2003; Ferraiolo and Kuhn 1992; Ferraiolo et al. 1995; Ferraiolo et al. 2001; Sandhu et al. 1996), we cannot cover them completely. We can only discuss the contributions that mainly concentrate on role mechanisms. We call the roles applied in RBAC RBAC-roles, which are proposed as a way to deal with the accessibilities of numerous objects or resources in a system to users. The central challenge is to construct an efficient data structure for RBAC-roles in order to express a user's rights to access such resources.

Ferraiolo and Kuhn (1992) pioneer in the research of RBAC. It is generally agreed that designing security for situations with many users and objects can be easily facilitated using a role-based design. RBAC aims to apply roles to simplify the tasks of security administrators in order to enforce access control policies. A role is described as a set of transactions that a user or set of users can perform within the context of an organization. Following their work (Ferraiolo and Kuhn 1992), RBAC has developed rapidly and received much attention in computer security and protection, and the literature has elaborated and examined the architectures and mechanisms of RBAC further (Ahn and Sandhu 2000; Ahn et al. 2003; Ferraiolo and Kuhn 1992; Ferraiolo et al. 1995; Ferraiolo et al. 2001; Sandhu et al. 1996). The mentioned research demonstrates that roles are excellent underlying mechanisms in dealing with access control and system security.

Sandhu et al. (1996) state that a role can represent specific task competency and embody the authority and responsibility of a user. Roles define both the specific individuals allowed to access resources and the extent to which resources can be accessed. In Ahn et al.'s work (2003), a role is defined as a set of actions and responsibilities associated with a particular activity. The persistence of roles in organizations is stressed. This property helps RBAC provide a powerful mechanism for reducing the complexity, cost, and potential for error in assigning permissions to users within an organization.

The Generalized RBAC (GRBAC) model extends traditional RBAC models (Covington et al. 2000; Moyer and Abamad 2001) by incorporating the notion of object roles and environment roles with the traditional notion of subject roles. From the viewpoint of security, subjects are entities used to access objects; objects are entities to be accessed; and environments are entities that help subjects access objects. Object roles capture various commonalities among the objects and emphasize the classification of them. Environment roles are used to capture security-relevant information about the environment and express common states of the objects and subjects.

The Temporal-RBAC (TRBAC) model supports periodic role enabling and disabling and temporal dependencies among permissions (Bertino et al. 2001) by introducing time into the access control infrastructure. The Generalized TRBAC (GTRBAC) model captures an exhaustive set of temporal constraint needs for access control (Joshi et al. 2005). It expresses periodic as well as durational constraints on roles, user-role assignments, and role-permission assignments. In an interval, the activation of a role can further be restricted as a result of numerous activation constraints including cardinality constraints and maximum active duration constraints. The GTRBAC model extends the syntactic structure of the TRBAC model and inherits the event and trigger expressions of TRBAC. Furthermore, it allows the expression of role hierarchies and separation of duty constraints for specifying fine-grained temporal semantics.

When discussing a proposed standard for RBAC, Ferraiolo et al. (2001) propose a role as a means for naming many-to-many relationships among individual users and permissions. A role is a function within the context of an organization with associated semantics regarding the authority and responsibility conferred on the user assigned to the role. The role characteristics in this category are the following (Ahn and Sandhu 2000; Ahn et al. 2003; Bertino et al. 2001; Covington et al. 2000; Ferraiolo and Kuhn 1992; Ferraiolo et al. 1995; Ferraiolo et al. 2001; Joshi et al. 2005; Nyanchama and Osborn 1999; Sandhu et al. 1996):

- Least Privilege: It requires that users be given no more privileges than necessary to perform their jobs.
- Separation of concerns: (i) a role can only be associated with an operation of a business function if the role is authorized and the role was not previously assigned to all of the other operations; (ii) a user is authorized as a member of a role only if that role is not mutually exclusive with any of the other roles for which the user already possesses membership; and (iii) a subject can only become active in a new role if the proposed role is not mutually exclusive with any of the roles in which the subject is currently active.
- Cardinality: The capacity of a role cannot be exceeded by an additional role member.
- Dependency constraints: There is a hierarchy of relationships among roles such as *contains*, *excludes*, and *transfers*.

Their work makes RBAC-roles have a mature, consistent, and standardized definition. The role concept in RBAC actually comes from similar ideas about the accessibility of resources used in operating systems. A role is a tag that can be used by a system to perform protection on resources in it. For protection, the system only grants users certain access rights to files based on their roles. In RBAC, a user is a client and a system is a server. RBAC-roles only emphasize roles in the aspect of rights because it is sufficient to meet the requirement of access control. Compared with the modeling-roles, RBAC-roles are practical and sufficient in many applications. On the other hand, RBAC takes roles as specific static identities to differentiate users. That is to say, a role is stated as a group of users, i.e. assigning access rights to a role is actually assigning the rights to the group of users who play this role. This allows the design of security strategies to be based on definite operations on resources. In general, a role-based information system requires that a role be dynamic. A user may bind to different roles from application to application and possibly during a single session within an application. Clearly, this is not captured by RBAC-roles.

RBAC has been researched actively for many years. A standard for RBAC has been proposed (Ferraiolo et al. 1995) and gradually applied to administrative work and modeling (Crampton and Loizou 2003). The basic ideas from RBAC are now applied to database management such as Oracle (2002), SQL server (2001), and Massachusetts Institute of Technology (MIT)'s Roles Database (Repa 1999). Although there are some trials on the research of fundamental and generalized role models (Nyanchama and Osborn 1999), the basic idea is still restricted to the administrative requirement. The basic view for roles in RBAC is well accepted, i.e. roles express the rights for users to access system objects. To generalize the role concepts of RBAC, many new contributions introduce the ideas of modeling roles, object technologies, and temporal data models into RBAC (Bertino et al. 2001; Covington et al. 2000; Joshi et al. 2005) to widen the applications.

In most practical database systems, roles are taken as a tool that allows administrators to collect users into a single unit to bestow permissions. Roles are successfully used in RDB (relational database) systems, called RDB-roles, based on RBAC-roles (Ahn et al. 2003; Chigrik 2001; Oracle 9i 2002; Repa 1999; Tolone et al. 2005). To meet more administration requirements, ARBAC97 (administrative RBAC97) is proposed in (Sandhu et al. 1999) to deal with role-based administration tasks such as user-role assignment, permission-role assignment, and role-role assignment. Moreover, ARBAC02 (Oh and Sandhu 2002) is proposed to improve ARBAC97 by removing the unnecessary integration of the user and permission pools and the role hierarchy. In ARBAC02, the organization structure is introduced as pools of users and permissions that are independent on roles or the role hierarchy. The motivation underlying these developments is to surpass the limitation of RBAC and apply RBAC-roles in system administration. We can call the above roles (Lupu and Sloman 1997; Oh and Sandhu 2002; Sandhu et al. 1999) system-roles. The system-roles consider more on such external properties as relationships among roles, users, and permissions.

2.5 Roles in CSCW Systems

People are more and more involved in Computer-Supported Cooperative Work (CSCW) because of the pressure from companies to improve their product-development and decision-making processes, and the convenience brought by the information super-highway (Ye et al. 2001). Due to the spread of coronavirus disease (COVID-19), working at home has become necessary. Working at home successfully largely depends on the achievements of the CSCW research and practice. Web conference platforms, such as Zoom (https://zoom.us/), Skype (https://www.skype.com/en/), GoToMeetings (https://www.gotomeeting.com/en-ca), Webex Meetings (https://www.webex.com/), WeChat (https://www.wechat.com/en), and Tencent Meeting (https://meeting.tencent.com/sg/en/) have rapidly established a large user base during this pandemic.

CSCW systems are computer-based tools that support collaborative activities and should meet the requirements of normal collaboration. They should not only support real/virtual face-to-face collaborative environments (Sugimoto et al. 2006) but also provide mechanisms to overcome the drawbacks of face-to-face collaboration, i.e. an aggressive person may dominate a meeting or discussion. It is natural for the CSCW systems to utilize the concept of roles (Zhu 2006) and we call such roles "CSCW-roles."

Even though roles were introduced by Leland et al. (1988) early into CSCW, the research on roles as underlying mechanisms of such systems has remained under-explored. The abovementioned web conference systems only incorporate specialized interfaces for different types of users, e.g. students, instructors, organizers, and attendees. Such applications still do not support flexible role negotiations, role specifications, and other role-related work. In past research, some systems consider roles as labels for users (Greenberg 1991; Guzdial et al. 2000; Turoff and Hiltz 1982; Zhu and Zhou 2006a). Based on these labels, the system designers use switch/case-like structures to define different working processes for different roles. In other words, systems use an enumeration list to express roles and give each role a specific function to perform. After they are released, the roles may not be changed (Zhu and Zhou 2006a). Some applications are built to support roles without clearly stating what roles are (Greenberg 1991; Smith et al. 1998). The assigned roles are based on people's intuitive decisions. Others use the role mechanism similar to that of RBAC, i.e. restricting the access rights to shared resources in collaboration (Edwards 1996; Smith et al. 1998). These systems can only support predefined or static roles. Users have no support facility to tune their roles in order to make collaboration more productive and efficient. Some systems, e.g. the Coordinator (Flores et al. 1988; Winograd 1988), are criticized as "naziware" because they over-restrict users. In other words, a user's operations are over-restricted by the roles they are playing. This criticism motivates the research and applications of dynamic roles in collaboration systems (Edwards 1996; Smith et al. 1998).

In the Electronic Information Exchange System (EIES) (Turoff and Hiltz 1982), roles are built out of a subset of primitive privileges (such as *append, link, assign,* and *use*) that are crucial to a human communication process. In Quilt (Leland et al. 1988), there are roles for writers (who are allowed to change their own work only), readers (who are not allowed to modify the document), and commentators (who can only add "margin notes" to it). Such CSCW systems have lost flexibility after introducing roles because they can provide only static role mechanisms, i.e. the user interface is predefined and cannot be changed. They lack flexible mechanisms for role specification, tuning, changing, and transitioning because a mechanism to express a role is simply missing.

Patterson (1991) emphasizes a role as an interface between objects. A user interface is simply an elaborate mechanism for moving messages back and forth between a user and objects. A role is an abstract user type. It is an object class of which a user is an instance. Given the roles of users, the associated messages are presented to users. In the RENDEZVOUS system (Patterson 1991), a role is taken as a filter of the input events and used to enable and disable an object's visibility. Patterson's work implicitly expresses the access rights of users by roles. Unfortunately, their work does not go far enough to provide a real abstract structure to support more flexible user interfaces based on roles.

Greenburg proposes the SHARE system to accommodate individual roles and group differences with roles (Greenberg 1991). Personalized groupware can lead to the wider acceptance of a product by offering a system that conforms to the individual needs of human users and groups. The roles in SHARE are static labels with built-in protocols to let the system respond to users in different ways.

Edwards introduces access control policies and roles to avoid chaos in collaborative applications (Edwards 1996). A role is described as a category of users within the user population of a given application; and all users of a certain role inherit a set of access control rights to objects within the application. By describing role membership, users can be relieved of some of the burdens of tracking, updating, and anticipating role membership explicitly. Dynamic roles are supported by binding a predicate function to a symbolic role name and mapping from roles to policies. The essential characteristics of roles are determined by policies in this way. A role's dynamic property can change by mapping a role's name to a policy (Edwards 1996; Leland et al. 1988). This mapping is accomplished by associating a predicate function with a role. A formal language INTERNEZZO (Edwards 1996) is developed to specify roles with attributes, predicates, and policies. The question arises: should additional methods be developed to further this research? Or, how can a policy be tuned, changed, or transferred to another role? Edwards uses roles as tools for access control and uses the INTERNEZZO language to tune, change, or restrict access control policies. Hence, it is still difficult for nonprofessional end users to do such tasks.

Smith et al. (1998) build the Kansas system and emphasize the importance of roles. People in a group play various roles even though they may not be well defined. Kansas intends to support roles by special treatments in multiuser interfaces. Roles are generally supported by the system's treatment of outputs and user inputs, which are similar to the view of incoming messages and outgoing messages at a more abstract level. Thus, Kansas can support flexible roles by managing the accessibility of resources such as inputs and outputs. This system applies a part of the RBAC-role implicitly because their accessibilities are similar to the permissions of RBAC-roles. "The physics underlying reality does not define roles, though it supports them" (Smith et al. 1998). This statement partially answers why there are so many different aspects to consider in using roles in information systems. Guzdial et al. (2000) inherit the idea of roles from Smith et al. (1998) and develop many roles that are typical in their collaborative tool CoWeb (Genilloud and Wegmann 2000) to accommodate a variety of users:

- They describe roles as specific concerns and activities associated with individuals who choose to use CoWeb; and
- They define a role as a human construct created and sustained by the interaction of minds. Their roles can be extended on- and off-line and are not associated with capabilities within the shared space.

In fact, CoWeb is another important practice similar to what Leland et al. (1988) tried before, i.e. roles are predesigned and static.

Zhu (2006a) and Zhu and Zhou (2006) completely review the role mechanisms applied in collaborative systems and propose a new role definition. It is composed of both responsibilities and rights. Their work establishes the requirements for Role-Based Collaboration (RBC); presents the concept, requirements, and principles of RBC; proposes a model Environment – Classes, Agents, Roles, Groups, and Objects (E-CARGO) for RBC; and describes the kernel mechanisms and their implementation to facilitate the development of RBC systems for industrial applications. The conclusion is that roles in collaborative systems are important and RBC requires more comprehensive research in order to achieve the mentioned goal to support and improve face-to-face collaboration. The role mechanism in E-CARGO extends the idea of Gaia in that roles specify requests which contain any kinds of outgoing messages, while Gaia's rights contain only specific access operations such as *read* and *write* (Wooldridge et al. 2000; Zigurs and Kozar 1994).

In the SPADE-1 environment, Bandinelli et al. (1996) introduce static roles such as project manager, system administrator, designer, and programmer. Users can perform operations on the global workspace according to their roles. Because their major task is to provide an environment for software engineering, their roles are intuitive and simple. They just use a string such as "ProjManager" to express the role of a project manager. In other words, their roles are used to specify the tasks or operations of a user, i.e. CSCW-roles.

Consequently, Zhu and Zhou (2006a) apply their E-CARGO model to build a software development tool. When considering software development as a collaborative activity, they discuss the importance of roles in software engineering and that of role-based software engineering; propose and describe a role-based software process; and implement a prototype tool for developing complex information systems with the help of role mechanisms.

In summary, in CSCW, there is neither a commonly accepted role concept nor methods to express roles. Some use roles as a commonly understood word (Guzdial et al. 2000; Leland et al. 1988; Turoff and Hiltz 1982); some support roles with a special user interface design framework (Greenberg 1991; Smith et al. 1998); and others propose a tool to specify roles dynamically (Edwards 1996; Stone and Veloso 1999). However, no matter what kinds of roles are considered, all the applications use roles to support human-computer and human-human interactions. Therefore, roles are considered to be interaction media in collaborative systems.

2.6 Roles in Social Psychology and Management

After investigating role concepts and mechanisms from different research areas in information systems, let us consider the role concepts from social psychology. Among the heavy literature, we are only selecting the most significant ones that discuss roles systematically and are highly relevant to information systems.

From Oxford English Dictionary (2004), we can find the common definitions of "role." A role is:

1) "The part or character which one has to play, undertake, or assume";
2) "The part played by a person in society or life"; or
3) "The typical or characteristic function performed by someone or something."

The term "role" derives from the theater and refers to the part played by an actor. Also, in Oxford 2004, a role is defined in a behavioral and psychological view. It is the behavior that an individual feels appropriate to assume in adapting to any form of social interaction; the behavior considered appropriate for the interaction demanded by a particular kind of work or social position. We call such defined roles social-roles.

The major concerns of social-roles are how to avoid conflicts and ambiguities among the people related to a position (Ashforth 2001; Biddle and Thomas 1966). The literature mainly discusses human resource management and staff recruitment. Thus, the basic question turns into how to present an accurate position description to recruit a person to fill this position.

Biddle and Thomas (1966) edit and publish the first book on role theory. They define roles as sets of prescriptions defining what the behavior of a positioning member should be. A role should be a collection of rights and duties. They conclude that role theory is realistic and possesses an identifiable domain of study, perspective, and language.

Bostrom (1980) perform a field study of role conflicts and ambiguities from the viewpoint of users and designers of Management Information Systems (MIS). A role is defined as a set of expectations about the behavior of a user in a particular position within a system. Generally, a role is a position occupied by a person in a social relationship. In this position, the person possesses special rights and responsibilities. It is concluded that an inadequate role-defining process is a key source of problems for users and designers in obtaining satisfaction when working with MIS.

Zigurs and Kozar (1994) conduct a field study on the requirements and usage of roles from the viewpoint of users of Group Support Systems (GSS). They review the literature on roles from the 1930s to the 1990s including (Biddle and Thomas 1966) and discussed the user behavior and influence on the role expectations of GSS users. The GSS used in their study has "only software that collects, manipulates, and aggregates users' inputs" and does not support role mechanisms. The roles of users are completely up to the organization of the user groups. Zigurs and Kozar define a role as "a dynamic set of recurring behaviors, both expected and enacted, within a particular group context." King and Sethi (1998) study the impact of socialization on role adjustments from information system professionals. Their work is motivated by the socialization process relevant to roles, such as role orientation, role ambiguity, and role conflict mentioned in the literature of sociology and psychology. It is concluded that institutionalized socialization tactics lead to a custodial role orientation; individual socialization tactics produce an innovative role orientation; and institutionalized tactics reduce role ambiguity and conflict among new employees. This conclusion demonstrates a clear requirement for an information system to be equipped with clear role mechanisms.

Ashforth (2001) does a comprehensive study and publish his book on role transitions. It is a good addition to the role theory literature in social psychology. Ashforth (2001) cites the role definition from Biddle and Thomas (1966). According to (Ashforth 2001), a role is defined as a position in a social structure. A position essentially means an institutionalized or commonly expected and understood designation in a given social structure such as accountant (company), mother (family), or church member (religious organization). All his discussions on roles are based on the identities of people.

Acuna and Juristo (2004) use role concepts intuitively in software project management. They concentrate on human resource management in software development and believe that the theory of psychology assists in software project management. They define roles as sets of responsibilities and capabilities required to carry out the activities of each subprocess. A capability defines the skills or attributes of a person. This method is capability-oriented and roles are taken as elements of a software engineering process.

For example, a development group and a data management group are faced with complaints that the results presented to customers are not based on the correct source data provided by the customers. Each group thinks that the other group is responsible for quality assurance. In building their role model, they can easily agree on the need for a quality assurance role. As they flesh out the responsibilities of that role, it becomes clear that it has to be filled by a team with representatives from both organizations. The team can then be further described in terms of the roles and responsibilities of its members.

In business management, role modeling (RM) is proposed as a business engineering technique (HLA 2001) used to provide a model of an organization in terms of roles, responsibilities, and collaboration among individuals and teams. RM also provides a process that allows both small- and large-scale enterprises to transform and advance. It acts as a vehicle for reengineering and process improvement. In RM, a role has a cohesive set of responsibilities and a purpose for its existence. A role may be played by a person, group, organization, team, or automated system. Each of these entities may fill many roles and a role may be filled by another role. RM draws a major part of its power by separating the "what" of a role from "who" is filling the role at any particular point in time. That separation allows teams to come together to discover the nature of the work that must be done to meet the demands placed upon them by their sponsors and figure out how the roles must be filled.

As discussed above, in social psychology and management, the major research is on how people behave when playing roles. The relevant problems about roles and human's behavior include role description or specification, expectation, conflicts, and transitions. All these problems largely affect the performances of a group of people (Ashforth 2001; Biddle and Thomas 1966; Bostrom 1980; King and Sethi 1998; Zigurs and Kozar 1994). In reality, it is difficult to describe roles clearly and strictly because natural languages are ambiguous to some extent. That is why different persons in the same position can make different contributions and human resource experts are required to describe a position as clearly as possible.

2.7 Convergence of Role Concepts

To fully evaluate and compare the role concepts from different research areas in information systems, we need to consider and learn the role concepts from the management and social psychology viewpoint because the role concepts in management and social psychology cover what a role should be. An information system is a virtual community that simulates, and certainly sometimes, surpasses a real one. Participants have two kinds of existence: service provider and client. When participants play a role, they provide specific services and have certain rights to ask for services. Keeping this in mind, a role then can be considered a person's view of the world. When they play a specific role, they have a special view of their surroundings. Their role in a working environment is actually a wrapper with a service interface (Genilloud and Wegmann 2000; Smith et al. 1998) and a request interface (Patterson 1991; Zhu 2006; Zhu and Zhou 2006).

From what we have discussed, we find that much work develops and applies the role concepts in an ad hoc manner. We also find that there is not much relevance among many papers that should be related. To develop a general understanding of the research and practice of roles, we compiled Table 2.1 to compare and contrast the various meanings and usages of roles in the relevant literature. We can conclude that *Agent-roles* and *System-roles*, expansions of RBAC-roles, are currently the most beneficial ones.

Let us review all role concepts discussed in this chapter. If all role concepts could be categorized with an inheritance hierarchy shown in Figure 2.2, there would be fewer arguments.

The rationality of Figure 2.2 comes from the classification of concepts and ideas of social organizations and major aims of different systems:

- Social-roles aim to describe human behavior in social lives. They should reflect explicitly or implicitly most of the properties of roles.
- Modeling-roles aim to describe the abstract mapping of a solution to a practical problem.
- Agent-roles aim to simulate the collaboration of agents in a system.
- System-roles are an expansion that accommodates more properties than RBAC-roles and aims to simplify the work of system administrators.
- CSCW-roles aim to support people to collaborate with the help of computers.

Figure 2.2 shows a classification, with a class hierarchy, of simple inheritance and whole inheritance from the viewpoint of object-orientation. CSCW, RBAC, and modeling-roles should inherit from social-roles.

We can also understand different role concepts from various viewpoints in information systems. They can be categorized into four unique views of roles according to the information system layers shown in Figure 2.3. From the viewpoint of operating systems, we agree that the idea of roles is similar to that of RBAC. From the viewpoint of modeling and programming methodologies, we have modeling-roles. Finally, from the viewpoint of applications and human users, we have CSCW-roles, agent-roles, and social-roles.

The reality of role concepts hitherto is described as an inheritance hierarchy in Figure 2.4. By solid lines, we mean that there is partial or intuitive inheritance. Solid rectangles in Figure 2.4 mean the role concepts are well accepted. A dashed

Table 2.1 Comparison among the different kinds of roles.

Aspects	Social-roles	Modeling-roles	CSCW-roles	RBAC-roles	System-roles	Agent-roles
The year initiated	1966	1977	1988	1992	1997	1999
Reasons why roles are applied	Classification, evolution, separation of concerns	Classification, evolution, separation of concerns	Interaction	Authorization	Classification	Classification, interaction, evolution, separation of concerns
Role players	Human	Object	User	User, operation	User, operation	Agent
Role content	Responsibility, right	Service	Request	Right	Right	Service, right
Role specification	Process, interface	Process, interface	Interface	Interface	Interface	Process, interface
Role assignment	Instantiation, allocation, constraint	Instantiation, allocation	Constraint	Constraint	Constraint	Allocation, constraints
Role relationship	Conflict, qualification, inheritance, place	Inheritance	No	Conflict, place	Conflict, place	Inheritance, place

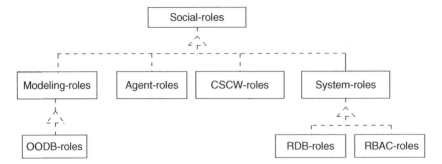

Figure 2.2 The ideal relationship hierarchy of role concepts.

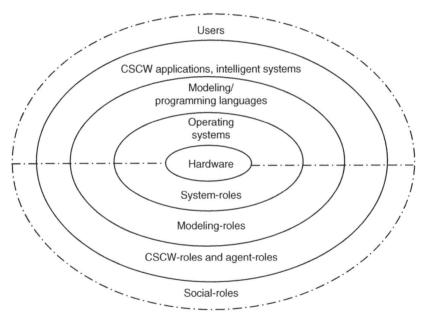

Figure 2.3 Different understandings of roles at different layers in an information system.

rectangle in Figure 2.4 means the role concept is not yet well accepted. Dashed lines mean that there is a partial or intuitive inheritance to a not well-accepted concept, i.e. social roles. For example, the modeling-roles and RBAC-roles possess only a part of those in social-roles (Figure 2.5). We can use Figure 2.5 to demonstrate the clear difference between that of modeling methodology and that of RBAC.

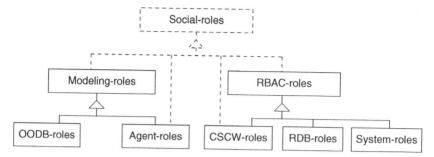

Figure 2.4 The relationships of the role concepts.

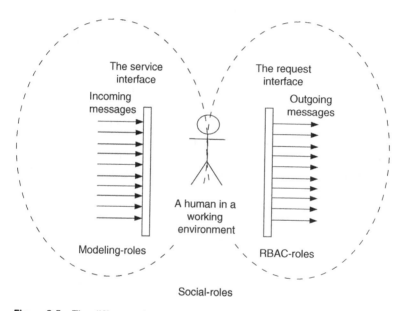

Figure 2.5 The difference between the modeling-roles and RBAC-roles.

Based on what we have discussed, we can use the inheritance relationship concepts from the object-oriented methodology to predict future work. There should be more specific applications applied to different branches of information systems. Additional studies are required regarding roles as the basic concepts in systems management and engineering.

Modeling-roles concentrate on services only, while RBAC-roles concentrate on special requests (*rights*) only. However, to support collaboration, we need to pay the same amount of attention to both services and requests. Roughly speaking, together, modeling-roles and RBAC-roles constitute social-roles. Agent-roles have considered *rights* as part of a role, and can thus be improved by accommodating more generalized *rights*, i.e. *requests*. As a simulation of intelligent systems, roles in agent systems, or agent-roles, are approaching the concept of social-roles.

2.8 Summary

Roles have been widely applied for many years. However, there are many more applications to come. For example, in ontological approaches (Albrecht et al. 2007), roles are fundamental concepts used to express structures and relationships (Guarino 1992); in agent systems (Brennan 2007; Cabri et al. 2005a; Liu et al. 2007; Odell et al. 2003), an agent's behavior can be adjusted and tuned by playing different roles or by tuning their current roles; in systems of systems, roles can be used to separate concerns in designing (Hipel et al. 2007); in exploiting the information web (Gregg and Walczak 2000), roles are good mechanisms to extract relevant information; in human–robot systems (Shah et al. 2007), roles can be used to regulate the interactions between humans and robots; and in Human Computer Interaction (HCI), personalized user interfaces present information in ways compatible with roles (Greenberg 1991; Shneiderman and Plaisant 1994). In CSCW, people can talk to others at a definite time if a role-playing schedule is set. By this schedule, interventions are blocked or arranged into special time slots. In management information systems, such as role-based workflow systems, one may do what the roles specify clearly by avoiding too much irrelevant information to improve productivity. This chapter concludes that:

1) Roles are important in information systems' design but are often introduced and applied intuitively.
2) The application of roles in different research fields has different motivations, aims, concentrations, and goals. Roles can be taken as modeling mechanisms, management strategies, interaction media, and behavior describers.
3) Obtaining the underlying commonalities of roles by referring to the ideas from management and social communities is required to promote the application of roles and it is an active trend in various fields of research.
4) There are still many challenges regarding the research and application of roles. We hope that more researchers and practitioners provide more efforts into role-related investigations.

In applying the role concepts, a key problem is how to specify a role and how to apply a role in a system. The traditional role mechanisms and role concepts are insufficient to accomplish these tasks. We need to develop new role mechanisms to support easy and clearer role specification. There are many essential topics that are still unexplored and require additional comprehensive research.

1) We should find more applications for RBAC-roles and modeling-roles, e.g. Computer-Aided Software Engineering (CASE) tools, personalized interfaces, business management, and Flexible Manufacture Systems (FMS).
2) A well-defined role model that accommodates generalized requirements in different aspects, such as role transfer, the dynamic structures of roles, and the roles for groups need to be developed.
3) The mechanism of group roles, i.e. how to have a group play a role, needs to be investigated.
4) A prototype system that fully supports the social roles, including agent-roles and CSCW-roles, needs to be built. CSCW-roles are required to concentrate on the interface aspect and agent-roles on the process aspect.
5) A role can be considered a set of incoming and outgoing messages. We still need to find a practical and efficient way to form user interfaces based on these messages. More research is needed to provide a role specification tool. We need a more structural and efficient way to bind messages to a role.

Roles as concepts in the information technology fields seem to have cooled down since 2010 and most researchers did not continue their research related to role concepts and mechanisms. The author contacted many researchers who did research on roles to seek the reasons why they did not continue such investigations, the answer is that it is hard to obtain research funds. None opposes the importance and significance of the research on roles. From experience, the author agrees that roles are a complex concept. It is difficult to define, specify, and implement as a well-defined and specified structure in computer-based systems. Many people adore buzzy words and feel negative about the terminology that has been used before. Perhaps all these are why most role-modeling researchers cease their effort. However, the author believes that after both the connotation and extension of the role concepts are completely specified, the next generation of programming languages will be role-based.

Please note that role theory and role concepts are embraced by the researchers in management and collaboration (Bowen 2016; Frydman 2016; Morris and Miller-Stevens 2016; Vandenberghe et al. 2017; Thies 2017). That means even though roles have not yet been established as a well-defined concept, they can be applied widely in management, system engineering, and modeling real-world problems.

We hope the readers could study the next chapters carefully and discover the benefits of the proposed E-CARGO model and apply relevant role concepts and mechanisms in dealing with complex problems.

Since the E-CARGO model was proposed by Zhu and Zhou (2006a), the research on RBC was started and opened a new breakthrough direction in applying roles in modeling and solving problems in collaboration systems (Zhu 2006; Zhu and Zhou 2006). The author hopes to use E-CARGO to advocate the application of role theory into the fast-developing knowledge and intelligence world.

References

Acuna, S.T. and Juristo, N. (2004). Assigning people to roles in software projects. *Software-Practice and Experience* 34 (7): 675–696.

Ahn, G.J. and Sandhu, R. (2000). Role-based authorization constraints specification. *ACM Transactions on Information and System Security (TISSEC)* 3 (4): 207–226.

Ahn, G., Zhang, L., Shin, D., and Chu, B. (2003). Authorization management for role-based collaboration. *Proceeding of IEEE International Conference on Systems, Man and Cybernetics*, vol. 5, Washington, DC, USA (5–8 October 2003), pp. 4128–4134.

Albano, A., Bergamini, R., Ghelli, G., and Orsini, R. (1993). An object data model with roles. *Proceedings of the International Conference on Very Large Databases*, Dublin, Ireland (24–27 August 1993), pp. 39–52.

Albrecht, C.C., Dean, D.L., and Hansen, J.V. (2007). An ontological approach to evaluating standards in E-commerce platforms. *IEEE Transactions on Systems, Man and Cybernetics, Part C: Applications and Reviews* 37 (5): 846–859.

Ashforth, B.E. (2001). *Role Transitions in Organizational Life: An Identity-Based Perspective*. Lawrence Erlbaum Associates, Inc.

Auramäki, E., Lehtinen, E., and Lyytinen, K. (1988). A speech-act-based office modeling approach. *ACM Transactions on Information Systems (TOIS)* 6 (2): 126–152.

Bachman, C.W. and Daya, M. (1977). The role concept in data models. *Proceedings of International Conference on Very Large Databases*, Tokyo, Japan (6–8 October 1977), pp. 464–476.

Bandinelli, S., di Nitto, E., and Fuggetta, A. (1996). Supporting cooperation in the SPADE-1 environment. *IEEE Transactions on Software Engineering* 22 (12): 841–865.

Bäumer, D., Riehle, D., Siberski, W., and Wulf, M. (2000). Role object. In: *Pattern Languages of Program Design 4* (eds. N. Harrison, B. Foote and H. Rohnert), 15–32. Addison-Wesley.

Becht, M., Gurzki, T., Klarmann, J., and Muscholl, M. (1999). ROPE: Role Oriented Programming Environment for multiagent systems. *Proceedings of Fourth IECIS International Conference on Cooperative Information Systems*, Edinburgh, UK (September 1999), pp. 325–333.

Bertino, E., Bonatti, P.A., and Ferrari, E. (2001). TRBAC: a temporal role-based access control model. *ACM Transactions on Information and System Security* 4 (3): 191–223.

Biddle, B.J. and Thomas, E.J. (eds.) (1966). *Role Theory: Concepts and Research*. John Willey & Sons, Inc.

Bostrom, R.P. (1980). Role conflict and ambiguity: critical variables in the MIS user-designer relationship. *Proceedings of the 17th Annual Computer Personnel Research Conference*, Miami, FL, USA (June 1980), pp. 88–115.

Botha, R.A. and Eloff, J.H.P. (2001). Designing role hierarchies for access control in workflow systems. *Proceedings of the 25th Annual International Computer Software and Applications Conference (COMPSAC'01)*, Chicago, IL, USA (October 2001), pp. 117–122.

Bowen, D.E. (2016). The changing role of employees in service theory and practice: an interdisciplinary view. *Human Resource Management Review* 26 (1): 4–13.

Brachman, R.J. and Schmolze, J.G. (1985). An overview of the KL-ONE knowledge representation system. *Cognitive Science* 9: 171–216.

Brennan, R.W. (2007). Toward real-time distributed intelligent control: a survey of research themes and applications. *IEEE Transactions on Systems, Man and Cybernetics, Part C: Applications and Reviews* 37 (5): 744–765.

Cabri, G., Ferrari, L., and Zambonelli, F. (2004). Role-based approaches for engineering interactions in large-scale multi-agent systems. In: *Software Engineering for Multi-Agent Systems II Research Issues and Practical Applications*, Lecture Notes in Computer Science, vol. 2940 (eds. C. Lucena, A. Garcia, A. Romanovsky, et al.), 243–263. Berlin, Heidelberg, Germany: Springer.

Cabri, G., Ferrari, L., and Leonardi, L. (2005). Injecting roles in Java agents through runtime bytecode manipulation. *IBM Systems Journal* 44 (1): 185–208.

Cabri, G., Ferrari, L., and Leonardi, L. (2005a). Supporting the development of multi-agent interactions via roles. *The 6th International Workshop on Agent-Oriented Software Engineering (AOSE) at AAMAS 2005*, Utrecht, The Netherlands (July 2005). Springer LNCS, vol. 3950, pp.154–166.

Chigrik, A. (2001). Understanding SQL server roles. http://www.databasejournal.com/features/mssql/article.php/1441261 (accessed 10 August 2020).

Covington, M. J., Moyer, M., and Ahamad, M. (2000). Generalized role-based access control for securing future applications. *23rd National Information Systems Security Conference*, Washington, DC, USA (April 2001). http://smartech.gatech.edu/dspace/bitstream/1853/6580/1/GIT-CC-00-02.pdf.

Crampton, J. and Loizou, G. (2003). Administrative scope: a foundation for role-based administrative models. *ACM Transactions on Information and System Security* 6 (2): 201–231.

Dafoulas, G.A. and Macaulay, L.A. (2001). Facilitating group formation and role allocation in software engineering groups. *Proceedings of the ACS/IEEE International Conference on Computer Systems and Applications (AICCSA'01)*, pp. 352–359.

Dahchour, M., Pirotte, A., and Zimányi, E. (2004). A role model and its metaclass implementation. *Information Systems* 29 (3): 235–270.

Demsky, B. and Rinard, M. (2002). Role-based exploration of object-oriented programs. *Proceedings of the 24th International Conference on Software Engineering (ICSE'02)*, Orlando, FL, USA (19–25 May 2002), pp. 313–324.

Depke, R., Heckel, R., and Kuster, J.M. (2001). Roles in agent-oriented modeling. *International Journal of Software Engineering and Knowledge Engineering* 11 (3): 281–302.

Edwards, W.K. (1996). Policies and roles in collaborative applications. *Proceedings of ACM 1996 Conference on Computer-Supported Cooperative Work (CSCW'96)*, Cambridge, USA (16–20 November 1996), pp. 11–20.

Ferber, J., Gutknecht, O., and Michel, F. (2004). From agents to organizations: an organizational view of multi-agent systems. In: *Agent-Oriented Software Engineering (AOSE) IV, Melbourne, July 2003*, Lecture Notes on Computer Science, vol. 2935 (eds. P. Giorgini, J. Müller and J. Odell), 214–230. Berlin, Heidelberg, Germany: Springer.

Ferraiolo, D.F. and Kuhn, D.R. (1992). Role-based access control. *Proceedings of the NIST-NSA National (USA) Computer Security Conference*, Baltimore, USA (13–16 October 1992), pp. 554–563.

Ferraiolo, D.F., Cugini, J.A., and Kuhn, D.R. (1995). Role-based access control (RBAC): features and motivations. *Proceedings of the 11th Annual Computer Security Applications Conference (CSAC '95)*, Los Alamitos, CA (11–15 December 1995), pp. 241–248.

Ferraiolo, D.F., Sandhu, R., Gavrila, S. et al. (2001). Proposed NIST standard: role-based access control. *ACM Transactions on Information and System Security* 4 (2): 224–274.

Flores, F., Graves, M., Hartfield, B., and Winograd, T. (1988). Computer systems and the design of organizational interaction. *ACM Transactions on Office Information Systems* 6 (2): 153–172.

Fortino, G. and Russo, W. (2012). ELDAMeth: an agent-oriented methodology for simulation-based prototyping of distributed agent systems. *Information and Software Technology* 54 (6): 608–624.

Fortino, G., Russo, W., and Zimeo, E. (2004). A statecharts-based software development process for mobile agents. *Information and Software Technology* 46 (13): 907–921.

Fortino, G., Garro, A., Mascillaro, S., and Russo, W. (2010). Using event-driven lightweight DSC-based agents for MAS modelling. *International Journal of Agent-Oriented Software Engineering* 4 (2): 113–140.

Frydman, J.S. (2016). Role theory and executive functioning: constructing cooperative paradigms of drama therapy and cognitive neuropsychology. *The Arts in Psychotherapy* 47: 41–47.

Genilloud, G. and Wegmann, A. (2000). A foundation for the concept of role in object modelling. *Proceedings of the 4th International Enterprise Distributed Object Computing Conference (EDOC 2000)*, Makuhari, Japan (September 2000), pp. 76–85.

Gottlob, G., Schrefl, M., and Röck, B. (1996). Extending object-oriented systems with roles. *ACM Transactions on Information Systems (TOIS)* 14 (3): 268–296.

Greenberg, S. (1991). Personalizable groupware: accommodating individual roles and group differences. *The European Conference of Computer Supported Cooperative Work (ECSCW '91)*, Amsterdam (September 1991), pp. 17–32.

Gregg, D.G. and Walczak, S. (2000). Exploiting the information web. *IEEE Transactions on Systems, Man and Cybernetics, Part C: Applications and Reviews* 37 (1): 109–125.

Guarino, N. (1992). Concepts, attributes and arbitrary relations: some linguistic and ontological criteria for structuring knowledge bases. *Data & Knowledge Engineering* 8: 249–261.

Guzdial, M., Rick, J., and Kerimbaev, B. (2000). Recognizing and supporting roles in CSCW. *The ACM 2000 Conference on Computer-Supported Cooperative Work (CSCW'00)*, Philadelphia, Pennsylvania, USA (December 2000), pp. 261–268.

Hawkins, D.I., Best, R.J., and Coney, K.A. (1983). *Consumer Behavior*. Plano, TX: Business Publications, Inc.

Hellriegel, D., Slocum, J.W. Jr., and Woodman, R.W. (1983). *Organizational Behavior*. St. Paul, Minnesota: West Publishing Co.

Hipel, K.W., Jamshidi, M.M., Tien, J.M., and White, C.C. III (2007). The future of systems, man, and cybernetics: application domains and research methods. *IEEE Transactions on Systems, Man and Cybernetics, Part C: Applications and Reviews* 37 (5): 726–743.

HLA Associates (2001). Role modeling. http://www.rolemodeling.com (accessed 10 August 2020).

Holt, A., Ramsey, H.R., and Grimes, J.D. (1983). Coordination system technology as the basis for a programming environment. *Electrical Communication* 57 (4): 307–314.

Joshi, J.B.D., Bertino, E., Latif, U., and Ghafoor, A. (2005). Generalized temporal role based access control model. *IEEE Transactions on Knowledge and Data Engineering* 7 (1): 4–23.

Kendall, E.A. (1999). Role model designs and implementations with aspect oriented programming. *ACM SIGPLAN Notices* 34 (10): 353–369.

King, R.C. and Sethi, V. (1998). The impact of socialization on the roles adjustment of information systems professionals. *Journal of Management Information Systems*, Spring 1998 15 (4): 195–217.

Kristensen, B.B. (1995). Object-oriented modeling with roles. *Proceeding of the 2nd International Conference on Object-Oriented Information Systems (OOIS'95)*, Dublin, Ireland (December 1995), pp. 57–71.

Kristensen, B.B. and Østerbye, K. (1996). Roles: conceptual abstraction theory & practical language issues. *Theory and Practice of Object Systems* 2 (3): 143–160.

Kühn, T., Stephan, B., Götz, S., et al. (2015). A combined formal model for relational context-dependent roles. *Software Language Engineering*, Pittsburgh, PA, USA (26–27 October 2015), Addison-Wesley Professional, pp. 141–160. ACM.

Kuncak, V., Lam, P., and Richard, M. (2002). Role analysis. *The 29th Annual ACM SIGPLAN - SIGACT Symposium on Principles of Programming Languages* (POPL'02) Portland, OR, USA (January 2002), pp. 17–31.

Leland, M.D.P., Fish, R.S., and Kraut, R.E. (1988). Collaborative document production using quilt. *Proceedings of the Conference on Computer-Supported Cooperative Work*, Portland, OR, USA (26–28 September 1988), pp. 206–215.

Liu, N., Abdelrahman, M.A., and Ramaswamy, S.R. (2007). A complete multiagent framework for robust and adaptable dynamic job shop scheduling. *IEEE Transactions on Systems, Man and Cybernetics, Part C: Applications and Reviews* 37 (5): 904–916.

Lupu, E. and Sloman, M. (1997). Towards a role based framework for distributed systems management. *Journal of Network and Systems Management* 5 (1): 5–30.

Morris, J.C. and Miller-Stevens, K. *Advancing Collaboration Theory: Models. Typologies, and Evidence.* New York, USA: Routledge.

Moyer, M. and Abamad, M. (2001). Generalized role-based access control. *Proceedings of the 21st International Conference on Distributed Computing Systems*, Mesa, AZ, USA (16–19 April 2001), pp. 391–398.

Murdoch, J. and McDermid, J.A. (2000). Modeling engineering design process with role activity diagrams. *Transactions of the Society for Design and Process Science(SDPS)* 4 (2): 45–65.

Nyanchama, M. and Osborn, S. (1999). The role graph model and conflict of interest. *ACM Transactions on Information and System Security* 2 (1): 3–33.

Odell, J. (2002). Objects and agents compared. *Journal of Object Technology* 1 (1): 41–53.

Odell, J., van Dyke Parunak, H., and Fleischer, M. (2003). The role of roles in designing effective agent organizations. In: *Software Engineering for Large-Scale Multi-Agent Systems*, Lecture Notes on Computer Science, vol. 2603 (eds. A. Garcia, C. Lucena, F. Zambonelli, et al.), 27–38. Berlin: Springer.

Odell, J., Nodine, M., and Levy, R. (2005). A metamodel for agents, roles, and groups. In: *Agent-Oriented Software Engineering (AOSE)*, Lecture Notes on Computer Science, vol. 3382 (eds. J. Odell, P. Giorgini and J. Müller), 78–92. Berlin: Springer.

Oh, S. and Sandhu, R. (2002). A model for role administration using organization structure. *Proceedings of the Seventh ACM Symposium on Access Control Models and Technologies*, Monterey, California, USA (3–4 June 2002), pp. 155–162.

Oracle (2002). Managing user privileges and roles. http://www.cise.ufl.edu/help/database/oracle-docs/server.920/a96521/privs.htm (accessed 10 August 2020).

Ould, M.A. (1995). *Business Processes: Modeling and Analysis for Re-Engineering and Improvement.* John Wiley & Sons.

Oxford University Press (2004). OED online. http://www.oed.com (accessed 10 August 2020).

Park, J.S., Sandhu, R., and Ahn, G.J. (2001). Role-based access control on the web. *ACM Transactions on Information and System Security* 4 (1): 37–71.

Partsakoulakis, I. and Vouros, G. (2004). Roles in MAS: managing the complexity of tasks and environments. In: *An Application Science for Multi-Agent Systems* (ed. T. Wagner), 133–154. Springer.

Patterson, J.F. (1991). Comparing the programming demands of single-user and multi-user application. *The fourth Symposium on User Interface Software and Technology*, Carolina, USA (11–13 November 1991), ACM Press, pp. 87–94.

Pernici, B. (1990). Objects with roles. *ACM SIGOIS Bulletin, The Conference on Office Information Systems* 11 (2-3): 205–215.

Reenskaug, T., Lehne, O.A., and Wold, P. (1995). *Working with Objects: The OOram Software Engineering Method.* Pentice Hall.

Reimer, U. (1985). A representation construct for roles. *Data & Knowledge Engineering* 1 (3): 233–251.

Repa, J. (1999). The roles database at the Massachusetts institute of technology. *EDUCAUSE Conference*, Long Beach, California (29 October 1999). http://web.mit. edu/rolesdb/www/educause/educause.html.

Riehle, D. and Gross, T. (1998). Role model based framework design and integration. *ACM SIGPLAN Notices, Proceedings of the Conference on Object-Oriented Programming, Systems, Languages, and Applications*, Vancouver, British Columbia, Canada (18–22 October, 1998), vol. 33, no. 10, pp. 117–133.

Riehle, D., Brudermann, R., Gross, T., and Mätzel, K.U. (2000). Pattern density and role modeling of an object transport service. *ACM Computing Surveys (CSUR)* 32 (1): 1–6.

Role Modellers Ltd. (2004). A better way to support collaboration. http://www. rolemodellers.com (accessed 10 August 2020).

Sandhu, R., Coyne, E., Feinstein, H., and Youman, C. (1996). Role-based access control models. *IEEE Computer* 29 (2): 38–47.

Sandhu, R., Bhamidipati, V., and Munawer, O. (1999). The ARBAC97 model for role-based administration of roles. *ACM Transactions on Information and System Security* 2 (1): 105–135.

Schrefl, M. and Thalhammer, T. (2004). Using roles in Java. *Software - Practice and Experience* 34: 449–464.

Shah, J.A., Saleh, J.H., and Hoffman, J.A. (2007). Review and synthesis of considerations in architecting heterogeneous teams of humans and robots for optimal space exploration. *IEEE Transactions on Systems, Man and Cybernetics, Part C: Applications and Reviews* 37 (5): 779–793.

Shneiderman, B. and Plaisant, C. (1994). The future of graphic user interfaces: personal role managers. *People and Computers IX, British Computer Society's HCI 94*, Glasgow, Scotland (August 1994), pp. 3–8.

Smith, R.B., Hixon, R., and Horan, B. (1998). Supporting flexible roles in a shared space. *The ACM 1998 Conference on Computer-Supported Cooperative Work (CSCW'98)*, Seattle, Washington, USA (14–18 November 1998), pp. 197–206.

Steimann, F. (2000a). A radical revision of UML's role concepts. *UML*, York, UK (2–6 October 2000) pp. 194–209.

Steimann, F. (2000b). On the representation of roles in object-oriented and conceptual modeling. *Data & Knowledge Engineering* 35: 83–106.

Steimann, F. (2001). Role = interface: a merge of concepts. *Journal of Object-Oriented Programming* 14 (4): 23–32.

Stone, P. and Veloso, M.M. (1999). Task decomposition, dynamic role assignment, and low-bandwidth communication for real-time strategic teamwork. *Artificial Intelligence* 110 (2): 241–273.

Sugimoto, M., Kusunoki, F., and Hashizume, H. (2006). A system for supporting group activities with a sensor-embedded board. *IEEE Transactions on Systems, Man, and Cybernetics-Part C* 36 (5): 693–700.

Thies, C.G. Role theory and foreign policy analysis in Latin America. *Foreign Policy Analysis* 13 (3): 662–681.

Tolone, W., Ahn, G., Pai, T., and Hong, S. (2005). Access control in collaborative systems. *ACM Computing Surveys* 37 (1): 29–41.

Turoff, M. and Hiltz, S.R. (1982). The electronic journal: a progress report. *Journal of the American Society for Information Science* 33 (4): 195–202.

Vandenberghe, C., Bentein, K., and Panaccio, A. (2017). Affective commitment to organizations and supervisors and turnover: a role theory perspective. *Journal of Management* 43 (7): 2090–2117.

VanHilst, M. and Notkin, D. (1996). Using role components to implement collaboration-based designs. *Proceedings of ACM 1996 Conference on Object-Oriented Programming, Systems, Languages and Applications (OOPSLA'96)*, San Jose, CA, USA (6–10 October 1996), pp. 359–369.

Winograd, T. (1988). A language/action perspective on the design of cooperative work. *Human-Computer Interaction* 3 (1): 3–30.

Wooldridge, M. and Jennings, N.R. (1995). Intelligent agents: theory and practice. *Knowledge Engineering Review* 10 (2): 115–152.

Wooldridge, M., Jennings, N.R., and Kinny, D. (2000). The Gaia methodology for agent-oriented analysis and design. *Journal of Autonomous Agents and Multi-Agent Systems* 3 (3): 285–312.

Ye, Y., Boies, S., Huang, P.Y., and Tsotsos, J.K. (2001). Agents-supported adaptive group awareness: smart distance and WWWaware. *IEEE Transactions on Systems, Man, and Cybernetics-Part A* 31 (5): 369–380.

Zambonelli, F., Jennings, N.R., and Wooldridge, M. (2003). Developing multiagent systems: the Gaia methodology. *ACM Transactions on Software Engineering and Methodology* 12 (3): 317–370.

Zhao, L. and Kendall, E. (2000). Role modeling for component design. *The Proceedings of the 33rd Hawaii International Conference on System Sciences*, Washington, DC, USA (June 2000), pp. 312–323.

Zhu, H. (2003). Some issues in role-based collaboration. *Proceedings of IEEE Canada Conference on Electrical and Computer Engineering (CCECE'03)*, Montreal, Canada (May 2003), vol. 2, pp. 687–690.

Zhu, H. (2003a). A role agent model for collaborative systems. *Proceedings of the 2003 Int'l Conference on Information and Knowledge Engineering (IKE'03)*, Las Vegas, Nevada (June 2003), pp. 438–444.

Zhu, H. (2006). Role mechanisms in collaborative systems. *International Journal of Production Research* 44 (1): 181–193.

Zhu, H. (2006a). Separating design from implementations: role-based software development. *Proceedings of the 5th IEEE International Conference on Cognitive Informatics*, Beijing, China (17–19 July 2006), pp. 141–148.

Zhu, H. and Zhou, M.C. (2006). Role-based collaboration and its kernel mechanisms. *IEEE Transactions on Systems, Man and Cybernetics, Part C* 36 (4): 578–589.

Zhu, H. and Zhou, M.C. (2006a). Supporting software development with roles. *IEEE Transactions on Systems, Man and Cybernetics, Part A* 36 (6): 1110–1123.

Zigurs, I. and Kozar, K.A. (1994). An exploratory study of roles in computer-supported groups. *MIS Quarterly* 18 (3): 277–297.

Exercises

1 Why are roles a complex concept? Why is it challenging to specify roles?

2 What do modeling-roles mean? What are the conceptual problems of modeling-roles?

3 What do roles express in RBAC?

4 What would you like to specify a role?

5 What are the differences between roles in agent systems and in CSCW systems?

6 Why should roles follow the ideas of social psychology?

7 Describe the differences between classes and roles.

8 Why is it inappropriate to let object play roles?

9 Do you agree that roles are independent entities on agents and objects? Why or why not?

10 Should roles be independent on other entities? Why?

11 Try to describe real-world roles with the role concept summarized in this chapter, i.e. with rights and responsibilities: professor, student, teacher, project manager, software engineer, and senior analyst.

Part II

Methodologies and Models

3

Role-Based Collaboration

3.1 Requirements for Role-Based Collaboration

Role-Based Collaboration (RBC) was initially proposed to support natural collaboration through computer-based systems. We discuss RBC ideas and concepts using the terms *agent*, *human user*, and *participant* and do not differentiate between them in this chapter.

Definition 3.1 *A methodology* is a system of models, methods, processes, and rules used in an area of study or activity. It includes a set of concepts, principles, structures, architectures, and algorithms. It is the systematic and theoretical analysis of the methods applied in a field of study. It comprises the theoretical analysis of the body of methods and principles associated with a branch of knowledge.

Definition 3.2 *RBC* is a computational methodology that takes roles as an underlying mechanism to facilitate collaborative activities, including abstraction, classification, planning, policymaking, cooperation, coordination, and decision making. It is a promising approach to discovering and solving problems in complex systems including collaboration.

Collaboration technologies have the potential to enhance the effectiveness of teamwork within and between organizations. The roles played by participants in a collaborative activity are important factors in achieving successful outcomes (Zigurs and Kozar 1994). The role concept is a key concern in the development of computer-based systems (Shneiderman and Plaisant 1994). The contributions of Singley and Singh (Singley et al. 2000) and Smith (1998) demonstrate the importance of roles in a collaborative system. Some traditional Computer-Supported Cooperative Work (CSCW) systems, or groupware, have indeed utilized the concept of roles (Barbuceanu et al. 1998; Botha and Eloff 2001; Edwards 1996; Greenberg 1991; Gutwin and Greenburg 1999; Guzdial et al. 2000; Leland et al. 1988;

E-CARGO and Role-Based Collaboration: Modeling and Solving Problems in the Complex World, First Edition. Haibin Zhu.
© 2022 by The Institute of Electrical and Electronics Engineers, Inc.
Published 2022 by John Wiley & Sons, Inc.

Sheng et al. 2014; Smith et al. 1998; Turoff 1993). However, these systems apply the concept of roles intuitively without defining them clearly and precisely. Smith et al. (1998) state that people in a group play various roles, which are supported by a system's treatment of outputs and user inputs.

Collaboration is a common activity in a society. It means to work together (Hellriegel et al. 1983). We can infer that collaboration or teamwork is required when one single person cannot complete a whole task on their own.

CSCW receives increased attention due to the competitive pressures on companies to improve product development and other decision-making processes, and to use information and communication technology (including Internet technologies) as effectively as possible (Guzdial et al. 2000; Leland et al. 1988; Singley et al. 2000; Smith et al. 1998).

CSCW seeks to enhance collaboration among people. However, the question of how to maximize productive collaborations by manipulating role assignments and the configuration of teams is still largely unexplored. RBC is a recent innovation that addresses these concerns. It is a new methodology to organize collaboration by providing role specification, assignment, transition, and negotiation mechanisms. With these mechanisms, roles are clearer to people in collaboration, thereby making the collaboration more productive.

Based on the discoveries and requirements of managerial, social, and psychological science (Ashforth and Mahwah 2001; Becht et al. 1999, Biddle and Thomas 1979; Bostrom 1980; Hellriegel et al. 1983), an RBC system allows users to enhance the collaboration by negotiating, tuning, and transferring relevant roles in the system. The proposed RBC theory has significant applications in group decision support systems (GDS) (Bostrom 1980; Turoff 1993; Zigurs and Kozar 1994), management (Luo and Tung 1999; Plaisant and Shneiderman 1995), personalized interfaces (Guzdial et al. 2000; Plaisant and Shneiderman 1995; Shneiderman and Plaisant 1994; Smith et al. 1998), agent systems (Barbuceanu et al. 1998; Cabri 2012; Campbell and Wu 2011; de Wilde et al. 2003; Ferber et al. 2004; Ferrari and Zhu 2012; Hou et al. 2011; Jameson 2003; Jennings et al. 1998; Jennings and Wooldridge 1996; Li et al. 2012; Maes 1994; Nair et al. 2003; Nwan 1996; Nwan et al. 1996; Odell et al. 2003a, b, 2005; Padgham et al. 2009; Picard and Gleizes 2003; Russell and Norvig 2003; van Splunter et al. 2003; Wooldridge and Jennings 1995; Wooldridge et al. 2000; Zambonelli et al. 2003), and software engineering (Dafoulas and Macaulay 2001; Luo and Tung 1999; Murdoch and McDermid 2000; Ould 1995; Riehle et al. 2000; Pressman 2007; VanHilst and Notkin 1996).

The requirements for RBC can be clarified in the context of CSCW systems, collaboration systems, or, more broadly, complex systems. Firstly, from the technological perspective, collaboration systems should provide people with an environment that reproduces as closely as possible the experience of face-to-face collaboration, which is not possible or too expensive for these people:

- Responsive Communication: collaborators require that information be exchanged efficiently and rapidly over the network the collaboration system is built on.
- Sharing: collaborators require resources to be shared to complete their common tasks. This sharing includes data sharing and view sharing for all the collaborators.

Secondly, from the perspective of usability, a collaboration system should have the objective of providing an environment that supports a particular collaborative activity more effectively than face-to-face collaboration, at least in some aspects.

- User satisfaction: collaborators should be satisfied with the results from their collaboration on a collaboration system. Roles can help people collaborate without direct contact if they do not want to.
- More productive collaboration and better collaboration performance: with a collaboration system, collaborators should be more productive and perform better than they would in a natural environment without a collaboration system. Roles are such candidate mechanisms because they can be used to avoid or reduce conflicts and ambiguity. Productivity can also be improved through better coordination of work using tools like workflow systems.
- Cooperation encouragement: collaborators should receive continuous encouragement while participating in collaboration via a collaboration system. Roles can meet this requirement because roles can be considered ranks and rewards for people's success in collaboration (Zhu and Zhou 2006b, Zhu et al. 2012).
- Intelligent help: collaborators should receive assistance in their collaborative activities. Agents, as representatives in a system, may help people respond to simple requests or record other requests. Agents and roles can assist collaborators in clarifying their rights and responsibilities.

The conventional CSCW systems mainly focus on the technological perspective, i.e. providing a face-to-face-like collaboration environment on networked computer systems. The users encountered many problems such as unsatisfactory communication, frustrating waiting time, difficult discussion environments, and complicated operations. Collaboration using traditional CSCW systems is often clumsy, impractical, and frustrating compared to face-to-face collaboration (Tao and Zhang 2013; Twidale and Nichols 1998; Zhu and Zhou 2006b; Zhu et al. 2012, 2015). Recent advancement of computing and communication technologies has evidently overcome these difficulties. However, there are still challenges for us to face from the usability perspective. To make RBC a reality, there is a need for improved methods and technologies that support the specification, design, and implementation of RBC systems.

The scenario of collaboration based on an RBC system can be described as follows:

- An RBC system is installed on a server or a cloud.
- Each user uses an interface such as a web browser to register. Corresponding agents are created. Only a default role is assigned to each user.
- Users log into the system on client computers with default roles.
- Some users create classes, objects, agents, roles, environments, and groups in the system.
- Users negotiate roles and may play many roles but only one at a time. They may play different roles at a different time based on the requirements of collaboration.
- A user can send messages relevant to his/her roles using the request (right) interfaces and receive messages through the service (responsibilities) interfaces.
- The agents accept the incoming messages when the users are logged out of the session.
- By timely negotiation, users can transfer roles and roles can be modified or tuned by other users with specific roles based on the requirements of collaboration.
- Through the roles, users access objects and their agents, contact other agents, join groups, and contribute to the collaboration.
- The result of the collaboration is reflected by the states of objects, roles, agents, environments, groups, and users in the system.

3.2 Architecture of an RBC System

To understand a system, we need to understand its architecture. The architecture of a system describes the components and their general relationships. At first, an RBC system should have the component of roles, which are the specifications of the rights and responsibilities for a player to take. The second component in an RBC system is agents, which are role players. The third component is groups, which are formed to conduct collaboration among role players. The fourth component is environments, which establish a platform for agents to play roles. The fifth component is objects, which are the resources for role players to utilize. The sixth component is classes, which classify objects and allow objects to be easily and effectively managed. Objects and roles are the fundamental components that form an environment. Messages are the seventh component, which establishes connections among role players. Human users are also regarded as components in RBC systems because RBC systems support collaboration among the users. In RBC systems, human users and their corresponding agents together play designated roles.

With an object-oriented paradigm (Budd 2001; Zhu and Zhou 2006a), we can easily adopt a role's incoming messages to express its services with class mechanisms. Therefore, we can use a special object, i.e. an agent, to express a user providing specific services in a collaboration system. This idea has been discussed many times in the field of object modeling and systems (Seo et al. 2011; Sousa et al. 2006). In other words, when an object plays a role, it provides specific services relevant to that role.

No facility currently exists to define the outgoing messages of a role as an expression of its rights. Role-Based Access Control (RBAC) has done much in dealing with users' permission to use certain operations, but ignores the varieties of users' outgoing messages. We consider a role as two special interfaces, one called the request interface, the interface between a user and the system, and the other called the service interface, the interface between the agent representing a user and the system. In this framework, roles function as the entities that allow users to communicate with the system.

Therefore, with a role, all the requests and services are defined for a user and form interfaces. These interfaces are composed mainly of messages. The outgoing messages in the request interface are used by a user to ask for services. The incoming messages in the service interface are used by an agent or a user to provide services. Moreover, users can use outgoing messages to access the states of their agents and provide services with the help of the agents (Figure 3.1). In

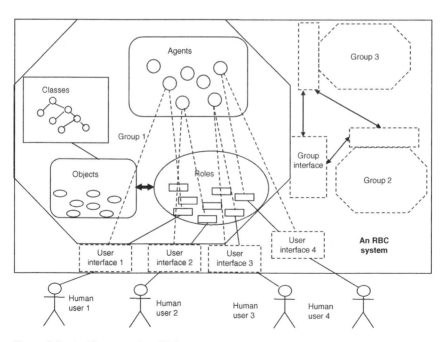

Figure 3.1 Architecture of an RBC system.

Figure 3.1, a dashed line between an agent and a human user means that the agent is a representative of the human user; solid arrows mean messages; and solid lines mean tightly coupled components. Environments, which are composed of objects and roles, are not shown in Figure 3.1 due to the difficulties of visualizing. The rectangles with curved corners are used to express both agents and objects because agents are also instances of a special class. Roles and classes are presented with different shapes to emphasize that roles should be at the same level of abstraction as classes.

In collaboration systems, when human users log into a system, they have two existences: their agents and themselves. The agents take the position of the users when they log out. With an RBC system, we can obtain a human–computer interaction scheme and a structure of RBC systems shown in Figure 3.1. Please note that in Figure 3.1, there are some agents without users linked to them. That is, the corresponding users are offline.

3.3 The Environment Established by Role-Based Collaboration

Collaboration advocates collectivism. We believe that the overall benefit of the group of participants is more important than the benefit of an individual participant. From this point of view, a collaboration system should encourage participants to contribute to the whole collaboration. RBC follows this principle. On the other hand, RBC does not oppose the consideration of individualism, i.e. it takes care of individual benefits by providing appropriate roles and structure for role promotion.

People cooperate in order to yield better or more productive results than individual work, and thus collaboration is inevitable (Hawkins et al. 1983; Hellriegel et al. 1983; Homans 1950; Miner 1992). Collaboration may produce both good and bad consequences (Cartwright and Lippitt 1957). As a tool to support collaboration, a collaboration system should help a group of people produce satisfactory results, and provide wealthy, attractive, and worthwhile experiences for the participants. However, the question of how to maximize productive collaboration by manipulating the configuration of teams is still largely unexplored (Turoff 1993). CSCW systems are virtual societies created by system developers. The design of such systems should reflect the rationalities of natural societies. We believe that roles are good mechanisms to support collaborative work with computers (Zhu 2003a, b, 2005, 2006; Zhu and Hou 2011; Zhu et al. 2012, 2015) because roles improve the management of groups and collaborative tasks (Ashforth and Mahwah 2001; Biddle and Thomas 1979; Zigurs and Kozar 1994) and roles facilitate coordination with groups of people and accomplishment of tasks

(Shneiderman and Plaisant 1994). An experience of collaboration can affect participants' satisfaction, motivation, and productivity. The design of the process of collaboration significantly affects the success of the collaboration (Hellriegel et al. 1983). During the collaboration process, some are active and aggressive, and some others are passive and peaceful.

To build a system that encourages participants' contributions in collaboration is interesting and challenging. RBC advocates the use of roles as flags to express a participant's progress and contributions in collaboration. To boost people's work ethic is a general aim for all the regulations or policies of societies or organizations. The initiatives of RBC are to encourage the participants to contribute as much as possible. The basic assumptions are based on Maslow's hierarchy (1970) of needs. We can say that diligent people hope to become great members of a great group. Those people that do not contribute to a team will eventually be discarded. Based on this assertion, we assume that most people would like to improve themselves by actively participating in the collaboration.

We can divide people into three types: (i) Diligent people who try to do as many jobs as possible. They hope to take on as many responsibilities as possible but do not care about the associated rights. (ii) Ordinary people who only want to complete the assigned jobs. They would like to take on responsibilities but require matching rights. (iii) Lazy people who try to grasp as many rights as possible but take on as few responsibilities as possible.

To restrain type (iii), encourage type (i), and support type (ii), roles should be composed of both rights and responsibilities. That is to say, more rights mean more responsibilities. If people hope to grasp more rights, they must play more or upper roles (Chapter 4), and thus take on more or more important responsibilities. For those who do not care about rights, there should still be enough rights for them to take on more responsibilities. This is a fundamental principle of RBC (Zhu 2003a, b, 2005, 2006; Zhu and Hou 2011; Zhu et al. 2012, 2015). With this, we propose to encourage participants' contributions to a group by well-designed role mechanisms.

Roles allow people to show their career development by the number and importance of the roles they play. That is to say, roles can act as a flag of advancement for people in collaboration. In an RBC team, the lazy people will eventually be discarded by the team and will have no roles to play, and the diligent people will be the majority and be promoted to play more important roles with more opportunities to contribute.

Logically, in RBC, we consider a human user and the corresponding agent the same entity. However, to build an RBC system, we still need to consider agents and human users separately. We can attach roles to an agent to express a human participant playing that role. In this system, human users contact each other through roles.

The promotion of roles can be in different routes. The roles in RBC form a similar hierarchy as shown in Figure 3.2, where MS means Master of Science, and BS

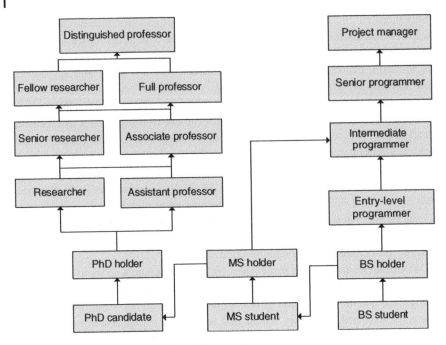

Figure 3.2 Examples of promotions of an agent.

means Bachelor of Science. RBC supports designing and implementing a role-based software engineering platform that tries to encourage software development team members to contribute to a project via roles. Our role transfer tool (Chapter 9) facilitates how to manage human resources in a team, which is different from the method of dealing with development processes with Role Activity Diagrams (RAD) (Murdoch and McDermid 2000).

When using role-based systems, human users upgrade themselves by accomplishing more and more services and improving the states of their agents. They will obtain more rights from more roles. The common tasks for human users are as follows:

- Task 1: provide services, to make a class of agents include all the implementations of the methods corresponding to the incoming messages of the roles.
- Task 2: ask for services, send messages to roles and obtain services from other agents.
- Task 3: accessing objects, to obtain the properties of objects authorized by the roles.
- Task 4: request a promotion, to play more or upper roles (Chapter 4).

Task 1 is the most important task for a human user. At least, a human user could make a class that implements all the required services by the role in the simplest forms (for example, return true or a simple response). Tasks 2–4 can be accomplished by human users interacting with the interface expressed by the roles, which also help human users accomplish task 1 more efficiently.

In RBC, a role is considered a template to evaluate agents. Agents can consider a role a model to simulate, learn, and behave. We can enhance a role by designing a role engine (Section 10.5, Chapter 10) to evaluate an agent and decide if the agent can play the role. In our method, we view playing an upper role as a promotion for an agent and its human user. Some authors take it as evolution (Alonso et al. 2003; Gottlob et al. 1996; Kristensen and Østerbye 1996; Steimann 2000, 2008). A role (instance) should be attached to an agent to represent a promotion for that agent. The above methods have a considerable effect on the promotion of human users, i.e. to play more roles. The former method should check if the agent can be attached to the role and return true or false as a result. The latter method should dispatch the message evenly and without bias. To determine if an agent is qualified to play a role, there are several criteria:

- The agent must provide methods to cover all the incoming messages for the role;
- The agent must have sufficient creditability, which is expressed by the credit points collected by playing roles;
- The agent must have enough abilities and resources including time and space;
- At least one of the roles that the agent was playing belongs to the lower roles of this new role if the new role has lower roles (Chapter 4); and
- Lastly, when a role is approved for an agent, the ability and resources of the agent including space and time should be adjusted based on the new role's requirement of resources necessary to play this role.

To manage credits, we assign a weight to each incoming message for a role. When we specify a role, we can also assign incoming messages to a role and set the specific weight for that role. Specifically, the same incoming messages might have different weights depending on different roles. When an agent executes a task relevant to the incoming message, it will collect the weight of the message as its credits.

In summary, RBC sets up a positive environment for agents to participate and motivates agents to work and contribute to collaboration.

3.4 The Process of Role-Based Collaboration

To understand the process of RBC, we need to clarify the categories of collaboration. Besides the categorization based on location and time, we can use another method to categorize different types of collaboration.

Collaboration can be divided into two types based on the synchronicity of collaboration:

- Synchronous collaboration: face-to-face collaboration or virtual face-to-face collaboration with the help of collaboration systems.
- Asynchronous collaboration: collaboration occurring at different times with the help of paper-based or computer-based communications.

Collaboration can also be divided into two categories based on the length of collaboration:

- Long-term collaboration: collaboration that takes a long time, such as a large project or a large venture such as an enterprise.
- Short-term collaboration: collaboration that takes a short time, such as a meeting or a conference.

Synchronous collaboration is normally used in short-term collaboration and asynchronous collaboration is typically used in long-term collaboration. Evidently, RBC is usually used to support asynchronous and long-term collaborations. However, RBC can also be used to support short-term synchronous collaboration. Based on the assumptions for collaboration, we conclude the following:

- A person's ability is limited. If they play too many roles, there is a chance that they cannot serve all the requests satisfactorily and lose credits.
- A person can collect his/her credit points by serving others.
- To assign new roles is to appraise a person's past work.
- A role might be disapproved based on the credit of the person.
- The dispatchers are fair and distribute messages evenly.

By RBC, we mean that people collaborate in an environment where their roles are clearly specified. The procedure for RBC is as follows (Figure 3.3):

Step 1: Negotiate roles. Agents, including administrators and team members, discuss or negotiate to specify the roles relevant to collaboration. If no agreement is reached, the collaboration aborts. There are two ways to negotiate roles. One is initiated by an agent and the other is initiated by a role facilitator, e.g. conventional human resource officers in administrations. The former happens when an agent wants to play a role and the latter happens when the role facilitator wants to nominate an agent to play a role. This step needs to clarify how many roles should be established in a group and what roles should be set up, i.e. role discovery; and what the roles' contents are, i.e. role specification.

Step 2: Assign roles. Every agent is assigned one or more roles. If no agreement is reached, the collaboration aborts. Both of the ways of negotiation in Step 1 will result in the same process, i.e. the human user should provide a class for their agent to cover all the services required by the role. In reality, there might be

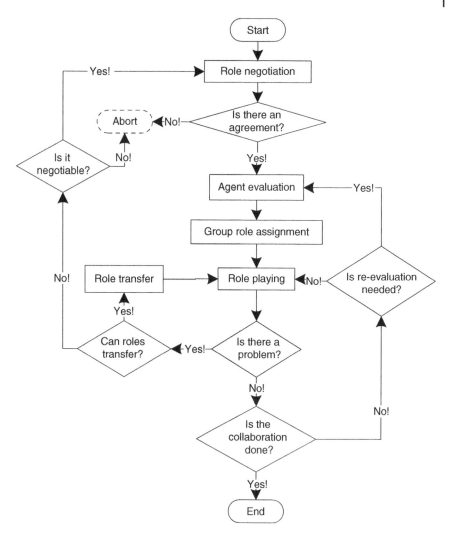

Figure 3.3 The life cycle or process of RBC.

declarations or working plans for a person to play a role. This step is divided into three substeps including agent evaluation, group role assignment, and role transfer according to dynamic requirements.

Step 3: Play roles. Agents work according to their roles until collaboration is successfully completed.

Step 3.1: Check incoming messages. Agents understand what they need to do at this time. The incoming messages are confined by the role responsibilities (the service interface). If there are conflicts or discontents, go to Step 1.

Step 3.2: Issue outgoing messages. People need to access and interact with the environment by sending messages or asking for others' services in order to provide their services. The messages are confined by their role's rights (the request interface). If there are conflicts or discontents, go to Step 1.

The general activities in an RBC system can be listed as follows:

1) To play a role. An agent wants to grasp rights to take responsibilities.
2) To serve others. An agent responds to incoming messages.
3) To evaluate an agent. The system evaluates an agent by checking the relevant information to see if the role is appropriate for the agent.
4) To disapprove a role. The system takes the role away from the agent.
5) To promote an agent. This system upgrades the agent by assigning it an upper role (Chapter 4).

In order to facilitate the above activities, we need to consider several important factors as follows:

- How can an agent collect credit points? We can set a weight number (credit points) for a message. When an agent serves this message, its weight is added to the credit points of the agent.
- How to dispatch messages? The dispatcher should have no bias. There should be a procedure to balance the dispatching, such as considering the factors of credit points, abilities, and the state of an agent.
- How to check if an agent is capable of fulfilling the role requirements? The only way is using the agent's current state including its abilities and credit points.

Example 3.1 Based on the above procedure, we can judge if a collaboration system is role-based or not. In fact, many traditional systems that apply role concepts cannot be called RBC systems because they only support some role views but not the underlying role mechanisms of collaboration systems.

A meeting is a common collaborative activity. There are roles in a typical meeting (Toastmasters Club 2001) such as toastmaster, introduction master, timer, speakers, table-topic master, and evaluator (Turoff 1993). A meeting requires the task of managing roles.

1) Negotiate roles. Before the meeting, people must negotiate and understand the roles in the meeting.
2) Assign roles. If there is an agreement, the roles are assigned, a meeting is scheduled, and the collaboration continues to Step 3, otherwise, it stops.
3) Play roles. People work according to their roles until the meeting is completed successfully or satisfied with a result.

   3.1) Check incoming messages. People understand what they need to do at this time. The incoming messages are confined by the role responsibilities (the service interface). The people collect credits by serving others, e.g. as a speaker or a chair. If there are conflicts, discontents, or promotion requests, go to 1).

3.2) Issue outgoing messages. People need to access and interact by sending messages or asking for others' services in order to provide their services, e.g. making a speech, or chair the meeting. If there are no incoming messages, people may issue messages as needed. The messages are confined by their roles' rights. If there are conflicts, discontents, and promotion requests, go to 1).

In fact, we practice collaboration based on roles even though we do not declare so. RBC is difficult to perform in natural environments because no role specification mechanism is available. Daily role negotiation cannot completely remove ambiguity. This is the major barrier to implementing RBC without the support of computing systems. RBC systems can provide clarity and enhanced support in situations where collaborative working is required.

3.5 Fundamental Principles of RBC

RBC has the following properties:

- Clear role specification: it is easy for human users to specify and understand their responsibilities and rights.
- Flexible role transition: it is flexible and simple for a human user to transfer from one role to another.
- Flexible role facilitation: it is easy for role facilitators to modify roles. Because collaborative activities are constantly evolving, even the existing roles might be required to be adjusted in correspondence with the development of the system.
- Flexible role negotiation: it is easy to negotiate a role's specification between a human user and a role facilitator.
- The interactions among collaborators are through roles.

From these properties, we find that roles are the key media for human users to interact and collaborate. The users are allowed to concentrate on interacting with roles rather than human users to make the collaboration more objective. The role specification and negotiation are the major tasks to form an environment for collaboration. The specification of the interactions among roles outlines the procedures of collaboration. Hence, role specification is a fundamental and key mechanism.

RBC means that human users apply an RBC system to cooperate and obtain an ideal collaborative result. To build an effective collaboration system, we must understand the fundamental principles with which we need to comply. Many principles we learned from modeling methodologies, software engineering, and collaboration systems development will serve as our basis.

With object-oriented principles, we can conceptually construct the underlying components of an RBC system, including classes, objects, messages, interfaces, agents, roles, groups, and human users. In this way, we can express a role's incoming messages in a class. Therefore, we can use a special object, i.e. an agent, to express a human user who provides services in a collaboration system. Here we need to emphasize that agents are the representatives of human users in a system and agents can do programmable jobs for human users.

It is a common idea used in object modeling methodologies that roles are specified from a server's viewpoint. Namely, when an object plays a role, it responds to incoming messages. Such roles only emphasize the server aspects, i.e. what services to provide. They do not mention the client aspects, i.e. what requests the player (an agent) can issue. This viewpoint considers roles with certain outgoing messages to facilitate the rights of a human user. Therefore, we need to introduce the ideas from roles in RBAC, which accommodates the rights of human users. It is a must to combine the role concepts of object modeling and those of RBAC to support the role models in collaboration systems. Hence, we need to comply with four sets of principles, i.e. object, agent, role, and group, when building an RBC system.

3.5.1 Object Principles

The object-oriented methodology is widely used in system modeling and software engineering. A collaboration system can also be considered from an object-oriented viewpoint. In an RBC system, we need to obey the following object-oriented principles (Zhu and Zhou 2006a; Zhu et al. 2015).

O1: Everything in the world is an object. Objects can be used to express everything in a collaboration system.

O2: Every system is composed of objects and a system is also an object.

O3: The evolution and development of a system are caused by the interactions among the objects inside and outside the system.

O4: The interactions among objects are expressed by sending messages that are requests to invoke objects' actions.

O5: A message is a way to activate the services of an object.

O6: Each object is an instance of a class that expresses the commonality of a group of objects.

O7: Each class (called a subclass) may inherit from another class (called a superclass).

O8: Classes can be taken as templates of objects.

O9: Each object can be created, modified, and destroyed.

3.5.2 Agent Principles

To express human users in collaboration, we need to separate these special objects called agents from ordinary ones. The agent principles are as follows (Zhu and Zhou 2006b).

A1: An agent is a special object, which is autonomous.

A2: An agent may represent the existence of a human user in a collaboration system.

A3: Each agent can be created (when a human user signs on), modified (when they want it to do more or better jobs), and deleted (when they sign out).

A4: An agent can help a human user do tasks, e.g. sending a notification message.

A5: A human user accesses his/her agent in collaboration.

A6: Collaboration may still advance despite the users logging out because their respective agents can perform tasks on their behalf.

A7: An agent may play one or more roles.

A8: An agent may join or leave a group (or groups).

3.5.3 Role Principles

A role can represent a special aspect of human users through the specification of incoming and outgoing messages. It provides an agent not only messages to serve others, but also messages to ask for services. We clarify the following principles of roles (Zhu and Zhou 2006b):

R1: A role is independent of agents or human users. We can define it separately. It is a common idea that a role is dependent on objects in object systems (Buckle 2006; Gottlob et al. 1996; Harrison and Ossher 1993; Kristensen and Østerbye 1996; Odell et al. 2005; Pernici 1991). In collaboration, however, collaborators may not care about a specific person. They only want to contact a person who plays a specific role. For example, professor X at Nipissing University may ask for services from the UTS (University Technology Service). X does not care which person comes to serve, as long as they play the role of technician at UTS.

R2: A role can be created, changed, and destroyed by a human user with a special role.

R3: A role includes both responsibilities (the service interface) when a player is viewed as a server, and rights (the request interface) when the player is viewed as a client. To specify a role means to specify both the responsibilities and rights of the role. A role does not accomplish tasks specified by the responsibilities. Only human users and their agents who are playing the role accomplish the tasks. As for the service interface, a role is actually a filter of messages sent to an agent. This filter only allows specified incoming messages to be sent to the player. As for the request interface, a role expresses or restricts the accessibility of an agent to the system.

R4: Roles are also a medium for interactions. Interactions among agents are based on their roles, i.e. a message to request collaboration with other human users is sent to the relevant roles. RBC emphasizes that the message receivers are mainly roles, not agents or human users unless each role has only one human user playing it, such as in a typical family. Human users are not required to identify a specific person to serve them, i.e. any human who plays the required role is acceptable, such as the client service department of a company.

R5: Playing a role means that the agent (and/or the human user) operates according to the requirement of the role. A role can be played by one or more human users or their agents at the same time.

R6: Roles can be used to support indirect and direct interactions. For direct interactions, each role has exactly one person playing it. For indirect interactions, each role has multiple persons playing it. In the former case, identifying a role means identifying a person. In the latter, identifying a role does not mean identifying a person.

3.5.4 Group Principles

In reality, human users work in a group and may hold multiple roles. Every work setting involves groups of individuals. To accomplish a task, the group members (human users) interact with each other. We should understand the following principles of groups (Zhu and Zhou 2006b):

G1: A group is a necessary structure in a collaboration system.

G2: A group can be created, changed, or destroyed.

G3: Before specifying a group, we must specify all the roles in the group.

G4: Forming a group is letting human users/agents join the group and play roles. They (human users/agents) are called members of this group.

G5: The state of a group dynamically changes.

G6: A group can be embedded, i.e. a group may be an object in another group.

G7: A group may overlap with other groups, i.e. a member may belong to two or more groups.

G8: A group can be public or private.

Public means that all the human users using the system can join the group. Private means that joining a group is controlled by a special human user who plays a special role (moderator).

G9: A group can be open or closed.

"Open" means that new human users can join the group and "closed" means that no new users can join the group.

3.6 Benefits of Role-Based Collaboration

Complex systems involve trust, dynamics, interaction, adaptation, coordination, and information sharing. Roles can be made to facilitate an organizational structure, provide orderly system behavior, and consolidate system security for both human and nonhuman entities that collaborate and coordinate their activities with or within systems (Sheng et al. 2014; Zhu 2003a, b, 2005, 2006; Zhu and Hou 2011; Zhu and Zhou 2006b; Zhu et al. 2012, 2015).

Compared with other role-based approaches (Al-Zaghameem and Alfraheed 2013; Caetano et al. 2009; Ferraiolo and Kuhn 1992; Ferraiolo et al. 1995, 2001; Kühn et al. 2015; Zhang 2008), RBC possesses the following properties that are considered advantages:

1) RBC has a specific and formalized model (Chapter 4);
2) RBC presents a clearly stated process of collaboration;
3) In RBC, objects and agents are clarified as different entities in collaboration, i.e. agents are role players but objects are not;
4) The formalized model provides more reusable "components" from its definitions rather than its "concepts" in other approaches;
5) With the formalized model, many real-world problems can be discovered, formalized, and then computer-based solutions can be specified or provided;
6) RBC provides a computational methodology for discovering a variety of problems: abstract or industrial;
7) RBC provides new ideas in collaboration research and assists in the application of collaboration technologies in other fields; and
8) RBC introduces a new perspective on collaboration, i.e. collectivism that emphasizes the team effort/performance other than the traditional individualism.

3.6.1 Establish Trust in Collaboration

The trust of an agent is an indicator of "being honest and reliable" that gives confidence for other agents to cooperate with the agent. A high degree of trust in an agent means that the agent will very likely be chosen as a collaboration partner.

If an agent has trust, it is easy for it to be assigned tasks, earn credits, and make money. Trust in collaboration systems means that an agent situates in an environment where each agent tries to choose the most reliable partner from potential agents and establish a strategy to adapt to the environment (Dasgupta 1998; Fan et al. 2011; Ramchurn et al. 2004).

Trust may increase with a positive experience and decrease with a negative experience or, over time, without any experience. This also makes it clear that trust is changeable. With roles, agents are regulated for their behaviors. Agents that lose trust will definitely lose credits by regular agent evaluation and dynamic role assignment. The most trustworthy agents hold the most important and valuable roles. Agents that gradually lose trust will finally be discarded by the system.

Roles are commonly accepted concepts that present enough information for trust. With roles, agents or people become role-players. The trust of a player is reflected by the roles they are playing. Playing a role means that the player possesses enough trust to do so. In collaboration with roles, agents do not have to care about the trust of other agents because they mainly interact with roles, which indicate trust. It is the system that ensures the trustworthiness of human users and their agents.

Roles are an equivalent mechanism to the reputation model (Sabater and Sierra 2001) used in multi-agent systems to establish trust (Dasgupta 1998; Fan et al. 2011; Ramchurn et al. 2004). In e-commerce systems, reputation enables buyers to choose the best sellers in the system. Moreover, reputation encourages sellers to behave well because they will be discarded by future buyers if their reputation goes down. Roles assist in building reputations in the following aspects:

- Providing methods to gather information regarding the trustworthiness of an agent, using their relationships with roles;
- Providing reliable evaluation methods to assess the qualifications of agents for specified roles;
- Providing mechanisms to endorse qualifications that describe the trustworthiness of an agent; and
- Building agents' beliefs, i.e. a role played by an agent a tells that a is capable of carrying out the delegated tasks as expected for the role.

In RBC, agents cooperate with other agents in an environment where roles dictate the agents' behaviors. Role mechanisms make agents not self-centered, and, therefore, prevent lying and collusion between agents. Compared with traditional trust models in multi-agent systems, roles apply a centralized evaluation methodology, thus enabling better aggregation and evaluation of qualifications to improve trust. Roles establish a system-level of trust that provides truth-eliciting interaction protocols; develop reputation mechanisms that foster trustworthy behavior; and set up security mechanisms that ensure new entrants can be trusted (Fan and Yen 2011; Ramchurn et al. 2004).

3.6.2 Establish Dynamics

In an agent community, dynamics is defined as a force or impetus for agents to be created, join, advance, evolve, and work in an environment (Zhu 2007, 2010). From this perspective, dynamics is considered an important property of both a multi-agent system and its agents. An agent is viewed as an entity that perceives its environment through sensors and acts upon that environment through actuators. A rational agent does the right things and should choose actions that are expected to maximize its performance according to its knowledge. Rationality is one of the dynamics of agent performance in a multi-agent system. However, the expression of rationality has not been comprehensively investigated. Roles are the rationalities for agents to live. Computer systems are added to the Internet to perform certain roles and to provide services on the Internet. In an RBC system, role factories are used to make roles match the requests and services in the role pool. If there are no more unsatisfied requests and unused services, the role factories should cease role production. There may be a situation where roles are not claimed by agents. After such a situation continues for a specific period, role production should stop.

Dynamics generate motivational or driving forces, physical or moral, in any field. The human world follows Maslow's hierarchy (Maslow 1970) of needs. A theory of dynamics should indicate the origin of the forces, the relationship among these driving forces, and the impacts of these forces on environments and agents. The principle of batteries (Buckle 2006) can be used as an analogue. Roles can be taken as cells and agents can act as electrons. In a battery, the power is stored when cells are charged or recharged. The power is used when the electrons move to fill the cells. When all the cells are filled with electrons, the power disappears. This situation is similar in RBC systems, i.e., roles attract agents. The greater the number of roles, the greater is the attraction on agents. Consider a factory that produces agents. The production of agents will be relative to the demand of roles. The force to produce an agent for a role is proportional to the cardinalities of that role, i.e. how many agents are required or allowed to play this role. The cardinality of a role is the magnitude of the force driving agent production. Driven by this force, the factory produces agents to adopt the roles. This is why American universities contributed many computer science graduates to the IT market in the years 1998–2002. Periodically, agent factories may produce a surplus of agents.

The agent factories should stop creating a kind of agent if the number of required agents is met. This is why the enrollment of computer science students dropped down in the years 2002–2006. This differs from the battery analogy, in that agent production has inertia in the production pipeline. This is why many computer science graduates in the United States could not find jobs in the years 2002–2006. In multi-agent systems, this inertia or surplus should be minimized. The possibility of minimizing such inertia is dedicated to timely records of the number of required agents for specific roles.

In RBC, if agents hope to grasp more rights, they must adopt more roles. As a result, they must take on additional responsibilities. For those who do not care about rights, there would still be enough rights for them to take additional responsibilities. Such an idea is a fundamental principle in RBC (Zhu and Zhou 2006b, Zhu et al. 2015). With this fundamental principle, role mechanisms can be taken as dynamics for agents to be proactive and mobile in an environment. By adopting roles, agents can progress by presenting how many significant roles they are playing. Roles can be taken as a flag of advancements for agents in collaboration. In an RBC system, lazy agents will be discarded and reclaimed by the agent factories. Diligent and regular agents will gradually become a majority in the system. Diligent agents will be promoted to play more important roles, providing them opportunities for greater contribution. In RBC:

- Roles can be specified with credits and services.
- Agents can earn credits by playing roles.
- Agents are trained to have abilities (to load more service implementations) to match the requirement of the role.
- The goal of agents is to earn the maximum number of credit points.
- To obtain this goal, agents must compete, coordinate, and collaborate.
- The goal of participating in collaboration is to collect more credits.

From the above discussion, an agent can be set a goal with definite credits and then started. The agents will automatically search out roles and adopt them.

After agents become proactive to pursue roles, a new problem may occur. A multi-agent system should have maximum cooperation and minimum competition. If agents are designed to adopt roles and collect as many credit points as possible, there must be competitions among agents. This may introduce negative factors into the system. Fortunately, agents are designed by people and they cannot grow spontaneously. The role distributions should be fair based on the abilities of agents. If an agent loses a bid for a role, it should seek out other roles. If it is inactive for more than a specified period, it should be reclaimed by the agent factories and rebuilt. Competition can be avoided in this way.

3.6.3 Facilitate Interaction

Interaction is essential for collaboration (Zhu 2008, Zhu and Hou 2011). People may fear interaction, as in society, misunderstanding can produce conflict, hate, and even war. That is to say, the major problem of interaction is a misunderstanding. Therefore, interaction requires common needs, backgrounds, views, and goals. All these constitute a shared model that is required for a successful interaction.

- A shared model makes interaction *concise* and *simple*. For example, even though mathematicians interact with one another when trying to solve a complicated mathematical problem, their interactions are succinct because they share similar knowledge of mathematical models.
- A shared model makes interaction *understandable*. Again, from the example of mathematicians, a difficult problem is comprehensible between two parties because they share the same models.

- A shared model makes interaction *comfortable*. For example, musicians interact with one another comfortably through musical symbols, i.e. a shared model. Mathematicians interact comfortably with formal systems, i.e. a shared model.
- A *successful outcome* may result from a shared model. For example, when one person asks another person for directions, it is easy to make this interaction successful if there is a map, i.e. the shared model. However, if there is no map at hand, the individuals involved lose the shared model and their interaction may be unsuccessful.
- Based on Norman's cognitive model (1986), there should be a general shared model in a computer system for human–computer interaction.

Ideal interactions should have a symmetric sharing model, i.e. both parties should share the same model and information flow should be easily regulated. In RBC, we hope to express an ideal way for two parties to interact, in which both parties have the same models not only for action specification and interpretation, but also for internal processing. For example, two people with the same five senses (which are physical models for perceptions, executions, actions, and interpretations) may fail in interaction because they have different models for processing the perceived knowledge based on each individual's unique prior training.

3.6.4 Support Adaptation

Change is an eternal phenomenon of the world. Adaptability is a special property that enables people to sustain themselves in dynamic environments. The often-quoted "survival of the fittest" is a phrase that captures this law of nature. In collaboration, the structures, members, relationships, and environments are always changing. To maintain good collaborative performance, adaptability is required.

After members join a group, their common goal is keeping the group in a workable state. In other words, the whole group must adapt to the new arrangement to pursue a higher performance. In reality, however, it is not the group members who do not want to adapt to the new arrangement. In many cases, it is the system that does not allow individuals to adapt since they must follow a set of static rules set by the system.

With the help of computers, teamwork and the adaptability of a team may be made easier to implement. That is, computer systems should assist people in adapting to the group in order to achieve higher group performance, and thus serving the interests of the individuals (Kerr and Tindale 2004). It is essential for a person to be adaptive within a group since optimal group performance can only be achieved when everybody within the group works with their highest potential. Ultimately, this will require individuals to adapt and change existing behavior. This is why traditional research on adaptation focuses on the adaptability of individual agents (Dahchour et al. 2004; de Wilde et al. 2003; Picard and Gleizes 2003;

van Splunter et al. 2003) and the adaptability of machines to individual users (Atterer et al. 2006; Botha and Eloff 2001; Fischer 2001; Gena 2005; HLA Associates 2001; Hou et al. 2007; Jennings and Wooldridge 1996; Kendal 1999; Pradel et al. 2012; Riehle et al. 2000; Riehle and Gross 1998). However, true adaptive collaboration concentrates on the adaptability of a group as a whole.

Group adaptation includes many aspects, such as group structure and the adaptabilities of individuals. It reflects the abilities of the group to work, live, socialize, and compete. The adaptability of a group is dependent on its organization, structure, culture, and regulations. There are many factors affecting collaborative performance in a group. For example, culture, interests, personalities, health, equipment, hardware resources, benefits, abilities, powers, motivations, and situations are all relevant factors. How to optimize group performance in a timely fashion is a big challenge for researchers of man–machine systems and collaborative technologies if we consider these factors.

The fundamental way to implement adaptive collaboration (Chapter 10) is to model the group and the factors affecting group performance using computer-based algorithms to timely recommend a group state, in order to obtain the highest performance.

3.6.5 Information Sharing

Information sharing is the foundation of collaboration and it must be swift and secure. Roles provide a new trade-off abstraction among classes, processes, and objects. Therefore, roles also provide a new way for information sharing. Information sharing in the context of collaboration is concerned with two major problems: protection and accessibility.

Database management systems are the most widely applied systems that support the sharing of data. Roles are used as a tool that allows administrators to collect users into a single unit against which permissions can be bestowed. In Role-Based Access Control (RBAC) (Ferraiolo and Kuhn 1992), accessibility and protection are implemented mainly through roles. RBAC aims to apply roles to simplify the tasks of security administrators in order to enforce access control policies. A role is described as a set of transactions that a user or set of users can perform within the context of an organization. Research on RBAC has developed rapidly since its inception, especially in the fields of computer security and protection. In particular, many researchers have discussed the architectures and mechanisms of RBAC (Ferraiolo and Kuhn 1992; Ferraiolo et al. 1995, 2001). Their research demonstrates that roles are excellent mechanisms in dealing with access control and system security. The effort on RBAC has produced a mature, consistent, and standardized definition of roles for data protection.

As for accessibility, conflict avoidance is a major concern. When there is an intention to access some objects, no conflict is the basic requirement for all successful accesses. Without roles, agents access objects directly with operations. Agents experience conflict when they access critical shared objects. With roles, agents are offered rights to access objects by their roles. With roles as message dispatchers, the accessing of objects is managed and handled through specific rules and policies, such as First Come First Served (FCFS) and Shortest Job First (SJF). Therefore, fewer conflicts will happen when roles are introduced.

3.6.6 Other Benefits

It is easy for a manager to distribute tasks using roles because it is simpler to find a technician than to find a specific person (Zhu and Zhou 2006b; Zhu et al. 2015). Based on roles, managers have ways to evaluate their staff. Roles provide a balanced means for anonymity and credibility. Managers can more easily find qualified players (people or agents) in collaborations. On the other hand, one person can play different roles and serve many clients in the same time period to save the human power of a company. Two or more people can play the same role to serve a client to improve the efficiency and quality of a specific service and make the client feel more comfortable and more satisfied. Roles can also help players:

- Identify the role player's "self" (Ashforth and Mahwah 2001);
- Avoid irrelevant interruptions that harm work performance;
- Enforce independence by hiding people under roles;
- Remove ambiguities to resolve expectation conflicts (Bostrom 1980; Hawkins et al. 1983; Hellriegel et al. 1983);
- Work with personalized user interfaces;
- Distribute tasks based on the overall requirements of a group;
- Decrease the workload of system administrators;
- Implement separation of concerns;
- Decrease the search space of knowledge; and
- Optimize the team's performance as a whole (Li et al. 2012; Rardin 1997).

3.7 Summary

Role-Based Collaboration (RBC) has developed into a computational methodology to guide the management, organization, and facilitation of collaboration. With continuous investigation, RBC can be widely applied to many new frontier cutting-edge fields (Armbrust et al. 2010; Fan and Yen 2011; Ferrari and Zhu 2012; Hou et al. 2007; Li et al. 2012; Lu and Käkölä 2014; Luo and Tung 1999; Shen and Sun 2011; Wang and Deng 1999; Wang et al. 2011; Weerawarana et al. 2008; Zhu and Zhou 2006b) to analyze, design, and implement related systems.

In the next chapter, the Environments-Classes, Agents, Roles, Groups, and Objects (E-CARGO) model will be described and specified. The E-CARGO model will extend the applications of RBC to complex systems to specify and solve more complex problems.

In the implementation of an RBC system, we should avoid using natural languages to describe what a role should be. We may use a group of messages or message patterns that have no or little ambiguity. By this strict definition, both role players and facilitators have a clear role specification without different role expectations to decrease role ambiguity. Certainly, there is still flexibility in how roles transmit the messages and how messages are processed by agents and human users.

In daily life, the roles of members in a group often change. However, various members may offer different contributions to the group even when they have the same role. Specifically, one member's role is different from another's role even if they have the same role name, such as assistant professor. The definition of roles seems to restrict the creative work of the members because a role is defined by certain message patterns. However, because message patterns are abstract, it is possible for message patterns to still support the creative work of users. If the right to create objects is granted to a role, then the users who play this role can conduct all kinds of creative work. At the same time, by interactive negotiations among users, the roles' responsibilities and rights can be modified to support the creative work of users.

The current situation is that CSCW systems and intelligent collaboration systems are two different fields. The former supports people to collaborate with computer systems but the latter tries to design and implement intelligent computer systems that emulate natural collaboration among people. Collaborative intelligent systems hope to process incoming messages by agents themselves, but in a CSCW system, people decide how to respond to a message.

These two kinds of systems can be combined. Our E-CARGO model is primarily aiming to support collaboration among people. With the development and the evolution of systems built with E-CARGO and Artificial Intelligence (AI) technologies, a collaboration system might become more and more intelligent and could support additional types of collaborative activities. RBC will become true after a role engine is built that automatically distributes messages to agents. With the support of the role engine (see Chapter 10), agents can play their assigned roles, respond to requests, and provide services.

References

Alonso, E., Kudenko, D., and Kazakov, D. (2003). *Adaptive Agents and Multi-Agent Systems*, LNAI, vol. 2636. Berlin, Germany: Springer.

Al-Zaghameem, A.O. and Alfraheed, M. (2013). An expressive role-based approach for improving distributed collaboration transparency. *International Journal of Computer Science Issues (IJCSI)* 10 (4–2): 61–67.

Armbrust, M., Fox, A., Griffith, R. et al. (2010). A view of cloud computing. *Communications of the ACM* 53 (4): 50–58.

Ashforth, B.E. and Mahwah, N.J. (2001). *Role Transitions in Organizational Life: An Identity-Based Perspective*. Lawrence Erlbaum Associates, Inc.

Atterer, R., Wnuk, M., and Schmidt, A. (2006). Knowing the user's every move: user activity tracking for website usability evaluation and implicit interaction. *Proceedings of the 15th International Conference on World Wide Web*, Edinburgh, UK (23–26 May 2006), pp. 203–212.

Barbuceanu, M., Gray, T., and Mankovski, S. (1998). Coordinating with obligations. *Proceedings of the Second International Conference on Autonomous Agents*, Minneapolis, Minnesota, United States (May 1998), pp. 62–69.

Becht, M., Gurzki, T., Klarmann, J., and Muscholl, M. (1999). ROPE: Role Oriented Programming Environment for multiagent systems. *Fourth IECIS International Conference on Cooperative Information Systems*, Edinburgh, UK (September 1999), pp. 325–333.

Biddle, B.J. and Thomas, E.J. (1979). *Role Theory: Concepts and Research*. New York: R. E. Krieger Publishing Co.

Bostrom, R.P. (1980). Role conflict and ambiguity: critical variables in the MIS user-designer relationship. *Proceedings of the 17th Annual Computer Personnel Research Conference*, New York, NY, USA (June 1980), pp. 88–115.

Botha, R.A. and Eloff, J.H.P. (2001). Designing role hierarchies for access control in workflow systems. *25th Annual International Computer Software and Applications Conference (COMPSAC'01)*, Chicago, IL, USA (8–12 October 2001), pp. 117–122.

Buckle, K. (2006). How do batteries store and discharge electricity? *Scientific American. com*. http://www.sciam.com/askexpert_question.cfm?articleID =0001A09D-0D79-1476-8D7983414B7F0000&catID =3&topicID=4, 2006 (accessed 10 August 2020).

Budd, T.A. (2001). *The Introduction to Object-Oriented Programming*, 3e. Boston, MA: Addison-Wesley.

Cabri, G. (2012). Agent roles for context-aware P2P systems. In: *Agents and Peer-to-Peer Computing*, LNCS, vol. 6573 (eds. S. Joseph, Z. Despotovic, G. Moro and S. Bergamaschi), 104–114. Berlin, Heidelberg, Germany: Springer.

Caetano, A., Silva, A.R., and Tribolet, J. (2009). A role-based enterprise architecture framework. *The 24th Annual ACM Symposium on Applied Computing*, Honolulu, Hawaii, USA (8–12 March 2009), pp. 253–258.

Campbell, A. and Wu, A.S. (2011). Multi-agent role allocation: issues, approaches, and multiple perspectives. *Autonomous Agents and Multi-Agent Systems* 22: 317–355.

Cartwright, D. and Lippitt, R. (1957). Group dynamics and the individual. *International Journal of Group Psychotherapy* 7: 86–102.

Dafoulas, G.A. and Macaulay, L.A. (2001). Facilitating group formation and role allocation in software engineering groups. *Proceedings of the ACS/IEEE International Conference on Computer Systems and Applications*, Beirut, Lebanon (25–29 June 2001), pp. 0352–0359.

Dahchour, M., Pirotte, A., and Zimányi, E. (2004). A role model and its metaclass implementation. *Information Systems* 29 (3): 235–270.

Dasgupta, P. (1998). Trust as a commodity. In: *Trust: Making and Breaking Cooperative Relations* (ed. D. Gambetta), 49–72. Blackwell.

Edwards, W.K. (1996). Policies and roles in collaborative applications. *Proceedings of ACM 1996 Conference on Computer-Supported Cooperative Work*, Boston, MA, USA (16–20 November 1996), pp. 11–20.

Fan, X. and Yen, J. (2011). Modeling cognitive loads for evolving shared mental models in human–agent collaboration. *IEEE Transactions on Systems, Man, and Cybernetics (SMC)(B)* 41 (2): 354–367.

Fan, Z.-P., Suo, W.-L., Feng, B., and Liu, Y. (2011). Trust estimation in a virtual team: a decision support method. *Expert Systems with Applications* 38 (8): 10240–10251.

Ferber, J., Gutknecht, O., and Michel, F. (2004). From agents to organizations: an organizational view of multi-agent systems. In: *Agent-Oriented Software Engineering (AOSE) IV*, LNCS, vol. 2935 (eds. P. Giorgini, J. Müller and J. Odell), 214–230. Berlin, Heidelberg, Germany: Springer.

Ferraiolo, D.F. and Kuhn, D.R. (1992). Role-based access control. *Proceedings of the NIST-NSA National (USA) Computer Security Conference*, Baltimore, USA (13–16 October 1992), pp. 554–563.

Ferraiolo, D.F., Cugini, J.A., and Kuhn, D.R. (1995). Role-Based Access Control (RBAC): features and motivations. *Proceedings of the 11th Annual Computer Security Applications Conference (CSAC '95)*, Los Alamitos, CA, (11–15 December 1995), pp. 241–248.

Ferraiolo, D.F., Sandhu, R., Gavrila, S. et al. (2001). Proposed NIST standard: role-based access control. *ACM Transactions on Information and System Security* 4 (2): 224–274.

Ferrari, L. and Zhu, H. (2012). Autonomous role discovery for collaborating agents. *Software: Practice and Experience* 42 (6): 707–731.

Fischer, G. (2001). User modeling in human–computer interaction. *User Modeling and User-Adapted Interaction* 11: 65–86.

Gena, C. (2005). Methods and techniques for the evaluation of user-adaptive systems. *The Knowledge Engineering Review* 20 (1): 1–37.

Gottlob, G., Schrefl, M., and Röck, B. (1996). Extending object-oriented systems with roles. *ACM Transactions on Information Systems* 14 (3): 268–296.

Greenberg, S. (1991). Personalizable groupware: Accommodating individual roles and group differences. *Proceedings of the Second European Conference on Computer Supported Cooperative Work*, Amsterdam, The Netherlands (September 1991), pp. 17–32.

Gutwin, C. and Greenburg, S. (1999). The effects of workspace awareness support on the usability of real-time distributed groupware. *ACM Transactions on Computer–Human Interaction (TOCHI)* 6 (3): 243–281.

Guzdial, M., Rick, J., and Kerimbaev, B. (2000). Recognizing and supporting roles in CSCW. *Proceeding of the ACM 2000 Conference on Computer-Supported Cooperative Work (CSCW'00)*, Philadelphia, Pennsylvania, USA (December 2000), pp. 261–268.

Harrison, W. and Ossher, H. (1993). Subject-oriented programming - a critique of pure objects. *Proceedings of OOPSLA*, Washington, DC, USA (26 September to 1 October 1993), pp. 411–428.

Hawkins, D.I., Best, R.J., and Coney, K.A. (1983). *Consumer Behavior*. Plano, TX: Business Publications, Inc.

Hellriegel, D., Slocum, J.W. Jr., and Woodman, R.W. (1983). *Organizational Behavior*. St. Paul: West Publishing Co.

HLA Associates (2001). Role modeling. http://www.rolemodeling.com (accessed 10 August 2020).

Homans, A. (1950). *The Human Group*. New York: Harcourt, Brace and Company.

Hou, M., Kobierski, R.D., and Brown, M. (2007). Intelligent adaptive interfaces for the control of multiple UAVs. *Journal of Cognitive Engineering and Decision Making* 1 (3): 327–362.

Hou, M., Zhu, H., Zhou, M., and Arrabito, G.R. (2011). Optimizing operator-agent interaction in intelligent adaptive interface design. *IEEE Transactions on SMC(C)* 41 (2): 161–178.

Jameson, A. (2003). Adaptive interface and agents. In: *The Human Computer Interaction Handbook* (eds. J.A. Jacko and A. Sears), 305–330. Erlbaum.

Jennings, N.R. and Wooldridge, M. (1996). Software agents. *IEE Review* 42 (1): 17–20.

Jennings, N.R., Sycara, K., and Wooldridge, M. (1998). A roadmap of agent research and development. *Autonomous Agents and Multi-Agent Systems* 42 (1): 7–38.

Kendal, E.A. (1999). Role model design and implementation with aspect-oriented programming. *ACM SIGPLAN Notices* 34 (10): 353–369.

Kerr, N.L. and Tindale, R.S. (2004). Group performance and decision making. *Annual Review of Psychology* 55: 623–655.

Kristensen, B.B. and Østerbye, K. (1996). Roles: conceptual abstraction theory & practical language issues. *Special Issue of Theory and Practice of Object Systems on Subjectivity in Object-Oriented Systems* 2 (3): 143–160.

Kühn, T., Stephan, B., Götz, S. et al. (2015). A combined formal model for relational context-dependent roles. Proceedings of Software Language Engineering, Pittsburg, USA (25-27 October 2015), pp. 141–160.

Leland, M.D.P., Fish, R.S., and Kraut, R.E. (1988). Collaborative document production using quilt. *Proceedings of CSCW'88*, Portland, Oregon, USA (26–28 September 1988), pp.206– 215.

Li, X., Chen, Y., and Dong, Z. (2012). Qualitative description and quantitative optimization of tactical reconnaissance agents system organization. *International Journal of Computational Intelligence Systems* 5 (4): 723–734.

Lu, Y. and Käkölä, T. (2014). A dynamic life-cycle model for the provisioning of software testing services. *Systems Science & Control Engineering: An Open Access Journal* 2 (1): 549–561.

Luo, W. and Tung, Y.A. (1999). A framework for selecting business process modeling methods. *Industrial Management & Data Systems* 1999: 312–319.

Maes, P. (1994). Modeling adaptive autonomous agents. *Artificial Life* 1 (1): 135–162.

Maslow, A. (1970). *Motivation and Personality*, 2e. Harper & Row.

Miner, J.B. (1992). *Industrial-Organizational Psychology*. McGraw- Hill, Inc.

Murdoch, J. and McDermid, J.A. (2000). Modeling engineering design process with role activity diagrams. *Transactions of the Society for Design and Process Science(SDPS)* 4 (2): 45–65.

Nair, R., Tambe, M., and Marsella, S. (2003). Role allocation and reallocation in multiagent teams: towards a practical analysis. *Proceedings of the second Int'l joint Conference on Autonomous agents and multiagent systems*, Melbourne, Australia (14 July 2003), ACM. http://teamcore.usc.edu/papers/2003/nair-aamas03.pdf.

Norman, D.A. (1986). Cognitive engineering. In: *User Centered System Design: New Perspectives on Human–Computer Interaction* (eds. D.A. Norman and S.W. Draper), 32–65. Hillsdale, NJ: Lawrence Erlbaum Associates.

Nwan, H.S. (1996). Software agents: an overview. *Knowledge Engineering Review* 11 (3): 205–244.

Nwan, H.S., Lee, L., and Jennings, N.R. (1996). Coordination in software agent systems. *BT Technology Journal* 14 (4): 79–89.

Odell, J.J., Van Dyke Parunak, H., Brueckner, S., and Sauter, J. (2003a). Changing roles. *Journal of Object Technology* 2 (5): 77–86.

Odell, J., van Dyke Parunak, H., and Fleischer, M. (2003b). The role of roles in designing effective agent organizations. In: *Software Engineering for Large-Scale Multi-Agent Systems*, Lecture Notes on Computer Science, vol. 2603 (eds. A. Garcia, C. Lucena, F. Zambonelli, et al.), 27–38. Berlin: Springer.

Odell, J., Nodine, M., and Levy, R. (2005). A metamodel for agents, roles, and groups. In: *Agent-Oriented Software Engineering (AOSE)*, Lecture Notes on Computer Science, vol. 3382 (eds. J. Odell, P. Giorgini and J. Müller), 78–92. Berlin: Springer.

Ould, M. (1995). *Business Processes – Modelling and Analysis for Re-engineering and Improvement*. Chichester: John Wiley & Sons.

Padgham, L., Winikoff, M., DeLoach, S., and Cossentino, M. (2009). A unified graphical notation for AOSE. In: *Agent-Oriented Software Eng. IX*, LNCS, vol. 5386 (eds. M. Luck and J.J. Gomez-Sanz), 116–130. Berlin, Heidelberg, Germany: Springer.

Pernici, B. (1991). Objects with roles. *ACM SIGOIS Bulletin* 11 (2–3): 205–215.

Picard, G. and Gleizes, M.-P. (2003). An agent architecture to design self-organizing collectives: principles and application. In: *Adaptive Agents and Multi-agent Systems*, LNAI 2636 (eds. E. Alonso, D. Kudenko and D. Kazakov), 110–124. Berlin: Springer.

Plaisant, C. and Shneiderman, B. (1995). Organization overviews and role management: inspiration for future desktop environments. *Proceedings of IEEE 4th Workshop on Enabling Technologies: Infrastructure for Collaborative Enterprises*, Berkeley Springs, WV, USA (April 1995), pp. 14–22.

Pradel, M., Henriksson, J., and Aßmann, U. (2012). A good role model for ontologies. In: *Enterprise Information Systems and Advancing Business Solutions* (ed. M. Tavana), 225–235. IGI Global.

Pressman, R. (2007). *Software Engineering: A Practitioner's Approach*. Boston, MA: McGraw-Hill.

Ramchurn, S.D., Huynh, D., and Jennings, N.R. (2004). Trust in multi-agent systems. *The Knowledge Engineering Review* 19 (1): 1–25.

Rardin, R.L. (1997). *Optimization in Operations Research*. Upper Saddle River, NJ: Prentice Hall.

Riehle, D. and Gross, T. (1998). Role model based framework design and integration. *Proceedings of ACM Conference on Object-Oriented Programming Languages, Systems, and Applications*, Vancouver, British Columbia, Canada (18–22 October 1998), pp. 117–133.

Riehle, D., Brudermann, R., Gross, T., and Mätzel, K.U. (2000). Pattern density and role modeling of an object transport service. *ACM Computing Surveys (CSUR)* 32 (1): 1–6.

Russell, D. and Norvig, P. (2003). *Artificial Intelligence: A Modern Approach*, 2e. Upper Saddle River, NJ: Pearson Education, Inc.

Sabater, J. and Sierra, C. (2001). REGRET: a reputation model for gregarious societies. *Proceedings of the 1st Int'l Joint Conference Autonomous Agents and Multi-Agent Systems*, Bologna, Italy (May 2001), pp. 475–482.

Seo, W., Yoon, J., Lee, J., and Kim, K. (2011). A state-driven modeling approach to human interactions for knowledge intensive services. *Expert Systems with Applications* 38 (3): 1917–1930.

Shen, H. and Sun, C. (2011). Achieving data consistency by contextualization in web-based collaborative applications. *ACM Transactions on Internet Technology* 10 (4): 13–37.

Sheng, Y., Zhu, H., Zhou, X., and Wang, Y. (2014). Effective approaches to group role assignment with a flexible formation. *The IEEE Int'l Conference on SMC*, San Diego, USA (5–8 October 2014), pp. 1445–1450.

Shneiderman, B. and Plaisant, C. (1994). The future of graphic user interfaces: personal role managers. *People and Computers IX*, British Computer Society, Glasgow, Scotland (August 1994), pp. 3–8.

Singley, M.K., Singh, M., Fairweather, P., et al. (2000). Algebra jam: supporting teamwork and managing roles in a collaborative learning environment. *Proceedings of CSCW'00*, Philadelphia, Pennsylvania, USA (December 2000), pp. 145–154.

Smith, R.B., Hixon, R., and Horan, B. (1998). Supporting flexible roles in a shared space. *Proceedings of CSCW'98*, Seattle, Washington, USA (26–28 September 1988), pp. 197–206.

Sousa, P., Caetano, A., Vasconcelos, A. et al. (2006). Enterprise architecture modelling with the Unified Modelling Language. In: *Enterprise Modelling and Computing with UML* (ed. P. Rittgen), 67–94. Idea Group Inc.

van Splunter, S., Wijngaards, N.J.E., and Brazier, F.M.T. (2003). Structuring agents for adaptation. In: *Adaptive Agents and Multi-agent Systems*, LNAI 2636 (eds. E. Alonso, D. Kudenko and D. Kazakov), 174–186. Berlin: Springer.

Steimann, F. (2000). On the representation of roles in object-oriented and conceptual modelling. *Data & Knowledge Engineering* 35 (1): 83–106.

Steimann, F. (2008). Role + counter-role = relationship + collaboration. *Workshop on Rel. and Asso. in Object-Oriented Languages*, Nashville, Tennessee, USA (19–23 October 2008), OOPSLA. http://deposit.fernuni-hagen.de/2201/1/ Role_Counter_Role%20_Relationship_Collaboration.pdf.

Tao, L. and Zhang, M. (2013). Understanding an online classroom system. *IEEE Transactions on Human-Machine Systems* 43 (5): 465–477.

Toastmasters Club (2001). Roles at typical meeting. http://www.geocities.com/ tictalkers/Roles.htm (accessed 10 August 2020).

Turoff, M. (1993). Distributed group support system. *MIS Quarterly* 17 (4): 399–417.

Twidale, M.B. and Nichols, D.M. (1998). A Survey of Applications of CSCW for Digital Libraries. *Technical Report CSEG/4/98*, Computing Department, Lancaster University, UK.

VanHilst, M. and Notkin, D. (1996). Using role components to implement collaboration-based designs. *Proceedings of OOPSLA'96*, San Jose, CA, USA (6–10 October 1996), pp. 359–369.

Wang, J. and Deng, Y. (1999). Incremental modeling and verification of flexible manufacturing systems. *Journal of Intelligent Manufacturing* 10 (6): 485–502.

Wang, S., Li, J., and Ma, S. (2011). Dynamic and secure business data exchange model for SaaS-based collaboration supporting platform of industrial chain. *International Journal of Advanced Pervasive and Ubiquitous Computing* 3 (3): 46–59.

Weerawarana, S., Curbera, F., Leymann, F. et al. (2008). *Web Services Platform Architecture*. Upper Saddle River, NJ: Prentice Hall.

de Wilde, P., Chli, M., Correia, L. et al. (2003). Adapting populations of agents. In: *Adaptive Agents and Multi-agent Systems*, LNAI 2636 (eds. E. Alonso, D. Kudenko and D. Kazakov), 110–124. Berlin: Springer.

Wooldridge, M. and Jennings, N.R. (1995). Intelligent agents: theory and practice. *The Knowledge Engineering Review* 10 (2): 115–152.

Wooldridge, M., Jennings, N.R., and Kinny, D. (2000). The Gaia methodology for agent-oriented analysis and design. *Autonomous Agents and Multi-Agent Systems* 3: 285–312.

Zambonelli, F., Jennings, N.R., and Wooldridge, M. (2003). Developing multiagent systems: the Gaia methodology. *ACM Transactions on Software Engineering and Methodology* 12 (3): 317–370.

Zhang, Y. (2008). A role-based approach in dynamic task delegation in agent team work. *Journal of Software* 3 (6): 9–20.

Zhu, H. (2003a), A Role Agent Model for Collaborative Systems, in Proc. of The 2003 Int'l Conf. on Info. & Knowledge Engineering (IKE'03), Las Vegas, Nevada, 2003, pp. 438–444.

Zhu, H. (2003b). Some issues in role-based collaboration. *Proceedings of IEEE Canada Conf. on Electrical and Computer Engineering (CCECE'03)*, Montreal, Canada (4–7 May 2003), vol. 2, pp.687–690.

Zhu, H. (2005). Encourage participants' contributions by roles. *Proceedings of the IEEE Int'l Conference on Systems, Man and Cybernetics*, Waikoloa, HI, USA (9–12 October 2005), pp. 1574–1579.

Zhu, H. (2006). Role mechanisms in collaborative systems. *International Journal of Production Research* 44 (1): 181–193.

Zhu, H. (2007). Role as dynamics of agents in multi-agent systems. *System and Informatics Science Notes* 1 (2): 165–171.

Zhu, H. (2008). Fundamental issues in the design of a role engine. *Proceedings of ACM/ IEEE Int'l Symp on CTS*, Irvine, CA, USA (19–23 May 2008), pp. 399–407.

Zhu, H. (2010). Role-based autonomic systems. *International Journal of Software Science and Computational Intelligence* 2 (3): 32–51.

Zhu, H. and Hou, M. (2011). Role-based human-computer interaction. *International Journal of Cognitive Informatics and Natural Intelligence* 5 (2): 37–57.

Zhu, H. and Zhou, M.C. (2006a). *Object-Oriented Programming with C++: A Project-Based Approach*. Beijing, China: Tsinghua University Press.

Zhu, H. and Zhou, M.C. (2006b). Role-based collaboration and its kernel mechanisms. *IEEE Transactions on Systems, Man and Cybernetics, Part C* 36 (4): 578–589.

Zhu, H., Zhou, M.C., and Alkins, R. (2012). Group role assignment via a Kuhn-Munkres algorithm-based solution. *IEEE Transactions on Systems, Man, and Cybernetics, Part A: Systems and Humans* 42 (3): 739–750.

Zhu, H., Zhou, M.C., and Hou, M. (2015). Support collaboration with roles. In: *Contemporary Issues in Systems Science and Engineering* (eds. M.C. Zhou, H.-X. Li and M. Weijnen), 575–598. IEEE Press.

Zigurs, I. and Kozar, K.A. (1994). An exploratory study of roles in computer supported groups. *MIS Quarterly* 18 (3): 277–297.

Exercises

1 What is a methodology? Describe RBC as a methodology.
2 What are the major components of an RBC system?
3 Discuss the interaction among agents in an RBC system.

4 Describe the architecture of an RBC system.

5 Discuss the dynamics of an RBC system. Do you agree with the idea that agents in an RBC system will be well behaved? Why?

6 Describe the process of RBC. Who will be the user of the RBC process presented in Figure 3.3?

7 Describe the four sets of principles of RBC, i.e. Object, Role, Agent, and Group principles.

8 Do you think the principles mentioned in Question 7 are the complete principles of RBC? Why?

9 Should we establish a set of environmental principles? If yes, please describe them.

10 What are the benefits of RBC?

4

The E-CARGO Model

4.1 First Class Components

To establish a formal model, we need to clarify the components at first. To specify components, we need fundamental concepts to be well defined. A well-defined model needs all its components to be well defined. We discuss the first class components in this section, where "first class components" mean the components that are independently defined by primitive concepts or components. By "primitive concepts," we mean those concepts that do not need more specifications or divisions.

Definition 4.1 *Identification* is to make differences between unique entities in a system. Externally, it means an identifier composed of letters and digits. Internally, it means the initial address of a memory unit. Identification should be unique. We use $\overline{D}$ to express the set of all the identifications, and $\overline{d}$ for a specific one, i.e. $\overline{d} \in \overline{D}$ and $\overline{d_r}$, $\overline{d_a}$, and $\overline{d_g}$ are identifications of a role, agent and group, respectively. We use identification to mean both internal and external unless otherwise stated.

Definition 4.2 A *data state or state* is a group of logical memory units that store data values. We use $\mathcal{D}$ to express the set of all the states in consideration and s for a specific state, i.e. $\mathit{s} \in \mathcal{D}$.

Definition 4.3 A *human user* is a person who has signed on to the system. It is shortened to "user" thereafter. We use $\mathcal{H}$ to express all the users and u for a specific one, i.e. $\mathit{u} \in \mathcal{H}$.

To help understand the definitions and discussions in this chapter, the principles discussed in Chapter 3 are immensely helpful. Readers should review Chapter 3 if necessary.

E-CARGO and Role-Based Collaboration: Modeling and Solving Problems in the Complex World,
First Edition. Haibin Zhu.
© 2022 by The Institute of Electrical and Electronics Engineers, Inc.
Published 2022 by John Wiley & Sons, Inc.

To understand this complex world, we need abstraction, which is a tool used to know, create, and understand a complex thing. Abstraction is also a general methodology to study the world. When we have many concrete objects to manage, we generally need to remember their common properties and each one's specialties. This is an abstraction process. The result of this process is an abstract concept.

By abstraction, we can understand other relevant concepts such as information hiding. We can say that abstraction is a way to implement information hiding, or by information hiding, we obtain abstraction.

Abstraction is also a difficult skill to master. As Confucius said more than 2000 years ago, "四十而不惑," read as "Sìshí ér bùhuò," i.e. "one begins to understand the world at 40." In other words, he mastered the abstraction ability and all the problems in the world had a model and a solution in his mind at that age.

Something should be forgotten to concentrate on other things. However, abstraction is not just random forgetting but an intentional one. That is to say, we must expose the nature of the thing we are thinking of. The loss of nature is a failure of abstraction. In this book, the whole world is abstracted into three categories of entities: agents, roles, and objects. This does not lose the nature of the world.

Based on the principles of object-orientation, that everything is an object is the highest level of abstraction. To help solve problems in collaboration, we need to add one new level of abstraction, i.e. introduce the two new concepts: agents and roles, both of which are special objects (Figure 4.1), where Figure 4.1a informs that the world is composed of objects, and Figure 4.1b illustrates that agents and roles are special objects.

Agents are special objects that are active, dynamic, automatic, autonomous, and possess self-awareness. Objects other than agents are natural or man-made entities that do not have dynamic and active properties, abilities, and activities. Roles are abstract intermediates between objects and agents. Roles are special objects other than agents. Without roles, agents and objects are not connected, i.e. agents are

Figure 4.1 The world of three types of entities. (a) Sets of objects, roles, and agents. (b) Classes of objects, roles, and agents.

isolated from objects, and objects are not accessible. Finally, agents are role players but objects are not. That is why we take roles as a fundamental concept for specifying systems.

Similar to abstraction, roles are an essential part of a collaboration. Roles are often understood in the real world as meaning "responsibilities." However, besides responsibilities, roles also specify rights. Why are roles emphasized as "responsibilities" traditionally? It is because the nature of humans is to pursue rights but avoid responsibilities, and usually roles imply rights. Therefore, people try hard to pursue and play roles to grasp hidden rights. This is the fundamental reason why responsibilities are emphasized.

Roles are abstractions that describe the actions that a player (i.e. an agent assigned with a role) can perform. Therefore, if roles are clarified, agents' behaviors are clarified too. After roles are specified, the leftover is how to assign roles to agents and how agents play roles.

4.1.1 Objects and Classes

To follow the fundamental principle of object orientation, i.e. everything in the world is an *object* (Kay 1993), and every object has a *class*, we should first formalize the concepts of objects and classes. An object must possess the following characteristics (Budd 2001; Kay 1993; Steimann 2000; Zhu and Zhou 2006a):

- "uniquely identified," i.e. any object should carry a unique identification;
- "created or destroyed";
- "communicative," i.e. an object can exchange messages with other objects;
- "nested," i.e. a complex object has other objects as its components (which, in turn, may have object components);
- "active and autonomous," i.e. an object may respond to messages without people's intervention; and
- "collaborative," i.e. collaborative relationships among objects arise when they exchange messages.

Classification comes from the natural requirements to solve problems in any domain of application and describe the commonalities among a collection of objects. This is the basic activity of people when organizing perceptions in their early mental development. Similarly, scientists organize knowledge within scientific disciplines, and application programmers organize domain knowledge and behaviors.

Therefore, classes are a mechanism of classification and a tool for abstraction. In other words, a class is either a classification, an abstraction, or both. A class is not only a concept, but also a practical mechanism in a program. Extracting classes from concrete objects is undoubtedly a process of abstraction.

Classes are a stepping-stone in the advancement of programming technologies. They make it possible to design object-oriented programs. Every object-oriented programming language must support the structure of classes. Classes are closely related to abstraction and classification (Budd 2001; Zhu and Zhou 2006a). Classes make it easy to abstract objects with formulated structures and a stepwise approach. In problem solving, if there are too many objects in a problem to handle, we may need to group them together, i.e. making a class.

In general, a class serves two distinct purposes. It is a definition of an implementation (methods and data structures) shared by a group of objects and it is a template from which objects can be created. It contains a definition of the state descriptors and methods to create an object. There are two perspectives on the concept of classes: narrow and broad. In general, a class provides a definition of the structure of its instances.

The narrow view of a class is that objects are instances of a class. Their state and behavior structures are determined by the class definition. A class is a template to create objects.

The broad view of a class is that it is an object warehouse and object factory. Object warehouse means that a class implicitly maintains a class extension, which consists of all instances of the class, like a set. Hence, an object can be "of" a certain group of objects. An apple, which is an instance of class apple, is of class fruit. A pear, which is an instance of class pear, is of class fruit. However, they may not be instances of class fruit, which may be abstract. Classes in this meaning are sets. Object factory means that there exists a constructor for each class. The purpose of this constructor is to generate new instances of that class.

A *class* is defined as a template for a set of objects that share a common structure and common behavior (Budd 2001; Zhu and Zhou 2006a). Considering the meaning and properties of objects, we can express a *class* as a quadruple.

Definition 4.4 *A class* is defined as a 4-tuple, $e :: = < \overline{d_c}, \mathcal{D}_e, \mathcal{F}_e, \mathcal{X}_e >$, where

- $\overline{d_c}$ is the identification of the class;
- $\mathcal{D}_e$ is a data structure description for storing the state of an object;
- $\mathcal{F}_e$ is a set of the function definitions or implementations; and
- $\mathcal{X}_e$ is a unified interface for all the objects of this class. It is a set of all the message patterns relevant to the functions of this class. A message pattern tells how to send a message to invoke a function.

We use e to express a specific class and C the set of all classes. Next, we define an *object* based on a class.

Definition 4.5 An *object* is defined as an instance of a class, i.e. $o:: = <\overline{d_o}, c, \textit{s}>$, where

- $\overline{d_o}$ is the identification of the object;
- c is the object's class identified by the class identification or name; and
- $\textit{s}$ is a set of values called attributes, properties, or states typed of the $\mathcal{D}_c$ of c.

We use o to express a specific object and $\mathcal{O}$ the set of all objects. The above two definitions comply with object principles O1–O9 stated in Chapter 3.

Example 4.1 A desk, a computer, a flower, a garden, a student, and a professor are all objects. The concept of *desk, computer, flower, garden, student, and professor* are all classes.

For class *Student*, we have:

- $\overline{d_c}$ = "Student";
- $\mathcal{D}_c$ = {"int SID," "string LastName," "string FirstName," "float GPA," "int Credit," "float TuitionBalance"};
- $\mathcal{F}_c$ = {"getSID(){...}," "getLastName(){...}," "getFirstName(){...}," "getName() {...}," "getGPA(){...}," "getCredit()," "getTB()," "setSID(int) {...}," "setLastName (string) {...}," "setFirstName(string) {...}," "setName(string, string) {...}," "setGPA(float) {...}," "setCredit(int) {...}," "setTB(float) {...}"}; and
- $\mathcal{X}_c$ = {"getSID()," "getName()," "getGPA()," "getCredit()," "getTB()," "setSID (int)," "setName(string, string)," "setGPA(float)," "setCredit(int)," "setTB (float)"}. Note, {...} means the implementation details of a function.

For object *student Chris Smith*:

- $\overline{d_o}$ = "Chris Smith";
- c = "Student"; and
- $\textit{s}$ = {05498, "Smith," "Chris," 3.8, 30, 5800}.

4.1.2 Roles and Environments

We discussed the role concept comprehensively in Chapter 2. In this chapter, roles are specified formally to form the Environments - Classes, Agents, Roles, Groups, and Objects (E-CARGO) model.

The term "role" is derived from theater and refers to the part played by an actor. A role represents a specific status that possesses certain rights and accompanying responsibilities. It can be defined as a set of regulations defining the behavior of a player. A role defines a set of responsibilities and capabilities needed to perform the activities relevant to these responsibilities (Al-Zaghameem and Alfraheed

2013; Ashforth and Mahwah 2001; Biddle and Thomas 1979; Bostrom 1980; Botha and Eloff 2001; Cabri 2012; Caetano et al. 2009; Colman and Han 2007; Dafoulas and Macaulay 2001; Dahchour et al. 2004; Gottlob et al. 1996; Guzdial et al. 2000; Kendal 1999; Kühn et al. 2014; Marwell and Hage 1970; Murdoch and McDermid 2000; Nair et al. 2003; Nyanchama and Osborn 1999; Odell et al. 2005; Pernici 1990; Pradel et al. 2012; Ren et al. 2007; Riehle and Gross 1998; Riehle et al. 2000; Shneiderman and Plaisant 1994; Singley et al. 2000; Smith et al. 1998; Steimann 2000). A role can also be defined as the prescribed pattern of behavior expected of a person in a given situation and a position (Ashforth and Mahwah 2001; Biddle and Thomas 1979). Roles can be more simply defined as a position in a social structure. A position is an institutionalized or commonly expected and recognized designation in a given social structure, such as an accountant, mother, or church member (Biddle and Thomas 1979). A role is a set of expectations regarding the behavior for a particular position within a society. Generally speaking, a role is a position occupied by a person in a social relationship. By occupying this position, one possesses specific rights and takes on specific responsibilities.

Roles are defined as a concept that is founded but not semantically rigid (Brachman and Schmolze 1985). "Founded" means that a concept can exist independently. "Semantically rigid" means that a concept contributes to the identity of its instances. This definition can help determine if a concept is a role or not. Roles allow not only for the representation of multiple views of the same phenomenon, but also for the representation of changes in time. Roles are the bridges between different levels of detail in an ontological structure.

Many researchers consider roles a fundamental concept in modeling (Becht et al. 1999; Cabri 2007, 2012; Dafoulas and Macaulay 2001; Dahchour et al. 2004; Gottlob et al. 1996; Guzdial et al. 2000; Kendal 1999; Kühn et al. 2014; Kristensen and Østerbye 1996; Marwell and Hage 1970; Murdoch and McDermid 2000; Nair et al. 2003; Nyanchama and Osborn 1999; Odell et al. 2005; Pernici 1990; Pradel et al. 2012; Ren et al. 2007; Riehle and Gross 1998; Riehle et al. 2000; Shneiderman and Plaisant 1994; Singley et al. 2000; Smith et al. 1998; Steimann 2000). Roles are entities that temporarily confine the behavior of objects. A role is considered as an abstraction and decomposition mechanism related to objects.

There are many different discussions concerning the concepts of roles. Some aspects of roles seem contradictory. The comprehensive studies on roles as fundamental concepts and mechanisms discussed in Chapter 2 are good guidelines for understanding roles in object-oriented systems. To support role-playing with object-oriented programming languages, we need to confine our role concepts in the sense of data modeling and programming. Based on the basic view of object orientation and the system modeling requirement, we can constitute the fundamental principles relevant to object evolution and separation of concerns based on roles.

Conventional role modeling (Dahchour et al. 2004; Kendal 1999; Kühn et al. 2014; Pradel et al. 2012; Riehle and Gross 1998) asserts that objects may play roles. When an object plays a role, it accepts messages and provides services related to its role. A role constitutes a subset of an object's behavior obtained by considering only the interactions of that role and hiding all other interactions. The following characteristics of roles are commonly accepted in the literature of modeling methodologies:

1) Roles are a concept or specialization of a concept.
2) Roles can be acquired and abandoned independently of each other.
3) Roles can be organized in hierarchies, generalized, or specialized.
4) Roles model a perspective on a phenomenon.
5) Roles are bound to an existing object.
6) Roles are used to emphasize how entities interact with each other.
7) Roles can be dynamic and involve sequencing, evolution, and role transfer.
8) An object may play several roles at a time.
9) Various roles of an object may share common structures and behavior.
10) The states and features of an object can be role-specific.

Roles, as states or tags, can also be used to express the evolution of objects (Budd 2001; Gottlob et al. 1996; Kristensen and Østerbye 1996; Odell et al. 2005; Pernici 1990; Steimann 2000, 2008). The major concern is how to organize the roles in a well-defined structure and let the objects easily attach their roles to express the objects' states.

From the viewpoint of modeling-roles, services are its key concerns. Every object, data item, or component has responsibilities to serve others and a role is a part of all the responsibilities of an object. That is, an object is considered a server. However, the thought that objects can play roles creates many difficulties in modeling complex systems. That is why we differentiate agents from objects in the E-CARGO model, where *agents are role players but objects are not* (Zhu and Zhou 2006b).

In system management, roles are utilized as a mechanism to organize human users to simplify the task of user management. This is the basic idea of Role-Based Access Control (RBAC) (Ferraiolo and Kuhn 1992; Ferraiolo et al. 1995, 2001) and database systems. The role concept from RBAC is now successfully applied in industries such as system security administration and database systems. In system management, human users are the major concerns when designing roles. It is human users, not objects, that play roles. Therefore, rights are the key concern for roles in RBAC. Namely, in a system with RBAC, a user has rights to access objects in the system and users play one role at a time when interacting with the system, i.e. possessing the rights relevant to their roles. System administrators can assign rights to roles but not to users because a role can be assigned to many people. Note that the number of roles is significantly less than the number of users. That is why roles can be used to group people by using object access rights.

In a social environment, roles are taken as a tool to specify human behavior. The design of collaboration systems should reflect the ideas of roles in a social environment. Current information systems have become a tool for collaboration. People generally interact with each other by firstly interacting with a computer or an information system (Figure 4.2).

If people collaborate through a system, the system should be viewed as both servers and clients. Therefore, the roles played by the system should represent both the right to issue requests and the ability to provide services.

By incorporating both requests and services, roles can be used to express different meanings or usages depending on objects, system components, systems, and people, e.g. interfaces, interactions, specifications, transitions, evolutions, and separation of concerns.

In reality, a role is actually a wrapper with a service interface and request interface. A person's role can be divided into two parts: the service interface including all the incoming messages, and the request interface including all the outgoing messages.

From Figure 4.3, we can see the concept of an interface role. Roles are abstract entities that express the interfaces between role players. Roles only specify what

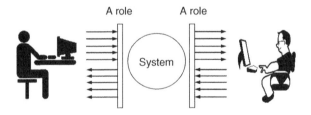

Figure 4.2 People-system-people interactions.

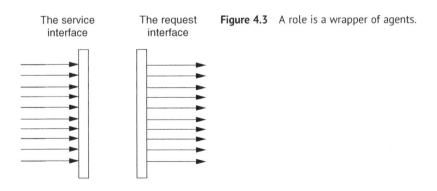

The service interface The request interface **Figure 4.3** A role is a wrapper of agents.

the services and requests are. How these services and requests are processed depends on their players. In modeling, system analysis, and system design, roles belong to this category.

A *role* can show the specialties of agents. It provides them with not only message patterns to serve others but also message patterns to access objects, classes, groups, and other roles. A role, as an object, is a dynamic entity. Initially, we first define the role as an interface for agents and concentrate on its data structures, i.e. what it is composed of. The operations/services/methods of roles will be discussed through relations among roles and agents. The major data structure of roles consists of several parts.

Definition 4.6 A *role* is defined as $r:: = < \overline{d_r}, \mathcal{M}_{in}, \mathcal{M}_{out}, V^r_p, V^r_r, \mathcal{A}_c, \mathcal{A}_p, \mathcal{A}_o >$, where

- $\overline{d_r}$ is the identification of the role;
- $\mathcal{M}_{in}$ expresses a set of incoming messages, which should be processed by a player;
- $\mathcal{M}_{out}$ expresses a set of outgoing messages, which can be sent by a player;
- V^r_p and V^r_r denote two sets of objects, where V^r_p expresses what will be provided to the agent that plays this role, called *supplies*, and V^r_r expresses the requirements for agents that are able to play this role, called *demands*; and
- $\mathcal{A}_c, \mathcal{A}_p$, and $\mathcal{A}_o$ are used to denote the current agents, potential agents, and previous agents, respectively, where current agents are those currently playing this role, potential agents are those qualified to play but not currently playing the role, and previous agents are those that played this role before.

This definition shows both the requests and services of agents. Incoming messages are used to express services and outgoing messages to express requests. This type of role is also called an *interface role*. Interface roles are dynamic. They use $\mathcal{A}_c, \mathcal{A}_p$, and $\mathcal{A}_o$ to emphasize the inter-relations among agents and roles. Initially, $\mathcal{A}_c, \mathcal{A}_p$, and $\mathcal{A}_o$ are all empty as the initial state of role r. Note that we separate agents from objects to emphasize that roles are the media to access agents. The only way to interact with other agents is by sending messages to roles. $\mathcal{R}$ is used to denote the set of all roles.

In software development, interface roles are used in project management, system analysis and design when architecture and structures are more important. The identification and specification of interface roles are the major tasks of project managers, system analysts, and systems designers.

From engineering practices, we find that only interface roles do not express all the meanings of roles. In reality, there are many roles that possess very concrete specifications of how to accomplish a task. For example, in a factory of electronics, the role of an assembly line worker is strictly specified as how to perform the operations on the line within a time limit. Similarly, in robotics, roles are specially designed processes that are stored in the memory of the robots. When a robot plays a role, it behaves according to the process specified by the role (Stone and Veloso 1999).

Process roles are concrete specifications that regulate the behaviors of their players. They specify not only what services and requests are, but also how they are processed. Roles in multi-agent systems (Becht et al. 1999; Cabri 2007; Nair et al. 2003; Wooldridge et al. 2000) and process management (Murdoch and McDermid 2000; Ould 1995) are process roles.

Definition 4.6.1 A *process role* specifies how a player implements both the requests and services, and is defined as $\tilde{r} :: = \ < \overline{d_r}, \mathcal{M}_{in}, \mathcal{M}_{out}, V_p^r, V_r^r,$ $\mathcal{A}_c, \mathcal{A}_p, \mathcal{A}_o, \mathcal{D}_p, \mathcal{P}_{in}, \mathcal{P}_{out} >$, where

- $\overline{d_r}$, $\mathcal{M}_{in}$, $\mathcal{M}_{out}$, V_p^r, V_r^r, $\mathcal{A}_c$, $\mathcal{A}_p$, and $\mathcal{A}_o$ have the same meanings of r in *Definition 4.6*;
- $\mathcal{D}_p$ is a space description for memory to accommodate the state of the role;
- $\mathcal{P}_{in}$ expresses a set of service implementations; and
- $\mathcal{P}_{out}$ expresses a set of request implementations.

In a process role, incoming messages are used to invoke services and outgoing messages are used to issue requests. A process role concentrates on how to deal with requests and services but ignores the inter-relations among agents and roles. Process roles are used in system implementations for agents to behave according to the specifications of the roles. The design and implementation of process roles are the major tasks for programmers.

In fact, process roles can be applied to business management. Namely, to play a role, a manager performs activities clearly specified by the role. Software engineering hopes to regulate such kinds of processes in the development of software. Therefore, process roles are commonly applied in process management and software engineering (Pressman 2007).

Example 4.2 *Programmer, professor, student, technician, or security officer* is a role. *Pipeline operator* is a process role because such a role regulates how a player operates on a pipeline. *President* is an interface role because no document clearly states how a person plays the role of president.

For role *programmer*, we have:

- $\overline{d_r} =$ "Programmer";
- $\mathcal{M}_{in} = \{coding, testing, design\}$;
- $\mathcal{M}_{out} = \{Accessing the workspace, Using the Internet/Intranet, Request for maintenance\}$;

- V_r^r =<BS of Computer Science, two years' experience of C++ programming>;
- V_p^r = <Salary, Benefit, Office>; and
- $A_c = A_{\not{p}} = A_o = \emptyset$, i.e. an empty set.

People collaborate in an environment. People normally build groups in an environment. We can mimic a play on a stage. The stage is the environment, the play or collaboration is performed by a group of actors. Therefore, we introduce a new concept to facilitate the definition of a group, i.e. *environment*.

Definition 4.7 An *environment* is defined as $e:: = < \overline{d_e}, B, \mathcal{D}_e >$, where

- $\overline{d_e}$ is the identification of e.
- B is a set of *role requirements*, which are tuples of role, number range and an object set, $B:: = \{ < r, q, \mathcal{D}_a > \}$. The number range q expressed by $[\ell, u]$ inform how many agents (at least by ℓ and at most by u) may play this role in the environment and q also determines the number of role instances (Section 4.3.1). For example, q might be $[1, 1]$, $[2, 2]$, $[1, 10]$, $[3, 50]$, ... $\mathcal{D}_a$ means the objects to be accessed only by the role player of r, e.g. offices, desks, shelves, computers, ...
- $\mathcal{D}_e$ is a set of objects that are shared by all the agents who are playing role in the environment, called *shared objects*.

$\mathcal{E}$ denotes the set of all the environments in a system, and e an environment. Based on this definition, before creating an environment, we need to first specify roles, i.e. set up a leaf role class (Section 4.3). At the same time, before we specify a role, we need to first create objects (classes) for the roles to access. The introduction of the environment concept is an innovation that establishes the model E-CARGO. It is the foundation for a group to be created and work properly.

Example 4.3 A computer science department environment can be expressed as

- $\overline{d_e}$ = "Department of Computer Science";
- $B = \{$<"Chair," $[1, 1]$, an office>, <"Faculty," $[3, 15]$, offices>, <"Secretary," $[1, 5]$, cubicles>, <"Computer administrator," $[1, 4]$, cubicles> $\}$; and
- $\mathcal{D}_e$ = {labs, libraries, equipment, meeting rooms}.

4.1.3 Agents and Groups

An agent is defined as an autonomous entity consisting of a set of provided services (Cabri 2007; Wooldridge et al. 2000). An *agent* is a special object that represents a user involved in the computer-supported collaboration. It is created when a new user signs onto the system. Agents are necessary because when a user is offline, the agent can still accept incoming messages and send out simple reply messages as needed. Here, we emphasize that agents cannot delegate all the outgoing messages of a role in order to let human users operate directly. If agents could delegate all the outgoing messages of a role, no human users would be required in the collaboration and the system would become autonomous multi-agent systems (Becht et al. 1999; Cabri 2007; Nair et al. 2003; Wooldridge et al. 2000).

In Role-Based Collaboration (RBC), an agent is a role player, which should be autonomous. We can provide special states and methods in the class of *Agent*. To manage an agent's progress, we need to introduce the concepts of abilities, credits, and workload. Time and space are used to express the ability of an agent. Time expresses how many units of free time the agent possesses and space expresses how many units (similar to the memory space of a computer) the agent has. Time and space can be reset based on the performance of an agent's services. Even though the time in one day is 24 hours, an agent may have a different amount of time and space based on their abilities including working efficiencies, attitudes, and goals. When a new agent is created, its ability can be initialized. This initial declaration is difficult to evaluate due to an improper self-appraisal, which is also a challenge for trust establishment. A wrong appraisal would result in detrimental effects on the agent. Therefore, in most cases, we may assume that the initial declaration is correct. Estimated time and space values may be conservative or aggressive. The time and space values could be adjusted and fine-tuned by the role facilitators.

We can use credit points for an agent to express its previous performance of playing other roles. Credit points can be used as a criterion to assign roles. The difference between abilities and credit points is that abilities are declared by the agent while credit points are earned by an agent's previous work performance. Based on this assumption, agents will usually try to play as many roles as possible. Similarly, agents try to complete as many services as possible to collect credit points. To avoid an agent having too much workload to accomplish services efficiently, each agent should have a state expressing its maximum workload. The state can be set by role facilitators or by human users. The combination of time, space, credit, and workload forms the state of an agent, i.e. s_a.

Definition 4.8 An *agent* is defined as $a :: = \ < \overline{d_a}, c_a, s_a, V_p^a, V_r^a, \mathcal{R}_c, \mathcal{R}_p, \mathcal{R}_o, \mathcal{D}_g >$, where

- $\overline{d_a}$ is the identification of agent a;
- c_a is a special class that describes the common properties of agents;
- s_a is the state value of agent a, including the time, space, credit points, and workload;
- V_p^a and V_r^a denote two sets of objects, where V_p^a expresses what services or qualifications a can provide, to meet the *demands* V_r^r of a role and V_r^a expresses what services or supplies a requests to check a role's *supplies* V_p^r;
- $\mathcal{R}_c$ is the identification of the roles that the agent is currently playing;
- $\mathcal{R}_p$ is a set of identifications of roles a is able to play ($\mathcal{R}_c \cap \mathcal{R}_p = \emptyset$);

- $\mathcal{R}_o$ is a set of identifications of roles played by a before; and
- $\mathcal{D}_g$ is a set of identifications of groups that a belongs to.

This definition complies with agent principles A1–A8 discussed in Chapter 3. c_a and s_a are used to reflect A4–A6. Human users should provide a special subclass of c_a to specify their specialties. $\mathcal{R}_c$, $\mathcal{R}_p$, $\mathcal{R}_o$, and $\mathcal{D}_g$ are used to reflect A7–A8. Agents are distinguished from objects by their different message responses. Objects respond to messages by initiating their class methods, but agents may respond to messages by showing the messages to their users or transferring the messages to other agents. We can also introduce special identifications to distinguish agents from objects. To support an agent's functionality, an agent should have knowledge about groups, classes, objects, and other agents in the system. $\mathcal{A}$ denotes the set of all agents.

Note that we use $\mathcal{R}_c$ instead of r_c (Zhu and Zhou 2006b) to mean that an agent may have more than one current role. We use $\mathcal{R}_r$ to represent the set of current and potential roles of a, i.e. $\mathcal{R}_r = \mathcal{R}_c \cup \mathcal{R}_p$. This definition leaves space for further extensions and applications.

V_p^r and V_r^r in *Definition* 4.8 and V_r^a and V_p^a in *Definition* 4.6 correspond with each other. That is to say, the V_r^a of an agent corresponds with the V_p^r of a role and the V_p^a is to V_r^r. They are used to check the suitability between agents and roles. $\mathcal{R}_c$, $\mathcal{R}_p$, $\mathcal{R}_o$, and $\mathcal{D}_g$ may be empty at the beginning. They are used to express the dynamic properties of an agent. For example, if $|\mathcal{R}_c| = 0$, the agent has not yet been assigned any role. When we issue a message $r.addCurrentAgent(a)$, we add the identification of agent a to role r, i.e. $\mathcal{R}_c := \mathcal{R}_c \cup \{r\}$, and $\mathcal{A}_c := \mathcal{A}_c \cup \{a\}$.

For an agent to collect credit points, it must address the problem of how to reply to the service. One method is synchronous responses, i.e. to reply directly to the role within a method in the agent class. The other is asynchronous responses, i.e. to reply by human users' operations to the role. The process for an agent to collect credit points is common, i.e. playing roles. The roles should be able to record the state of message dispatching, reply to the earlier request, and let corresponding agents collect credits. Another problem is how to evaluate the performance of a human user or an agent. A human user may only provide a simple reply method implementation for an incoming message to collect the credit points relevant to the service without doing actual work. To address this problem, we need to verify the quality of the reply and set a misconduct record in the role engine design. Another way is to let human users evaluate the quality of a reply and the role engine can evaluate the corresponding human users/agents. We may need to learn from our daily lives in which misconduct can be uncovered by tracing the work history of people because if the initial task was satisfactory, the following ones should also be satisfactory. Misconduct records may suggest that there is something wrong with previous work. Sufficient misconduct records will result in the removal of related people or agents from collaboration. Role facilitators will also be able to judge and reset the credit points of agents as needed.

Example 4.4 A robot, drone, person, computer, or agency is an agent.
For a *drone* agent, we have:

- $\overline{d_a}$ = "Drone X1";
- c_a = class Drone;
- ω_a = {time:120 minutes, load capacity: 5 kg, credit: 10, current load: 2 kg};
- V_p^a = {flying, positioning, weight lifting, communicating};
- V_r^a = {Recharge in two hours};
- $\mathcal{R}_c = \mathcal{R}_p = \mathcal{R}_o = \emptyset$; and
- $\mathcal{D}_g$ = {Rescuing Group G1}.

Users work in groups and hold roles. Every workplace involves groups of individuals (Ashforth and Mahwah 2001; Dafoulas and Macaulay 2001; Odell et al. 2005; Turoff 1993; Zigurs and Kozar 1994). In a group, to accomplish a task, the group members (users) interact with each other. From conventional thought, without considering roles, all researchers agree that a group is composed of agents who interact with each other. There are numerous connections between agents.

Definition 4.9 A *group* is defined as a set of agents playing roles in an environment, $g = <\overline{d_g}, e, \mathcal{J}>$, where

- $\overline{d_g}$ is the identification of the group;
- e is an environment for the group to work; and
- $\mathcal{J}$ is a set of tuples of identifications of an agent and role, i.e. $\mathcal{J} = \{<\overline{d_a}, \overline{d_r}>\}$.

Definition 4.9 complies with group principles G1–G5 outlined in Chapter 3, where e and $\mathcal{J}$ are used to reflect principles G4–G5. Initially, $\mathcal{J} = \emptyset$.

The above definition of a group states the fact that without the users' participation, no collaboration could occur. We use g to express a specific group and $\mathcal{G}$ to specify the set of all groups.

Example 4.5 A robot team, basketball team, hockey team, company, school, university, organization, meeting, or conference is a group.
The computer science department of Nipissing University is a group where

- $\overline{d_g}$ = "CS Department of Nipissing University";
- e = "Department of Computer Science"; and
- $\mathcal{J}$ = {<"HZ," "Chair">, <"MW," "Faculty">, <"BJS," "Faculty">, <"TH," "Secretary">, <"KC," "Computer Administrator">,... }.

Suppose there are m agents in the group, n roles in the environment e of group $\mathcal{G}$, and one agent plays exactly one role in the group. We use $\mathcal{G}_j$ to mean the role number range for the jth role, where $\mathcal{G}_j.u$ means the upper limit of the number of the agents playing the jth role, and $\mathcal{G}_j.\ell$ means the lower limit. We have the following inequality for a group: $\sum_{j=0}^{n-1} \mathcal{G}_j.\ell \leq m \leq \sum_{j=0}^{n-1} \mathcal{G}_j.u$.

The inequality suggests that every agent must play a role in a group. An agent could join the group only when there are available vacancies for a role, and there must be a sufficient number of agents to play relevant roles. It also fulfills the principle G9 by creating an open and functioning group. If $m = \sum_{j=0}^{n-1} \mathcal{G}_j.u$, the group is closed and does not accept new members.

If we want the resources in an environment to be used completely and without wastes, the following equality should be enforced, for each $\mathcal{B}_j \in \mathcal{B}$ in group $\mathcal{G}. e$,

$$\sum_{j=0}^{n-1} |\mathcal{B}_j.\mathcal{D}_o| = |\mathcal{J}|.$$

$\sum_{j=0}^{n-1} |\mathcal{B}_j.\mathcal{D}_o| > |\mathcal{J}|$ means that there are more resources than required and

$\sum_{j=0}^{n-1} |\mathcal{B}_j.\mathcal{D}_o| < |\mathcal{J}|$ means that there are not enough resources.

For principle G6, we can state that $\mathcal{G}_1$ is embedded in $\mathcal{G}_2$ if $\{\overline{d_r} \mid \exists \mathcal{G}, \mathcal{D}_o(<\overline{d_r}, \mathcal{G}, \mathcal{D}_o> \in \mathcal{G}_1.e.\mathcal{B})\} \subset \{\overline{d_r} \mid \exists \mathcal{G}, \mathcal{D}_o(<\overline{d_r}, \mathcal{G}, \mathcal{D}_o> \in \mathcal{G}_2.e.\mathcal{B})\}$, and $\{\overline{d_a} \mid \exists \overline{d_r}(< \overline{d_a}, \overline{d_r} > \in \mathcal{G}_1.\mathcal{J}\} \subset \{\overline{d_a} \mid \exists \overline{d_r}(< \overline{d_a}, \overline{d_r} > \in \mathcal{G}_2.\mathcal{J}\}$. That is to say, all the roles, agents, and role assignments in $\mathcal{G}_1$ are also in $\mathcal{G}_2$.

For G7, we can state that $\mathcal{G}_1$ is overlapped with $\mathcal{G}_2$ if $\{\overline{d_r} \mid \exists \mathcal{G}, \mathcal{D}_o(<\overline{d_r}, \mathcal{G}, \mathcal{D}_o > \in \mathcal{G}_1.e.\mathcal{B})\} \cap \{\overline{d_r} \mid \exists \mathcal{G}, \mathcal{D}_o (<\overline{d_r}, \mathcal{G}, \mathcal{D}_o > \in \mathcal{G}_2.e.\mathcal{B})\} \neq \phi$ and $\{\overline{d_a} \mid \exists \overline{d_r} (< \overline{d_a}, \overline{d_r} > \in \mathcal{G}_1.\mathcal{J} \} \cap \{\overline{d_a} \mid \exists \overline{d_r}(< \overline{d_a}, \overline{d_r} > \in \mathcal{G}_2.\mathcal{J} \} \neq \phi$, i.e. some of the roles, agents, and role assignments in $\mathcal{G}_1$ are in $\mathcal{G}_2$.

For G8, we can state that $\mathcal{G}$ is public if $\mathcal{G}$ is in all the roles' object sets, i.e. $\forall \imath(\imath \in \mathcal{R} \rightarrow \mathcal{G}.\overline{d_g} \in \imath.V_p^r)$ and $\mathcal{G}$ is private if $\mathcal{G}$ is in some roles' object sets, i.e. $\exists \imath(\imath \in \mathcal{R} \rightarrow \mathcal{G}.\overline{d_g} \notin \imath.V_p^r)$. In plain descriptions, a public group $\mathcal{G}$ is accessible by all the roles in consideration.

Now we can describe the relationships among the first-class components with UML (Unified Modeling Language) class diagrams, which are presented in Figure 4.4. Simply, as an object, a class is composed of one or more (1..*) objects; a group is composed of one or more (1..*) agents; an environment is composed of one or more (1..*) roles; and a group is built on an environment (1..1). Class, Agent, Group, Role, and Environment all inherit from Object.

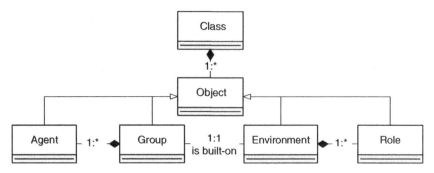

Figure 4.4 The relationships among first class components of E-CARGO.

4.2 Second Class Components

4.2.1 Users or Human Users

To model a computer-based collaboration system, we cannot avoid people as users of the system. In the E-CARGO model, agents are representatives of people or users. All the formal and expressible properties such as profiles, simple automatic message processing, and well-accepted user models are expressed by agents. In an RBC system including human users or users, i.e. $\mathcal{H}$, as in *Definition* 4.3, the inexpressible properties in collaboration such as natural intelligence are applied.

After a group is built, users can log into the system, join the group by playing a role, access relevant objects, and interact with each other. To support a group of people to work together is basically the routine function of a collaboration system. In this definition, group g, role $\imath$, agent a and object o have such a relationship: $<a,\imath>\in g.\mathcal{J} \rightarrow ((a.\overline{d_a} \in \imath.\mathcal{A}_e) \wedge (\imath.\overline{d_r} \in a.\mathcal{R}_e))$, i.e. joining a group means playing a role.

4.2.2 Message

Interaction is necessary for collaboration. Messages are necessary to facilitate interactions among roles and emphasize the principles O4–O5 discussed in Chapter 3.

Definition 4.10 *A message* (Zhu and Zhou 2006b, Zhu 2015) is defined as $\natural :: = <\overline{d_m},\imath_m,\mathit{f}, \mathcal{P}_m, \ell_m, \alpha, \beta>$ where

- $\overline{d_m}$ is the identification of the message;
- $\imath_m$ is null or the receiver (role) of the message expressed by an identification of the role;
- f is the pattern of a message (function), specifying the message name, types, sequence, and number of parameters;
- $\mathcal{P}_m$ is a list of objects taken as parameters for the message pattern f, where $\mathcal{P}_m \subset O$;

- ℓ_m is a label that expresses any, some, or all-message corresponding to unicast, multi-cast and broadcast, respectively;
- α is the space limit for the message to be processed; and
- β is the time limit for the message to be processed.

We use $\flat$ to express a specific message and $\mathcal{M}$ a set of all the messages. A message is a way to facilitate interactions among the components of the E-CARGO model.

Example 4.6   A letter, email, flashing light, assignment, flyer, or radio news is a message. The assignment posted on the blackboard is an *all-message* to the role of *student*. The flyers put onto the open table of a shopping mall are a *some-message* to the role of *potential buyer*. The radio news is an *all-message* to the role of *listener*. The email sent to the help desk is, in fact, an *any-message* to the role of *technician*.

For COSC 1557 Assignment 1, we have:

- $\overline{d_m}$ = "COSC 1557 Assignment 1";
- r_m = "COSC 1557 Student";
- $\flat$ = <A1, Objectives, Descriptions, Requirements>;
- $\mathcal{P}_m$ = <"Objectives:...," "Descriptions:...," "Requirements:...">;
- ℓ_m = all;
- α = a computer in A120; and
- β = one week.

With *Definition* 4.10, we can understand more about a role's message sets, i.e. $\mathcal{M}_{in} \subset \mathcal{M}, \mathcal{M}_{out} \subset \mathcal{M}$, where $\mathcal{M}$ is the set of all messages in the world. This definition follows role principles R1–R6 outlined in Chapter 3. $\mathcal{M}_{in}$ and $\mathcal{M}_{out}$ are used to reflect R3. r_m is used to reflect R8–10. $\mathcal{A}_c$, $\mathcal{A}_p$, and $\mathcal{A}_o$ are used to specify the dynamic state of a role. In a run-time system, $\mathcal{M}_{in}$ is a subset of $\mathcal{X}$ (interface) of e_a of agent a that plays this role. The elements of $\mathcal{M}_{out}$ are constructed with the subsets of $\mathcal{M}_{in}$ of other roles. Suppose that we have at least one agent currently playing a role, for roles r_1 and r_2, $\forall m_{in}(m_{in} \in r_1 \cdot \mathcal{M}_{in}) \rightarrow \exists a \ (m_{in} \in a.e_a.\mathcal{X} \wedge a. \overline{d_a} \in r_1 \cdot \mathcal{A}_c)$ and $\forall m_{out}(m_{out} \in r_1 \cdot \mathcal{M}_{out}) \rightarrow \exists r_2 (\exists m_{in}(m_{in} \in r_2 \cdot \mathcal{M}_{in}) \wedge m_{in} \cdot \flat = m_{out} \cdot \flat)$.

Note that in RBC, we emphasize that messages are sent to roles. Messages are exchanged among roles. A role is also a message dispatcher to the agents that are playing this role. Therefore, in E-CARGO, objects are passive entities that are accessed by agents through roles. Objects cannot process messages. Role players, i.e. agents, are the only components that process messages. This idea clearly clarifies the differences between agents, roles, and objects.

To dispatch a message, when a role receives a message, it checks the tag of this message to differentiate *all-*, *any-* or *some-*messages, it decides if the message is sent to all, any or some agents. If it is an *any-message*, it evaluates all agents' workloads and selects the agent with the lowest workload and sends the message to them.

In a traditional object model, we concentrate mainly on objects and their classes because executable programs run automatically with little interaction with users. A traditional object-oriented paradigm emphasizes the messages accepted by a class of objects. However, it does not consider much about the messages an object may send out. In the E-CARGO model, we emphasize that a role is a message receiver and a user sends messages by playing roles. In this definition, if r_m is null, this message is an incoming one to be dispatched by the role to an agent with f and P_m. If r_m is an identification of a role, it is an outgoing message that should be dispatched by the role specified by the message. Outgoing messages should be filled with f, P_m, ℓ_m, α, and β. We divide the outgoing messages into three categories by ℓ_m, *any-message*, *some-message* and *all-message*. By any-message we mean that the message may be sent to any agent who plays the role. Some-message means the messages should be sent to some agents and all-message means that the messages should be sent to all the agents who play the role.

4.2.3 System

A system $\sum$ can now be described as a 9-tuple.

Definition 4.11 A *system* is defined as a 9-tuple, i.e. $\sum :: = \; < C, O, A, M, R, E, G, s_0, H >$ where

- C is a finite set of classes;
- O is a finite set of objects;
- A is a finite set of agents;
- M is a finite set of messages;
- R is a finite set of roles;
- E is a finite set of environments;
- G is a finite set of groups;
- s_0 is the initial state of the system; and
- H is a finite set of users.

The initial state s_0 is expressed by the initial values of all the components C, O, A, M, R, and E, such as, built-in classes, initial objects, initial agents, primitive roles, primitive messages, and primitive environments. By "initial" and "primitive," we mean those entities that should be created before a system starts to work.

With the participation of users H, such as logging onto a collaboration system $\sum$, accessing objects of the system, sending messages through roles, forming a group in an environment, $\sum$ evolves, develops, and functions. The results of the collaboration are a new state of $\sum$ that is expressed by the values of C, O, A, M, E, G, and H. We include H to express that users might be involved in and affected by collaboration. Please note that without the participation of users, the system can only do what the agents can do. This is why users are an essential part of a system that is not autonomous.

To conduct collaboration with E-CARGO, we specify roles, build an environment, and let users play their roles in the environment to create effective collaboration. In other words, when users want to work in a group, they need to first play a role. They are then confined by the role to specific responsibilities and rights. If users are not satisfied with their current role, they can ask for tuning of their role or transferring to another role. Without a role specification mechanism, we would have no effective way of performing role transitions and tuning E-CARGO establishes the foundation to support all the properties of RBC (Chapter 3).

4.3 Fundamental Relationships in E-CARGO

A country or a society requires a government to manage, supervise, and guide its people. A society requires a board of governors to regulate the recruiting, behaviors, and collaborations of the members. Every system requires self-control and self-supervision. Similarly, a collaboration system requires self-management and self-control. Furthermore, a center (called role engine, see Chapter 10) to manage and control the components of E-CARGO is required in RBC systems.

An organization can be expressed in terms of a set of roles that determine the social position of agents within the organization and their relations to other members of the organization. Agents should be designed in such a way that enables them to reason about the social position they gain by adopting a specific role, and what the correct behavior in such a role would be (Ashforth and Mahwah 2001; Marwell and Hage 1970; Skarmeas 1995).

In social life, roles are abstract and implicit. It is the people, not the roles, that interact with each other in collaboration. People transfer roles when they are collaborating. This role transfer is implicit and there is no evidence for others to easily and clearly view if the people do not declare the transfer. In some cases, even though people declare that they play roles, others may not confidently know what the roles are. Some external objects such as chairs, tables, rooms, and titles are very weak flags to express roles. People may be in one place for one role and another place for another role. This demonstrates the requirement of role specification methodologies and tools.

In RBC, a role engine is required to deal with the management of roles and role players (agents), and control messages sent from roles to roles. In computer-based systems, roles can be made into facilities with clear and explicit flags for other agents to know what roles the agents are playing.

Roles have three functions: to specify special behavior (as interface roles), to form the behavior of an agent (as process roles), and to set a place for an agent in a group (or define the inter-relations among agents). Although many mention the relations among roles, there is no complete and comprehensive discussion on role relations.

Role specification includes two aspects: one is the content of a role and the other is its relation with other roles. Role relation specification significantly affects almost all the aspects of RBC systems, such as *role modification, role assignment* (Zhu et al. 2012), *role transfer, role execution*, and *role interactions*. The clear specification of role relations is fundamental to system analysis, system design, and system construction.

A role engine is a platform to support RBC. The basic functions of a role engine and its fundamental relations in RBC are clarified and formally defined below. These definitions can be used to discover more properties and analyze the design and construction of an RBC system. An RBC system is very complicated. To check and keep it consistent is difficult. Formal definitions of relations would help check its consistency.

To fully accept the RBC principles, one needs to accept that, in a society, it is the system structure that controls the people's behavior and the structure is formed by its roles. A role engine is a system to control the agents' behavior based on roles. To construct a role engine, it is necessary to answer the following questions: What are the relations among roles? What are the relations between roles and agents? To answer such questions, we need to concentrate on binary relations among roles, agents, and between agents and roles. For a binary relation $\mathbb{R}$ over a set $\mathbb{S}$, if $x, y \in \mathbb{S}$, we use $<x, y> \in \mathbb{R}$ to express *x is related to y*.

4.3.1 The Relations Among Roles

The relations among people are complicated in societies (Marwell and Hage 1970). Role relations are more abstract than people's relations (Zhu 2008). Based on the principle of information hiding in the abstraction process, role relations should be easier to understand and handle than people's relations. The classification and clarification of role relations are the foundations to implement a role engine.

4.3.1.1 Role Classes and Instances
To specify role relations, it is required that we specify what a role is. When roles are discussed, arguments about roles always occur. What are roles? Are roles classes? Are roles instances? What is a message to a role?

From the viewpoint of modeling-roles (Chapter 2), there are many role instances of a role class. Role instances are attached to an object that is playing them. These kinds of roles can be applied to express the evolution of objects. The state of a modeling-role is a part of the state of the object playing the role. The requests are implicitly described in the implementation, i.e. the methods of the modeling-role class.

In RBC, we state that roles are instances of τ in *Definition 4.6* with properties of classes, and τ is actually a meta-class in the view of object-orientation. As an instance, a role has its concrete agents and messages. As a class, its cardinalities $(|\mathcal{A}_c|, |\mathcal{A}_g|, |\mathcal{A}_o|)$ express how many instances are created for agents to play it. To classify this, we define role classes and role instances.

Definition 4.12   A *role class* is an instance of τ, i.e. *Definition* 4.6.

From this definition, we understand that *Definition* 4.6 provides a meta-class definition for all the role classes.

Definition 4.13   A *role instance* is an instance of a role class.

In *Definition* 4.6, τ is a meta-class for all role classes. When a new role is defined, it is an instance of τ. When we say an agent plays a role, the agent plays a role instance linked with a role class. In RBC and E-CARGO, role classes and concrete role instances exist in an environment (*Definition* 4.7). Role instances are implemented for agents to play them. The τ in the tuples of $\mathcal{B}$ is a role class and q expresses the number of role instances. A role instance should express the concrete objects in V_p^r. The concrete objects described by V_r^r should be linked to an agent, i.e. the player. The number of role instances is restricted by the cardinalities of a role, i.e. q. The following discussions will focus on the relations between role classes. In RBC, a message is first sent to a *leaf role* class *(Section 4.3.1.2)* in an environment. The role class then dispatches the message to agents that are players of an instance of a leaf role, which is in an environment.

4.3.1.2 Inheritance Relation

Roles themselves can be a classification tool if they are related to a classification hierarchy.

Definition 4.14   Role τ_2 is a *super role* of role τ_1 if τ_1 possesses all the properties of τ_2. Vice versa, τ_1 is a *sub role* of τ_2. An *inheritance relation*, denoted as Ω, is a set of tuples of roles $<\tau_1, \tau_2 >$, where τ_2 is a super role of τ_1 and τ_1 is a sub role of τ_2. Here, "inheritance" means the simple and whole inheritance from the view of object-orientation (Zhu and Zhou 2006a).

Figure 4.5 shows the inheritance relations among a group of role classes.

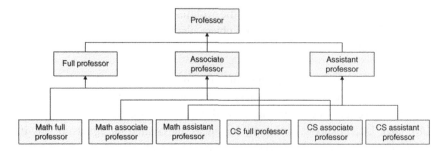

Figure 4.5 Super roles and sub roles. Note: CS means Computer Science.

Definition 4.15 Role r is called a *root role* if it has no super roles. Role r is called a *leaf role* if it has no sub roles.

In Figure 4.5, *Professor* is a super role of roles *Full Professor, Associate Professor*, and *Assistant Professor. Professor* is also a root role, and *Math Full Professor, Math Associate Professor, ...,* and *CS Assistant Professor* are leaf roles. In such a hierarchy, it is necessary to clarify that an agent should be assigned with a leaf role because leaf roles are the only roles executable by agents.

Please note that this role relation is not a partial order (Skiena 1990), nor are other role relations. The reflexive property is not held for this relation. For example, it is meaningless to say that a *professor* inherits from a *professor*.

Another note is that in the tree of inheritance, a role instance can only be created based on a leaf role class. That is, all the non-leaf role classes are abstract role classes. It is also understandable that all the other classes are between leaf role classes.

Definition 4.16 In E-CARGO, a message can be of three types: *any, all*, and *some*. Suppose r is a non-leaf role, $r_0, r_{1,}, ...,$ and $r_{u-1,}$ are leaf roles whose super role is r. *Any-messages* to r should be sent to an agent that plays any of the roles $r_0, r_{1,}, ...,$ and r_{u-1}. *All-messages* should be sent to all the agents that play $r_0, r_{1,}, ...,$ and r_{u-1}. *Some-messages* should be sent to some agents that play $r_0, r_{1,}, ...,$ and r_{u-1}.

Definition 4.17 When a message is sent to a role, the message is *successfully* sent if the role matches the message pattern and the role finds a player to process this message.

We need to check if this message is included in the service set of the role or one of its sub roles. If the message is in the sets and required agents are ready to respond to the message, the message sending is successful. In RBC systems, a role accepts and dispatches messages. A role cannot respond to messages directly.

4.3.1.3 Promotion Relations

In a well-designed community, there are predesigned role promotion relations. This relation encourages the members of the community to work hard to contribute. It means that an agent could play an upper-level role only when it ever played a lower-level role before.

Definition 4.18 Role r_2 is an *upper role* of role r_1 if r_1 must be played before r_2 is assigned. Vice versa, r_1 is called a *lower role* of r_2. A *promotion relation*, denoted as Λ, is a set of tuples of roles $<r_1, r_2>$, where r_1 is a lower role of r_2 and r_2 an upper role of r_1.

Role promotion is not natural but man-made. It is set in a role engine to act as a policy for role assignment. It is set during initialization and may be adjusted during collaboration.

Figure 4.6 gives two examples of promotion relations in the education and information technology fields.

4.3.1.4 Report-to Relations

A role provides many services. However, not all other roles can request its services and obtain instant responses. Some regulations are introduced to control the accessibilities of the services. For example, a team leader can tell a team member to complete a task but a team member cannot tell a team leader what to do (Colman and Han 2007). This leads to a report-to relation.

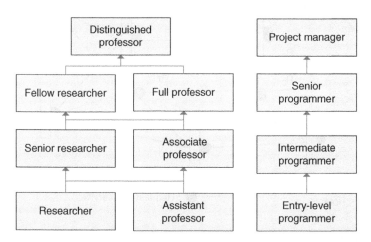

Figure 4.6 Examples of promotion relations.

Definition 4.19 *Report-to relation, supervisor role, and supervisee role.* Role r_2 is a *supervisor role* of role r_1 if role r_1 must respond to the requests from role r_2 in a time limit set by r_2. Vice versa, r_1 is a *supervisee role* of r_2. A *report-to relation* denoted as Δ is a set of tuples of roles $<r_1,r_2>$, where r_1 is a supervisee role of r_2 and r_2 is a supervisor role of r_1.

Based on this definition, a rule can be designed in a role engine: an agent playing a supervisee role obtains repercussions if it does not respond to its supervisor's requests on time, i.e. its credit point record in the agent's state s_a is decreased. A report-to relation is different from the promotion relation in that an agent playing a supervisee role may not be promoted to the supervisor role and a lower role may not always be supervised by its upper role.

4.3.1.5 Request Relations

Everybody serves others and everybody requests others' services. Therefore, a request role is an evident relation between roles.

Definition 4.20 Role r_1 is called a *request role* of role r_2 if r_2 provides the services requested by r_1, i.e. $r_1.\mathcal{M}_{out} \subseteq r_2.\mathcal{M}_{in}$ and r_2 is called a *service role* of r_1. A *request relation* denoted as Θ is a set of tuples $<r_1, r_2>$, where r_1 is a request role of r_2 and r_2 is a service role of r_1, i.e. $< r_1, r_2 > \in \Theta$ if $r_1.\mathcal{M}_{out} \subseteq r_2.\mathcal{M}_{in}$.

The request relation also holds the irreflexive property (Figure 4.7). This property guarantees that a role does not issue a request to itself.

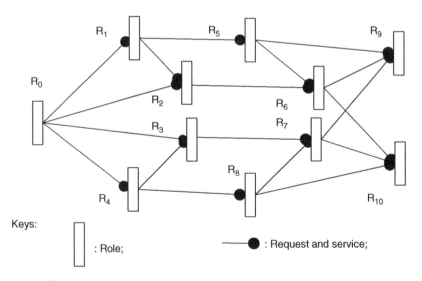

Figure 4.7 A request relation.

4.3.1.6 Derived Relations

The following relations are derived from the above basic relations. They are not independent.

Definition 4.21 Role r_1 is called a *competitor role* of role r_2 if roles r_1 and r_2 have the same request role or the same upper role. A *competition relation* denoted as $\in$ is a set of tuples of roles $<r_1, r_2>$, where r_1 and r_2 are competitor roles of each other. More exactly, $< r_1, r_2 > \in \in$ if $\exists r_3 \ni (< r_3, r_1 > \in \Theta \wedge < r_3, r_2 > \in \Theta)\vee (< r_1, r_3 > \in \Lambda \wedge < r_2, r_3 > \in \Lambda)$.

For example, roles *software consultants* and *software developers* are competitor roles because they provide the same services, i.e. software development and role *client* may request the same service from them; in Figure 4.6, roles *researcher* and *assistant professor* are competitor roles but roles *programmer* (including *entry-level programmer*, *intermediate programmer*, and *senior programmer*) and *researcher* are not competitor roles. The competition relations are used to establish regulations for assigning roles and dispatching messages.

Definition 4.22 Role r_1 is a *peer role* of role r_2 if r_1 and r_2 have the same supervisor role. A *peer relation* denoted as $\Diamond$ is a set of tuples of roles $<r_1, r_2>$, where r_1 and r_2 are peer roles of each other. More exactly, $< r_1, r_2 > \in \Diamond$ if $\exists r_3 \ni (< r_1, r_3 > \in \Delta \wedge < r_2, r_3 > \in \Delta)$.

4.3.1.7 Conflict Relations

In RBAC, Nyanchama and Osborn (1999) point out that there are many conflicts when authorization of roles is concerned: user-user/group-group/user-group conflicts, role-role conflicts, privilege-privilege conflicts, user-role assignment conflicts, and role-privilege assignment conflicts. From the viewpoint of RBC, roles may be in conflict when they are assigned to agents.

Definition 4.23 Roles r_1 and r_2 are *conflicting* if one agent cannot take r_1 and r_2 at the same time or during a period of time. r_1 is called a *conflicting role* of r_2 and vice versa. A *conflict relation* denoted as Ξ is a set of tuples $<r_1, r_2>$, where r_1 and r_2 are conflicting roles of each other. More exactly, $< r_1, r_2 > \in \Xi \rightarrow \forall a \in A, r_1, r_2 \in R (\neg(r_1 \in a.R_e \wedge r_2 \in a.R_e))$.

In an RBC system, the conflict relation is actually predefined in order to protect conflicting roles from being assigned to the same agent.

As a summary, Γ is used to express all the relations among roles in an RBC system. Suppose $r_1, r_2 \in R$ and $r_1 \neq r_2$, $\Gamma ::= \Lambda \cup \Delta \cup \Theta \cup \in \cup \Diamond \cup \Xi$, where

- Λ is a promotion relation and $<\!\imath_1,\imath_2\!>\in\Lambda$ if $\imath_1$ is a lower role of $\imath_2$ and $\imath_2$ is an upper role of $\imath_1$.
- Δ is a report-to relation and $<\!\imath_1,\imath_2\!>\in\Delta$ if $\imath_1$ is a supervisee role of $\imath_2$ and $\imath_2$ is a supervisor role of $\imath_1$.
- Θ is a request relation and $<\!\imath_1,\imath_2\!>\in\Theta$ if $\imath_1.\mathcal{M}_{oct}\subseteq\imath_2.\mathcal{M}_{in}$.
- ϵ is a competition relation and $<\imath_1,\imath_2>\in\epsilon$ if $\exists\imath_3\ni(<\imath_1,\imath_3>\in\Theta$ $\wedge<\imath_2,\imath_3>\in\Theta)\vee(<\imath_1,\imath_3>\in\Lambda\wedge<\imath_2,\imath_3>\in\Lambda))$.
- $\Diamond$ is a peer relation and $<\imath_1,\imath_2>\in\Diamond$ if $\exists\imath_3\ni(<\imath_1,\imath_3>\in\Delta\wedge<\imath_2,\imath_3>\in\Delta)$.
- Ξ is a conflict relation and $<\imath_1,\imath_2>\in\Xi$ if $\imath_1$ and $\imath_2$ are conflicting.

Note that we denote Ω the inheritance relation especially, and $<\!\imath_1,\imath_2\!>\in\Omega$ if $\imath_1$ inherits from $\imath_2$. As discussed above, inheritance is special compared with other relations. In the following statements, by roles, we mean *leaf role classes* if we do not emphasize inheritance.

As we know, to understand relations, the properties of relations should be clarified. The most important properties in specifying relations including (Schmidt 2010) *Reflexivity, Transitivity, Symmetricity*, and *Circularity*.

Definition 4.24 A binary relation $\mathbb{R}$ over a set $\mathbb{S}$ is *reflexive* if $\forall x\in\mathbb{S}<x,x>\in\mathbb{R}$, $\mathbb{R}$ is called *irreflexive*, or *anti-reflexive*, if $\forall x\in\mathbb{S}<x,x>\notin\mathbb{R}$. *Reflexivity* is a typical property of the equivalence relation.

Definition 4.25 A binary relation $\mathbb{R}$ over a set $\mathbb{S}$ is *transitive* if $\forall x, y, z\in\mathbb{S}<x, y>\in\mathbb{R}$ and $<y, z>\in\mathbb{R}$, then $<x, z>\in\mathbb{R}$. *Transitivity* is a key property of both partial orders and equivalence relations.

Definition 4.26 A binary relation $\mathbb{R}$ over a set $\mathbb{S}$ is *symmetrical* if $\forall x, y\in\mathbb{S}<x, y>\in\mathbb{R}$, then $<y, x>\in\mathbb{R}$.

Definition 4.27 A binary relation $\mathbb{R}$ over a set $\mathbb{S}$ is *circular* if $\exists x_0, x_1, x_2, ..., x_n\in\mathbb{S}$, $<x_i, x_{i+1}>(x_i\neq x_{i+1})\in\mathbb{R}$ $(i=0, 1, ..., n-1)\ni(x_n=x_0)$. $\mathbb{R}$ is *noncircular* if $\neg\exists x_0, x_1, x_2, ..., x_n\in\mathbb{S}$, $<x_i, x_{i+1}>(x_i\neq x_{i+1})\in\mathbb{R}$ $(i=0, 1, ..., n-1)\ni(x_n=x_0)$.

For example, the equivalence relation has all the above four properties.

Definition 4.28 A *role net* is a directed graph formed by all the relations of Γ, denoted as $<\mathcal{R},\Gamma>$, where $\mathcal{R}$ is the node set and Γ is the edge set.

Definition 4.29 Roles $\imath_1$ and $\imath_2$ are *interrelated* if $<\!\imath_1,\imath_2\!>$ belongs to a role net, i.e. $<\!\imath_1,\imath_2\!>\in\Gamma$.

4.3.2 The Relations Between Roles and Agents

After roles and role relations have been set, a collaborative platform is built. Agents (actors) can collaborate (perform) on this platform (stage) called a *role engine*. Agents should be assigned roles, behave on a role engine, transfer roles, and, finally, leave the engine, i.e. the relations between agents and roles should be established. Some relations are specified and some are established by executing predicates (operations).

Definition 4.30 Role r is the *current role* of agent a if a is currently playing r, i.e. $r \in a.\mathcal{R}_c$; at the same time, a is called a current agent of role r, i.e. $a \in r.\mathcal{A}_c$.

Definition 4.31 Role r is a *potential role* of agent a if a is qualified to play but not currently playing this role, i.e. $r \in a.\mathcal{R}_p$; at the same time, a is called a potential agent of r, i.e. $a \in r.\mathcal{A}_p$.

From these definitions, one agent can play many current and potential roles, and one role may have many current and potential agents.

Definition 4.32 Role r is a *previous role* of agent a if a used to play r but r is not a current role of a, i.e. $(r \in a.\mathcal{R}_o) \bigwedge (r \notin a.\mathcal{R}_c)$; at the same time, a is called a previous agent of r, i.e. $a \in r.\mathcal{A}_o$.

Definitions 4.33–4.38 belong to a role engine dealing with the relations between agents and roles.

Definition 4.33 *Apply-for* (a,r) is a service belonging to the role class of r. The message is sent by the agent class when agent a applies for role r.

After the message *apply-for* (a,r) is received by the system, the system should check if a is qualified to play r based on the qualifications of a.

Definition 4.34 *Approve* (a,r) is an event that occurs when *apply-for* (a,r) has been issued and the agent is evaluated to be qualified to play role r. After this predicate is executed, r is added to the potential role set of agent a, i.e. $r \in a.\mathcal{R}_p$.

Many agents may bid for one role. There must be a set of rules to select (*approve*) the best agent to play the role. It means that the approved agent should be the most efficient, least costly, and best fit to accomplish the tasks specified by the role. Some system restrictions should be enforced when approving an agent: the agent can play a direct upper role, a leaf role, a lower role, a peer role, and a previous role of its repository roles, but it cannot play a conflicting role of its repository roles.

Definition 4.35 *Disapprove* (a,r) is a predicate that disapproves a role from an agent.

This predicate is executed in two situations: when an agent is promoted to play an upper role, the lower role is disapproved; when an agent is checked and found that its credits are below the required credit of the role, the agent is disapproved of the role. After it is executed, r is deleted from the repository set and put into the previous role set of agent a, i.e. $r \notin a.R_r \wedge r \in R_o$.

Definition 4.36 *Transfer* (a, r_1, r_2) is a predicate for an agent (a) to transfer the current role from r_1 to r_2. After it is executed, r_2 is assigned as a current role of the agent and r_1 is put back to the potential role set. It can be executed only when the system is still workable after this role transfer occurs, i.e. all the roles have enough current agents.

Definition 4.37 *Dispatch* (r, a, m) is a predicate for a role (r) to dispatch message m to agent a. After it is executed, a does what m asks for.

Definition 4.38 *Reply* (a, r, p). It is a predicate for an agent (a) to reply an object p to role r. After it is executed, r completes one service and responds to its request role with object p.

4.3.3 The Relations Between Agents

Many agents are working together in collaboration. By playing different roles, they build different workflows and accomplish different tasks, information transmissions, and productions. In collaboration, competition is inevitable. Therefore, agents may collaborate or compete. Agents may be collaborators or competitors.

Definition 4.39 Agents a_1 and a_2 are *collaborators* if their roles are interrelated, i.e. $\exists r_1, r_2 \in a_1.R_r \cap a_2.R_r \ni \ <r_1, r_2> \ \in \Gamma$.

Definition 4.40 Agents a_1 and a_2 are *current collaborators* if they are currently playing interrelated roles or $a_1.R_e$ and $a_2.R_e$ are interrelated, i.e. $\exists r_1, r_2 \in a_1.R_e \cap a_2.R_e \ni < r_1, r_2 > \ \in \Gamma$.

Definition 4.41 Agents a_1 and a_2 are *potential collaborators* if they are not current collaborators and share potential interrelated, i.e. $a_1.R_e \cap a_2.R_e = \emptyset \wedge \exists r_1, r_2 \in a_1.R_r \cap a_2.R_r \ni \ <r_1, r_2> \ \in \Gamma$.

Definition 4.42 Agents a_1 and a_2 are *potentially previous collaborators* if they are not current collaborators and previously played interrelated roles, i.e. $a_1.\mathcal{R}_c \cap a_2.\mathcal{R}_c = \emptyset \wedge \exists r_1, r_2 \in a_1.\mathcal{R}_o \cap a_2.\mathcal{R}_o \ni \; <r_1, r_2> \; \in \Gamma$.

Definition 4.43 Agents a_1 and a_2 are *peers* if they play the same role, i.e. $a_1.\mathcal{R}_r$ $\cap a_2.\mathcal{R}_r \neq \emptyset$.

Definition 4.44 Agents a_1 and a_2 are *current peers* if they have the same current roles, i.e. $a_1.\mathcal{R}_c \cap a_2.\mathcal{R}_c \neq \emptyset$.

Definition 4.45 Agents a_1 and a_2 are *potential peers* if $a.\mathcal{R}_p \cap a_2.\mathcal{R}_p \neq \emptyset \wedge a_1.\mathcal{R}_c \cap a_2.\mathcal{R}_c = \emptyset$.

Definition 4.46 Agents a_1 and a_2 are *potentially previous peers* if $a.\mathcal{R}_o \cap a_2.\mathcal{R}_o \neq \emptyset \wedge a_1.\mathcal{R}_c \cap a_2.\mathcal{R}_c = \emptyset$.

Definition 4.47 Agents a_1 and a_2 are *competitors* if their repository role sets contain competition roles, i.e. $\exists r_1 \in a_1.\mathcal{R}_r, r_2 \in a_2.\mathcal{R}_r \ni (<r_1, r_2> \; \in \mathcal{E})$.

Definition 4.48 Agents a_1 and a_2 are *current competitors* if their current roles are competitor roles, i.e. $\exists r_1 \in a_1.\mathcal{R}_c, r_2 \in a_2.\mathcal{R}_c \ni \; <r_1, r_2> \; \in \mathcal{E}$.

Definition 4.49 Agents a_1 and a_2 are *potential competitors* if $\exists r_1, r_2 \in a_1.\mathcal{R}_p \cap a_2.\mathcal{R}_p \ni (<r_1, r_2> \; \in \mathcal{E}) \wedge \neg \exists r_1 \in a_1.\mathcal{R}_c, r_2 \in a_2.\mathcal{R}_c \ni \; <r_1, r_2> \; \in \mathcal{E}$.

Definition 4.50 Agents a_1 and a_2 are *potentially previous competitors* if $\exists r_1, r_2 \in a_1.\mathcal{R}_o \cap a_2.\mathcal{R}_o \ni (<r_1, r_2> \; \in \mathcal{E}) \wedge \neg \exists r_1 \in a_1.\mathcal{R}_c, r_2 \in a_2.\mathcal{R}_c \ni \; <r_1, r_2> \; \in \mathcal{E}$.

Definition 4.51 Agent a_1 is called a *client* of agent a_2 and a_2 is called a *server* of a_1 if $\exists r_1, r_2 \in a_1.\mathcal{R}_r \cap a_2.\mathcal{R}_r \ni (<r_1, r_2> \; \in \Theta)$.

Definition 4.52 Agent a_1 is called a *current client* of agent a_2 and a_2 is called a *current server* of a_1 if a_1 is currently playing a request role of the role a_2 is currently playing, i.e. $\exists r_1 \in a_1.\mathcal{R}_c, r_2 \in a_2.\mathcal{R}_c) \ni (<r_1, r_2> \; \in \Theta)$.

Definition 4.53 Agent a_1 is called a *potentially potential client* of agent a_2 and a_2 is called a *potential server* of a_1 if $\exists r_1, r_2 \in a_1.\mathcal{R}_p \cap a_2.\mathcal{R}_p \ni (<r_1, r_2> \; \in \Theta) \wedge \exists \neg r_1 \in a_1.\mathcal{R}_c, r_2 \in a_2.\mathcal{R}_c) \ni (<r_1, r_2> \; \in \Theta)$.

Definition 4.54 Agent a_1 is called a *previous client* of agent a_2 and a_2 is called a *previous server* of a_1 if $\exists r_1, r_2 \in a_1.\mathcal{R}_o \cap a_2.\mathcal{R}_o \ni (<r_1, r_2> \in \Theta) \wedge \exists \neg r_1 \in a_1.\mathcal{R}_e, r_2 \in a_2.\mathcal{R}_e) \ni (<r_1, r_2> \in \Theta).$

4.3.4 Properties of an RBC System

After the above definitions are clarified, the following properties should be kept in an RBC system.

Property 1 *Irreflexive.*

- $\forall r \in \mathcal{R} \ (<r,r> \notin \Omega) \wedge (<r,r> \notin \Delta) \wedge (<r,r> \notin \Lambda) \wedge (<r,r> \notin \Theta) \wedge (<r,r> \notin \epsilon) \wedge (<r,r> \notin \Diamond) \wedge (<r,r> \notin \Xi).$

Property 2 *Transitive.*

- $\forall r_1, r_2, r_3 \in \mathcal{R} \ (<r_1, r_2> \in \Omega \wedge <r_2, r_3> \in \Omega \rightarrow <r_1, r_3> \in \Omega);$
- $\forall r_1, r_2, r_3 \in \mathcal{R} \ (<r_1, r_2> \in \Lambda \wedge <r_2, r_3> \in \Lambda \rightarrow <r_1, r_3> \in \Lambda);$
- $\forall r_1, r_2, r_3 \in \mathcal{R} \ (<r_1, r_2> \in \Delta \wedge <r_2, r_3> \in \Delta \rightarrow <r_1, r_3> \in \Delta);$
- $\forall r_1, r_2, r_3 \in \mathcal{R} \ (<r_1, r_2> \in \epsilon \wedge <r_2, r_3> \in \epsilon \rightarrow <r_1, r_3> \in \epsilon);$ and
- $\forall r_1, r_2, r_3 \in \mathcal{R} \ (<r_1, r_2> \in \Diamond \wedge <r_2, r_3> \in \Diamond \rightarrow <r_1, r_3> \in \Diamond).$

Note that relations Θ and Ξ do not hold the transitive property. Even though role r_1 requests (or is in conflict with) r_2 and r_2 requests (or is in conflict with) r_3, r_1 does not have to request (or be in conflict with) r_3.

Property 3 *Symmetrical.*

- $\forall r_1, r_2 \in \mathcal{R} \ (<r_1, r_2> \in \epsilon) \rightarrow (<r_2, r_1> \in \epsilon);$
- $\forall r_1, r_2 \in \mathcal{R} \ (<r_1, r_2> \in \Diamond) \rightarrow (<r_2, r_1> \in \Diamond);$ and
- $\forall r_1, r_2 \in \mathcal{R} \ (<r_1, r_2> \in \Xi) \rightarrow (<r_2, r_1> \in \Xi).$

Property 4 *Noncircular.* There are no circles in inheritance, request, promotion, and report-to relations, i.e. $\neg \exists r_0, r_1, r_2, ..., r_n \in \mathcal{R}, <r_j, r_{j+1}> (r_j \neq r_{j+1}) \in \mathcal{X}$ $(\mathcal{X} = \Omega, \Lambda, \Delta \text{ or } \Theta) \ (j = 0, 1, ..., n-1) \ni (r_n = r_0).$

From the *Definitions* 4.39–4.50, interesting facts are discovered: (i) agents playing conflicting roles are collaborators; (ii) peers may be either collaborators or competitors; (iii) peer role players are collaborators; (iv) clients and servers are collaborators; and (v) collaborators may be also competitors.

For (i), in sales, roles *price setter* and *buyer* for a specific item are in conflict but two persons playing these roles are collaborators. In fact, two different agents are required to play different conflicting roles in collaborations. For (ii), in a review process of scientific research, peers cannot be reviewers. To be fair, reviewers should be agents who are neither collaborators nor competitors. For (iii), the people in an office collaborate to complete a task ordered by their common supervisor. For (iv), clients and servers collaborate to complete a service transaction in the service community. For (v), two soccer teams are competitors due to *the relation of promotion*, and also collaborators due to *the role player* of a game.

Definition 4.55 A system is *redundant* if some services are not requested, i.e. $\exists(r \in \mathcal{R}) \ni (r.\mathcal{M}_{in} - \bigcup_{j=0}^{n-1} r_j.\mathcal{M}_{out}(r_j \neq r)) \neq \emptyset$, where $n = |\mathcal{R}|$.

Definition 4.56 A system is *insufficient* if some requests are not served, i.e. $\exists(r \in \mathcal{R}) \ni (r.\mathcal{M}_{out} - \bigcup_{j=0}^{n-1} r_j.\mathcal{M}_{in}(r_j \neq r)) \neq \emptyset$, where $n = |\mathcal{R}|$.

Definition 4.57 A system is *consistent* if all the relations are consistent, i.e. it satisfies **Properties 1–4**.

One purpose of specifying role relations is to regulate message dispatches. Some regulations are:

- A message sent to a super role should be dispatched to its sub roles.
- A message accepted by an agent can be repackaged and forwarded to the supervisee roles of the role currently played by this agent.
- A message accepted by an agent can be repackaged and forwarded to the service roles of the role currently played by this agent.
- A message dispatched to an agent can be re-dispatched to the peers of this agent.

The relations and the properties specified in this chapter have been applied in a prototype of an RBC chatting system (Zhu 2007). The success of the prototype verified the discussed relations.

4.4 Related Work

Although roles as concepts have been discussed for many years, the literature lacks a well-accepted role model and systematic descriptions of the relations among roles and their players. Most of the literature did not present role relationship specification and classification. References discussed in Chapter 2 are highly related to the literature. The work on role relationships inspired the work described in this chapter.

A role is defined by Hawkins et al. (1983) as a prescribed pattern of behavioral expectations of a person in a given situation. A position means a commonly expected and understood designation in a given social structure such as an accountant, mother, and church member (Ashforth and Mahwah 2001). Bostrom (1980) defines a role as a set of expectations about behavior for a particular position within a society. A role is a position occupied by a person in a social relationship. In this position, the person possesses specific rights and takes specific responsibilities.

Marwell and Hage (1970) present 100 role relationships in different societies including economic, political, health and welfare, science and education, family, religion, art, and leisure sectors. Each role relationship involves occupants, activities, locations, and occurrences. All the relationships are classified in the senses of sociology, i.e. *gemeinschaft* and *gesellschaft*. Their analysis lacks abstraction in the sense of collaboration and management systems and is not very useful in such fields.

In RBAC, Simon and Zurko (1997) mention the role conflict relations when they discuss conflicts of interest relevant to complex tasks in a workflow management system. Conflicts are common in business transactions that require two signatures before a check is issued in order to avoid fraud. Nyanchama and Osborn (1999) discuss the role-role conflict relations by role graphs. A role-role conflict means that the two roles should never appear together. This implies they should never be assigned to a single user, which is checked on user-role assignment. Their discussion concentrates on the conflict of permission authorizations when accessing system resources.

Skarmeas (1995) proposes a role model to deal with the tree hierarchy of roles in organizations. Roles may contain other roles and/or elementary roles. In their proposed tree structures, elementary roles are leaves. In describing organizational relations, he introduces virtual roles to deal with the supervision relations among roles. Virtual roles (similar to the *supervisee role* in this chapter) are used to assign supervised roles to individuals. He mentions the role consistency problem, which is the major concern when role transfers occur. However, the tasks of a role and the roles themselves are not clearly distinguished in his work (Skarmeas 1995).

Olarnsakul and Batanov (2004) emphasize that roles have relationships (i.e. interactions and dependencies) with one another in order to fulfill assigned responsibilities. They present that roles and relationships are the building blocks of the organizational structure. In their model, a role relationship is a set of protocols that represent collaboration tasks or business processes.

Odell et al. (2005) describe a meta-model for agents, roles, and groups that are incorporated with Unified Modelling Language (UML) notations. They believe that roles provide the building blocks for agent social systems. Roles meet the requirements of describing interactions among agents. They use associations to describe the relationships among roles. Classifier classes are introduced to illustrate the relationships among agents and roles. Their meta-model is a good reference to analyze and design systems with roles.

Ren et al. (2007) present a coordination model, i.e. the Actor, Role, and Coordinator (ARC) model. Similar to the role dynamics mentioned in Section 3.6.2, roles are taken as a key driver in the ARC model. In their model, behaviors of actors, roles, and coordinators are formally defined and applied to support the reconfigurability and fault localization of open, distributed, embedded software systems. The specified behaviors regulate some relations in RBC systems, such as memberships and configurations.

4.5 Summary

RBC is a well-specified methodology that mainly uses roles as underlying components to facilitate collaboration activities. Its fundamental E-CARGO model can be extended and instantiated in different ways to discover challenges and solve complex collaboration problems. Research and investigations on RBC and E-CARGO will be valuable in promoting the analysis, design, and development of collaboration systems. As a computational model of a complex system, E-CARGO is both flexible and robust.

In the E-CARGO model, we notice that a role has no knowledge of its environments and groups because environments and groups are derived components based on roles. As for the roles of individuals, we should recognize that they can only be in one place at one time and they are limited in their ability to move from place to place. In the above definitions, we note that roles are defined in the systems' scope. Environments and groups are built after roles are specified. Such a scheme truly reflects the reality of an organization.

The agents and roles are the mediums in which users operate the system and interact with others. Users can contribute to collaboration by:

- Creating an agent class to help them automatically access objects, send outgoing messages and reply to incoming messages without their direct interference;
- Negotiating roles with role facilitators and negotiators;
- Adding, deleting, and modifying the methods of an agent class and having the agent play roles; and
- Playing roles, i.e. accessing objects, sending outgoing messages, and replying to incoming messages.

To understand the significance of roles in a collaborative system, let us remove roles from the system $\sum$. As a result, $\mathcal{E}$ and $\mathcal{G}$ vanish because they depend on roles. It now consists of only $\mathcal{C}, \mathcal{O}, \mathcal{A}, \mathcal{M}, \triangleleft_0$, and $\mathcal{H}$. $\sum$ becomes a traditional multi-agent system. We lose the mechanism to organize collaboration and the media of interactions between the users and other components of the system. That is why traditional CSCW or agent systems always argue about their multiple user interface design because the interfaces depend entirely on the implementations. Based on the E-CARGO model, the interfaces of a system are completely determined by the roles.

Future work (see Chapter 10) can be conducted by investigating the application of RBC/E-CARGO in other research fields and by scrutinizing each step in the life-cycle of RBC and each component of E-CARGO.

References

Al-Zaghameem, A.O. and Alfraheed, M. (2013). An expressive role-based approach for improving distributed collaboration transparency. *International Journal of Computer Science Issues (IJCSI)* 10 (4–2): 61–67.

Ashforth, B.E. and Mahwah, N.J. (2001). *Role Transitions in Organizational Life: An Identity-Based Perspective*. Lawrence Erlbaum Associates, Inc.

Becht, M., Gurzki, T., Klarmann, J., and Muscholl, M. (1999). ROPE: Role Oriented Programming Environment for multiagent systems. *Proceedings of Fourth IECIS International Conference on Cooperative Information Systems*, Edinburgh, UK (September 1999), pp. 325–333.

Biddle, B.J. and Thomas, E.J. (1979). *Role Theory: Concepts and Research*. New York: R. E. Krieger Publishing Co.

Bostrom, R.P. (1980). Role conflict and ambiguity: critical variables in the MIS user-designer relationship. *Proceedings of the 17th Annual Computer Personnel Research Conference*, New York, NY, USA (June 1980), pp. 88–115.

Botha, R.A. and Eloff, J.H.P. (2001). Designing role hierarchies for access control in workflow systems. *25th Annual International Computer Software and Applications Conference (COMPSAC'01)*, Chicago, IL, USA (8–12 October 2001), pp. 117–122.

Brachman, R.J. and Schmolze, J.G. (1985). An overview of the KL-ONE knowledge representation system. *Cognitive Science* 9 (2): 171–216.

Budd, T.A. (2001). *The Introduction to Object-Oriented Programming*, 3e. Boston, MA: Addison-Wesley.

Cabri, G. (2007). Environment-supported roles to develop complex systems. In: *Engineering Environment-Mediated Multi-Agent Systems*, Lecture Notes in Computer Science (LNCS), vol. 5049 (eds. D. Weyns, S.A. Brueckner and Y. Demazeau), 284–295. Berlin, Heidelberg, Germany: Springer.

Cabri, G. (2012). Agent roles for context-aware P2P systems. In: *Agents and Peer-to-Peer Computing*, LNCS, vol. 6573 (eds. S. Joseph, Z. Despotovic, G. Moro and S. Bergamaschi), 104–114. Berlin, Heidelberg, Germany: Springer.

Caetano, A., Silva, A.R., and Tribolet, J. (2009). A role-based enterprise architecture framework. *The 24th Annual ACM Symposium on Applied Computing*, Honolulu, Hawaii, USA (8–12 March 2009), pp. 253–258.

Colman, A. and Han, J. (2007). Using role-based coordination to achieve software adaptability. *Science of Computer Programming* 64: 223–245.

Dafoulas, G.A. and Macaulay, L.A. (2001). Facilitating group formation and role allocation in software engineering groups. *Proceedings of the ACS/IEEE International Conference on Computer Systems and Applications*, Beirut, Lebanon (25–29 June 2001) pp. 0352–0359.

Dahchour, M., Pirotte, A., and Zimányi, E. (2004). A role model and its metaclass implementation. *Information Systems* 29 (3): 235–270.

Ferraiolo, D.F. and Kuhn, D.R. (1992). Role-based access control. *Proceedings of the NIST-NSA National (USA) Computer Security Conference*, Baltimore, USA (13–16 October 1992), pp. 554–563.

Ferraiolo, D.F., Cugini, J.A., and Kuhn, D.R. (1995). Role-Based Access Control (RBAC): features and motivations. *Proceedings of the 11th Annual Computer Security Applications Conference (CSAC '95)*, Los Alamitos, CA (11–15 December 1995), pp. 241–248.

Ferraiolo, D.F., Sandhu, R., Gavrila, S. et al. (2001). Proposed NIST standard: role-based access control. *ACM Transactions on Information and System Security* 4 (2): 224–274.

Gottlob, G., Schrefl, M., and Rock, B. (1996). Extending object-oriented systems with roles. *ACM Transactions on Information Systems* 14 (3): 268–296.

Guzdial, M., Rick, J., and Kerimbaev, B. (2000). Recognizing and supporting roles in CSCW. *Proceeding of the ACM 2000 Conference on Computer-Supported Cooperative Work (CSCW'00)*, Philadelphia, Pennsylvania, USA (December 2000), pp. 261–268.

Hawkins, D.I., Best, R.J., and Coney, K.A. (1983). *Consumer Behavior*. Plano, TX, USA: Business Publications, Inc.

Kay, A.C. (1993). The early history of Smalltalk. *Proceedings of the 2nd ACM SIGPLAN Conference on History of Programming Languages (HOPL-II)*, Cambridge, Massachusetts, USA (20–23 April 1993), pp. 69–95.

Kendal, E.A. (1999). Role model design and implementation with aspect-oriented programming. *ACM SIGPLAN Notices* 34 (10): 353–369.

Kristensen, B.B. and Østerbye, K. (1996). Roles: conceptual abstraction theory & practical language issues. *Special Issue of Theory and Practice of Object Systems on Subjectivity in Object-Oriented Systems* 2 (3): 143–160.

Kühn, T., Leuthäuser, M., Götz, S. et al. (2014). A metamodel family for role-based modeling and programming languages. In: *Software Language Engineering*, LNCS, vol. 8706 (eds. B. Combemale, D.J. Pearce, O. Barais and J.J. Vinju), 141–160. Cham, New York, NY, USA: Springer.

Marwell, G. and Hage, J. (1970). The organization of role-relationships: a systematic description. *American Sociological Review* 35 (5): 884–900.

Murdoch, J. and McDermid, J.A. (2000). Modeling engineering design process with role activity diagrams. *Transactions of the Society for Design and Process Science (SDPS)* 4 (2): 45–65.

Nair, R., Tambe, M., and Marsella, S. (2003). Role allocation and reallocation in multiagent teams: towards a practical analysis. *Proceedings of the Second Int'l Joint Conference on Autonomous Agents and Multiagent Systems*, Melbourne, Australia (14 July 2003), http://teamcore.usc.edu/papers/2003/nair-aamas03.pdf.

Nyanchama, M. and Osborn, S. (1999). The role graph model and conflict of interest. *ACM Transactions on Information and System Security* 2 (1): 3–33.

Odell, J., Nodine, M., and Levy, R. (2005). A metamodel for agents, roles, and groups. In: *Agent-Oriented Software Engineering (AOSE)*, Lecture Notes on Computer Science, vol. 3382 (eds. J. Odell, P. Giorgini and J. Müller), 78–92. Berlin: Springer.

Olarnsakul, M. and Batanov, D.N. (2004). Customizing component-based software using component coordination model. *International Journal of Software Engineering and Knowledge Engineering* 14 (2): 103–140.

Ould, M.A. (1995). *Business Processes: Modeling and Analysis for Re-Engineering and Improvement*. John Wiley & Sons.

Pernici, B. (1990). Objects with roles. *ACM SIGOIS Bulletin, The Conference on Office Information Systems* 11 (2–3): 205–215.

Pradel, M., Henriksson, J., and Aßmann, U. (2012). A good role model for ontologies. In: *Enterprise Information Systems and Advancing Business Solutions* (ed. M. Tavana), 225–235. IGI Global.

Pressman, R. (2007). *Software Engineering: A Practitioner's Approach*. Boston, MA: McGraw-Hill.

Ren, S., Yu, Y., Chen, N. et al. (2007). The role of roles in supporting reconfigurability and fault localizations for open distributed and embedded systems. *ACM Transactions on Autonomous & Adaptive Systems (TAAS)* 2 (3): 10:1–10:27.

Riehle, D. and Gross, T. (1998). Role model based framework design and integration. *Proceedings of ACM Conf. on Object-Oriented Programming Languages, Systems, and Applications*, Vancouver, British Columbia, Canada, October 18–22, 1998, pp. 117–133.

Riehle, D., Brudermann, R., Gross, T., and Mätzel, K.U. (2000). Pattern density and role modeling of an object transport service. *ACM Computing Surveys (CSUR)* 32 (1): 1–6.

Schmidt, G. (2010). *Relational Mathematics*. Cambridge, UK: Cambridge University Press.

Shneiderman, B. and Plaisant, C. (1994). The future of graphic user interfaces: personal role managers. *People and Computers IX, British Computer Society*, Glasgow, Scotland (August 1994), pp. 3–8.

Simon, R. and Zurko, M.E. (1997). Separation of duty in role based access control environments. *Proceedings of the 10th IEEE Workshop on Computer Security Foundations*, Rockport, MA (June 1997). Los Alamitos, CA, USA, pp. 183–194.

Singley, M.K., Singh, M., Fairweather, P., et al. (2000). Algebra jam: supporting teamwork and managing roles in a collaborative learning environment. *Proceedings of CSCW'00*, Philadelphia, Pennsylvania, USA (December 2000), pp. 145–154.

Skarmeas, N. (1995). Organizations through roles and agents. *Int'l Workshop on the Design of Cooperative Systems (Coop' 95)*, Antibes-Juan-Les-Pins, France (January 1995). http://citeseer.ist.psu.edu/174774.html.

Skiena, S. (1990). *Implementing Discrete Mathematics: Combinatorics and Graph Theory with Mathematica*. Reading, MA: Addison-Wesley.

Smith, R.B., Hixon, R., and Horan, B. (1998). Supporting flexible roles in a shared space. *Proceedings of CSCW'98*, Seattle, Washington, USA (26–28 September 1988), pp. 197–206.

Steimann, F. (2000). On the representation of roles in object-oriented and conceptual modelling. *Data & Knowledge Engineering* 35 (1): 83–106.

Steimann, F. (2008). Role + counter-role = relationship + collaboration. *Workshop on Rel. and Asso. in Object-Oriented Languages*, Nashville, Tennessee, USA (19–23 October 2008), OOPSLA. http://deposit.fernuni-hagen.de/2201/1/Role_Counter_Role_Relationship_Collaboration.pdf.

Stone, P. and Veloso, M. (1999). Task decomposition, dynamic role assignment, and low-bandwidth communication for real-time strategic teamwork. *Artificial Intelligence* 110: 241–273.

Turoff, M. (1993). Distributed group support system. *MIS Quarterly* 17 (4): 399–417.

Wooldridge, M., Jennings, N.R., and Kinny, D. (2000). The Gaia methodology for agent-oriented analysis and design. *Autonomous Agents and Multi-Agent Systems* 3: 285–312.

Zhu, H. (2007). Role as dynamics of agents in multi-agent systems. *System and Informatics Science Notes* 1 (2): 165–171.

Zhu, H. (2008). Fundamental issues in the design of a role engine. *Proceedings of ACM/IEEE Int'l Symp on CTS*, CA Irvine, CA, USA (19–23 May 2008), pp. 399–407.

Zhu, H. (2015). Role-based collaboration and the E-CARGO: revisiting the developments of the last decade. *IEEE Systems, Man, and Cybernetics Magazine* 1 (3): 27–35.

Zhu, H. and Zhou, M. (2006a). *Object-Oriented Programming with C++: A Project-Based Approach*. Beijing, China: Tsinghua University Press.

Zhu, H. and Zhou, M. (2006b). Role-based collaboration and its kernel mechanisms. *IEEE Transactions on Systems, Man, and Cybernetics, Part C: Applications and Reviews* 36 (4): 578–589.

Zhu, H., Zhou, M.C., and Alkins, R. (2012). Group role assignment via a Kuhn-Munkres algorithm-based solution. *IEEE Transactions on SMC(A)* 42 (3): 739–750.

Zigurs, I. and Kozar, K.A. (1994). An exploratory study of roles in computer-supported groups. *MIS Quarterly* 18 (3): 277–297.

Exercises

1 What are the major components of the E-CARGO model?

2 Does the E-CARGO model have insufficient/sufficient/duplicate components? Why?

3 Are all the elements in the E-CARGO model necessary? Why?

4 What are the relationships between object and classes, roles and environments, and agents and groups?

5 Describe examples from the real world concerning the concepts of roles, role classes, and role instances.

6 What are the relations between roles?

7 What are the relations between agents and roles?

8 What are the relations between agents?

9 Can you find some agents' relations that cannot be described by their roles?

10 What are the properties of an RBC system?

11 What are primitive roles? Can you think of some examples?

12 Are role assignments required to form a group? Explain your argument.

13 Choose an organization or team you are familiar with and specify this organization with the E-CARGO model, i.e. specify all the constituents of the organization with the components of E-CARGO.

14 Design a template for roles to help extract roles from hiring websites.

15 Is friendship a relation between agents? or between roles? Why? Is a *friend* a role? Why?

16 Is there an equivalent relation between two roles in a system? If yes, how do you define this relation? If no, why?

17 Suppose that Peter tells you that he is a colleague of Ann. Can you formalize their relation with the definitions stated in this chapter?

18 Suppose that Bob and you are coauthors of a paper. Explain why you cannot review his manuscript submitted to a journal with the relations presented in this chapter.

5

Group Role Assignment (GRA)

5.1 Role Assignment

As discussed in Chapter 3, Role-Based Collaboration (RBC) is an emerging methodology to build an organizational structure, provide orderly system behavior, and consolidate system security for human and nonhuman entities that collaborate and coordinate their activities with or within systems (Zhu 2008; Zhu and Zhou 2006). The life cycle of RBC includes three major phases: *role negotiation, role assignment*, and *role execution* (Zhu 2016; Zhu and Zhou 2006).

Role assignment affects the efficiency of collaboration and the degree of satisfaction among members in the collaboration. Role assignment can be categorized into three steps: *agent evaluation, Group Role Assignment (GRA)* (Zhu and Alkins 2009a, b; Zhu and Alkins 2009a; Zhu et al. 2012)*, and *role transfer* (Zhu and Zhou 2008). Qualifications are the basic requirements for possible role-related activities (Black 1988). *Agent evaluation* rates the qualification of an agent for a role. This requires the capabilities, experiences, and credits of agents based on role specifications to be checked. Agent evaluation is a fundamental, yet difficult, problem that requires advanced methodologies, such as classification, mining, searching, and matching. GRA constructs a group by assigning roles to its members or agents to achieve its highest performance. *Role transfer* (also called *dynamic role assignment*) reassigns roles to agents or transfers agent roles (Zhu and Zhou 2008; Zhu and Zhou 2008, 2009) to meet the demand of system changes.

This chapter concentrates on the second step of role assignment, i.e. GRA. The assumption is that agent evaluation is done and a set of agent qualification values for roles has been obtained. Based on these qualification values, the GRA problem can be further divided into three cases: (i) agents are rated as only "yes" or "no" for roles and all the roles have the same importance; (ii) agents are rated between 0 and 1 for roles and roles are all equally important; and (iii) agents are rated between 0 and 1 for roles and roles have varying importance.

E-CARGO and Role-Based Collaboration: Modeling and Solving Problems in the Complex World, First Edition. Haibin Zhu.
© 2022 by The Institute of Electrical and Electronics Engineers, Inc.
Published 2022 by John Wiley & Sons, Inc.

This chapter provides an effective solution to the GRA problem using the Kuhn-Munkres (K-M) algorithm, advancing the field of role assignment significantly.

5.2 A Real-World Problem

To understand the problem of GRA, a scenario of a soccer team is considered. In a soccer team (Figure 5.1), there are 20 players ($a_0 - a_{19}$) in total. In the field, there are 4 roles and a total of 11 players for the 4-3-3 formation: 1 goalkeeper (r_0), 4 backs (r_1), 3 midfielders (r_2), and 3 forwards (r_3). Before each game, the most important task for the coach is to choose 11 players to be on the field. Players' states are affected by moods, emotions, health, fatigue, and past performance. In the problem description, we use "group performance" to express the sum of all the evaluated qualifications of the assigned agents in their designated roles. We will define "group performance" formally in Section 5.3.

Suppose that the coach has the data shown in Figure 5.2a that expresses the evaluated qualifications of the players with respect to each role, where rows represent players and columns represent roles. These values reflect the individual performance of each player relevant to a specific position. The coach needs to optimize the team's performance by role assignment. To simplify this process, the team performance is assumed to be a sum of the selected players' performance on their designated roles. However, the coach finds it difficult to find an exact solution to the role assignment problem and attempts two strategies.

In the first strategy, *the coach selects the best players from r_0 to r_3 to play in the field if they have not yet been selected.* This is indicated by the underlined numbers. The total group performance is 9.23 as shown in the first row of Table 5.1.

In the second strategy, *the coach selects the best players from r_3 to r_0, if they have not yet been selected.* This is indicated by the rectangles. The total group performance is 8.91 as shown in the last row of Table 5.1.

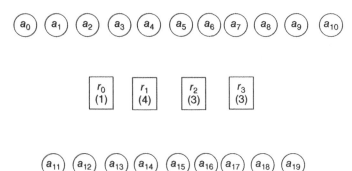

Figure 5.1 A soccer team.

Figure 5.2 The evaluation values of agents for roles and the assignment matrix. (a) A matrix of evaluation values. (b) An assignment matrix.

(a)

$$\begin{bmatrix} 0.65 & 0.98 & 0.96 & 0.90 \\ 0.26 & 0.33 & 0.59 & 0.19 \\ 0.72 & 0.61 & 0.19 & 0.63 \\ 0.06 & 0.48 & 0.43 & 0.90 \\ 0.87 & 0.35 & 0.06 & 0.25 \\ 0.72 & 0.15 & 0.28 & 0.01 \\ 0.33 & 0.59 & 0.37 & 0.67 \\ 0.75 & 0.59 & 0.25 & 0.45 \\ 0.12 & 0.10 & 0.01 & 0.51 \\ 0.84 & 0.13 & 0.96 & 0.63 \\ 0.01 & 0.29 & 0.82 & 0.12 \\ 0.07 & 0.52 & 0.36 & 0.95 \\ 0.97 & 0.90 & 0.88 & 0.54 \\ 0.14 & 0.54 & 0.51 & 0.26 \\ 0.04 & 0.03 & 0.83 & 0.70 \\ 0.44 & 0.70 & 0.16 & 0.39 \\ 0.12 & 0.48 & 0.04 & 0.76 \\ 0.30 & 0.14 & 0.52 & 0.08 \\ 0.91 & 0.50 & 0.96 & 0.21 \\ 0.53 & 0.06 & 0.85 & 0.85 \end{bmatrix}$$

(b)

$$\begin{bmatrix} 0 & 1 & 0 & 0 \\ 0 & 0 & 0 & 0 \\ 0 & 1 & 0 & 0 \\ 0 & 0 & 0 & 1 \\ 1 & 0 & 0 & 0 \\ 0 & 0 & 0 & 0 \\ 0 & 0 & 0 & 0 \\ 0 & 0 & 0 & 0 \\ 0 & 0 & 0 & 0 \\ 0 & 0 & 1 & 0 \\ 0 & 0 & 0 & 0 \\ 0 & 0 & 0 & 1 \\ 0 & 1 & 0 & 0 \\ 0 & 0 & 0 & 0 \\ 0 & 0 & 1 & 0 \\ 0 & 1 & 0 & 0 \\ 0 & 0 & 0 & 0 \\ 0 & 0 & 0 & 0 \\ 0 & 0 & 1 & 0 \\ 0 & 0 & 0 & 1 \end{bmatrix}$$

Table 5.1 Comparisons among assignment strategies.

Strategy	Assignment for $\{r_0\}\{r_1\}\{r_2\}\{r_3\}$	Group performance
(r_0, r_1, r_2, r_3)	{12}{0, 2, 6, 15}{9, 18, 19}{3, 11, 16}	9.23
(r_0, r_1, r_3, r_2)	{12}{0, 2, 6, 15}{9, 14, 18}{3, 11, 19}	9.30
(r_0, r_2, r_1, r_3)	{12}{2, 6, 7, 15}{0, 9, 18}{3, 11, 19}	9.04
(r_0, r_2, r_3, r_1)	{12}{2, 6, 7, 15}{0, 9, 18}{3, 11, 19}	9.04
(r_0, r_3, r_1, r_2)	{12}{2, 6, 7, 15}{9, 18, 19}{0, 3, 11}	8.98
(r_0, r_3, r_2, r_1)	{12}{2, 6, 7, 15}{9, 18, 19}{0, 3, 11}	8.98
(r_1, r_0, r_2, r_3)	{18}{0, 2, 12, 15}{9, 14, 19}{3, 11, 16}	9.35
(r_1, r_0, r_3, r_2)	{18}{0, 2, 12, 15}{9, 10, 14}{3, 11, 19}	9.41
(r_1, r_2, r_0, r_3)	{4}{0, 2, 12, 15}{9, 18, 19}{3, 11, 16}	9.44
(r_1, r_2, r_3, r_0)	{4}{0, 2, 12, 15}{9, 18, 19}{3, 11, 16}	9.44
(r_1, r_3, r_0, r_2)	{18}{0, 2, 12, 15}{9, 10, 14}{3, 11, 19}	9.41
(r_1, r_3, r_2, r_0)	**{4}{0, 2, 12, 15}{9, 14, 18}{3, 11, 19}**	**9.51**
(r_2, r_0, r_1, r_3)	{12}{2, 6, 7, 15}{0, 9, 18}{3, 11, 19}	9.04

(Continued)

Table 5.1 (Continued)

Strategy	Assignment for $\{r_0\}\{r_1\}\{r_2\}\{r_3\}$	Group performance
(r_2, r_0, r_3, r_1)	{12}{2, 6, 7, 15}{0, 9, 18}{3, 11, 19}	9.04
(r_2, r_1, r_0, r_3)	{4}{2, 6, 12, 15}{0, 9, 18}{3, 11, 19}	9.25
(r_2, r_1, r_3, r_0)	{4}{2, 6, 12, 15}{0, 9, 18}{3, 11, 19}	9.25
(r_2, r_3, r_0, r_2)	{12}{0, 9, 18}{1, 10, 14}{3, 11, 19}	8.79
(r_2, r_3, r_2, r_0)	{4}{0, 9, 18}{10, 12, 14}{3, 11, 19}	8.98
(r_3, r_0, r_1, r_2)	{12}{2, 6, 7, 15}{9, 18, 19}{0, 3, 11}	8.98
(r_3, r_0, r_2, r_1)	{12}{2, 6, 7, 15}{9, 18, 19}{0, 3, 11}	8.98
(r_3, r_1, r_0, r_2)	{18}{2, 6, 12, 15}{9, 14, 19}{0, 3, 11}	9.10
(r_3, r_1, r_2, r_0)	{4}{2, 6, 12, 15}{9, 18, 19}{0, 3, 11}	9.19
(r_3, r_2, r_0, r_1)	{4}{2, 6, 7, 15}{9, 12, 18}{0, 3, 11}	8.91
(r_3, r_2, r_1, r_0)	{4}{2, 6, 7, 15}{9, 12, 18}{0, 3, 11}	<u>8.91</u>

In the curly braces are the player numbers. The underlined line is an initial solution. The bold line is the result of the problem.

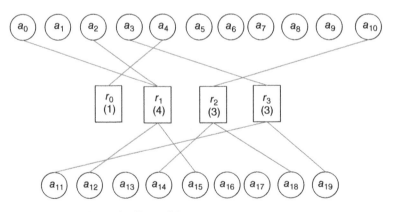

Figure 5.3 A solution for Figure 5.1.

In addition, we evaluate all other possible permutations of position assignment, i.e. (r_0, r_1, r_3, r_2), (r_0, r_2, r_1, r_3), ..., for a total of 22 permutations.

Table 5.1 shows the group performances of all 24 permutations. Among the 24 strategies, *the optimum one is* (r_1, r_3, r_2, r_0), as indicated by the bold row. The solution is also shown in Figure 5.2a using circles, in Figure 5.2b as an assignment matrix, and in Figure 5.3 as a graph. It is noted that the optimal solution is not intuitive and requires enumerations and comparisons, e.g. the most qualified goalkeeper a_{12} is not assigned to the goalkeeper (r_0) role because he contributes more to the team when he plays a back (r_1) role instead of the goalkeeper. The coach arranges the players according to the optimal role assignment and obtains the best group performance of 9.51.

5.3 Extended Expression of the E-CARGO Model

With the E-CARGO model (Zhu 2008; Zhu and Zhou 2006, 2008, 2008; Zhu et al. 2012), a tuple of agent a and role r, i.e. $<a, r>$ is called a role assignment (also called an agent assignment). A role r is workable if it is assigned sufficient (expressed by lower range ℓ of range q of base $\mathcal{B}$ of role r in environment e of group g, i.e. $g.e.\mathcal{B}[r].q.\ell$), current agents to play it (Zhu and Alkins 2009a). For example, the roles represented by the columns in Figure 5.2b are workable for the soccer team with the 4-3-3 formation shown in Figure 5.1.

In formalizing GRA problems (GRAPs), only agents and roles are considered. In the following discussions, current agents and roles are our focus; environments and groups are simplified into vectors and matrices, respectively; m ($=|\mathcal{A}|$) expresses the size of the agent set $\mathcal{A}$; n ($=|\mathcal{R}|$) the size of the role set $\mathcal{R}$, i the index of an agent, and j the index of a role.

Definition 5.1 A *role range vector* is a vector of the lower ranges of roles in environment e of group g. The role range vector is denoted as L, and $L[j] \in \mathcal{N}$, where $\mathcal{N}$ is the set of non-negative numbers and $0 \leq j < n$. Suppose that roles in $g.e$ are numbered as j ($0 \leq j < n$) and $\mathcal{B}[j]$ means the tuple for role j, then $L[j] = g.e.\mathcal{B}[j].q.\ell$.

For example, $L = [1\ 4\ 3\ 3]$ for the soccer team in Figure 5.1.

Definition 5.2 A *qualification matrix Q* of group g is an $m \times n$ matrix of values in $[0, 1]$, where $Q[i, j]$ expresses the qualification value of agent i for role j. It is denoted as $Q[i, j] \in [0,1] (0 \leq i \in \mathcal{N} < m; 0 \leq j < n)$.

For example, Figure 5.2a shows a qualification matrix for the soccer team from Figure 5.1.

Definition 5.3 A *role assignment matrix* is an $m \times n$ matrix of values in $\{0,1\}$. If $T[i, j] = 1$, agent i is assigned to role j and agent i is called an *assigned agent*. It is denoted as $T[i, j] \in \{0,1\}$ ($0 \leq i < m; 0 \leq j < n$).

For example, Figure 5.2b shows an assignment matrix for the soccer team from Figure 5.1.

Definition 5.4 A *group performance σ* is defined as the sum of the assigned agents' qualifications, i.e. $\sigma = \sum\limits_{i=0}^{m-1} \sum\limits_{j=0}^{n-1} Q[i,j] \times T[i,j]$.

For example, the group performance of Figure 5.2 is 9.51.

Definition 5.5 A *role weight vector* is a vector with the weights of the roles in the environment e of group g. The role weight vector is denoted as W, and $W[j] \in [0, 1](0 \leq j < n)$. Suppose that roles in g. e are numbered as j $(0 \leq j < n)$ and $B[j]$ means the tuple for role j, then $W[j] = g.e.B[j].w$, where w belongs to $\mathcal{D}_o$.

For example, for the soccer team in Figure 5.1, if we emphasize attack, we may set $W = [0.6\ 0.7\ 0.8\ 0.9]$; but if we emphasize defense, we may set $W = [0.9\ 0.8\ 0.7\ 0.6]$.

Definition 5.6 A *weighted group performance* σ_w is defined as the weighted sum of assigned agents' qualifications, i.e.

$$\sigma_w = \sum_{j=0}^{n-1} W[j] \times \sum_{i=0}^{m-1} Q[i,j] \times T[i,j]$$

For example, for the assignment in Figure 5.2, if $W = [0.6\ 0.7\ 0.8\ 0.9]$, the weighted group performance is 7.38.

Definition 5.7 Group g is *workable* if all its roles are workable, i.e. group g expressed by T is *workable* if $\left(\sum_{i=0}^{m-1} T[i,j] \geq L[j])(0 \leq j < n \right)$.

For example, Figure 5.2b forms a workable group for the team in Figure 5.1. However, a team with a red card does not because $L = [1\ 4\ 3\ 3]$. Note that "*not workable*" here only means that this assignment cannot implement the 4-3-3 formation but does not mean that the team cannot compete on the field.

5.4 Group Role Assignment Problems

GRAPs can be categorized into different types based on the conditions: simple, rated, and weighted role assignment (Zhu and Alkins 2009a, b; Zhu et al. 2012).

5.4.1 Simple Role Assignment

Definition 5.8 Given m, n, $Q(Q[i, j] \in \{0,1\}\ (0 \leq i < m; 0 \leq j < n))$, and L, a *Simple Group Role Assignment Problem (SimGRAP)* is to find a role assignment matrix T that makes g workable (Zhu and Alkins 2009b), where each agent is assigned at most one role.

For example, suppose that Q is shown as Figure 5.4a and $L = [1\ 4\ 3\ 3]$. One solution T is shown in Figure 5.4b.

Figure 5.4 A qualification matrix and solution. (a) The qualification matrix. (b) The assignment matrix.

(a)

$$\begin{bmatrix} 1 & 0 & 0 & 0 \\ 0 & 1 & 0 & 0 \\ 0 & 1 & 0 & 0 \\ 0 & 1 & 0 & 0 \\ 0 & 1 & 1 & 0 \\ 0 & 1 & 1 & 0 \\ 0 & 1 & 1 & 0 \\ 0 & 0 & 1 & 1 \\ 0 & 0 & 1 & 1 \\ 0 & 0 & 0 & 1 \\ 0 & 0 & 0 & 1 \end{bmatrix}$$

(b)

$$\begin{bmatrix} 1 & 0 & 0 & 0 \\ 0 & 1 & 0 & 0 \\ 0 & 1 & 0 & 0 \\ 0 & 1 & 0 & 0 \\ 0 & 1 & 0 & 0 \\ 0 & 0 & 1 & 0 \\ 0 & 0 & 1 & 0 \\ 0 & 0 & 1 & 0 \\ 0 & 0 & 0 & 1 \\ 0 & 0 & 0 & 1 \\ 0 & 0 & 0 & 1 \end{bmatrix}$$

5.4.2 Rated Group Role Assignment

Suppose that the qualification values of agents are rated between 0 and 1, where the value 0 represents disqualification and the value 1 represents a perfect qualification. In this situation, GRA seeks to find a solution with the maximum sum of qualification values for the group.

Definition 5.9 Given m, n, $Q(Q[i, j] \in [0,1]$ $(0 \le i < m, 0 \le j < n))$, and L, the *Group Role Assignment (GRA)* problem (also called *Rated Group Role Assignment Problem (RGRAP)*) is to find a matrix T that maximizes the group performance of group g, i.e.

$$\max \sigma = \sum_{i=0}^{m-1} \sum_{j=0}^{n-1} Q[i,j] \times T[i,j]$$

subject to

$$T[i,j] \in \{0, 1\} \quad (0 \le i < m, 0 \le j < n), \tag{5.1}$$

$$\sum_{i=0}^{m-1} T[i,j] = L[j] \quad (0 \le j < n), \tag{5.2}$$

$$\sum_{j=0}^{n-1} T[i,j] \le 1 \quad (0 \le i < m), \tag{5.3}$$

where expression (5.1) is a 0 or 1 constraint; (5.2) makes the group workable, and (5.3) means that an agent can be assigned at most one role, i.e. for each agent a, $|a.\mathcal{R}_c| \le 1$.

We usually use T^* to express the T obtained from Definition 5.9. For example, suppose that Figure 5.2a is Q and $L = [1\ 4\ 3\ 3]$. The solution T is shown in Figure 5.2b with a group performance of 9.51.

(a)

$$
\begin{bmatrix}
0.71 & 0.6 & 0.0 & 0.22 \\
0.29 & 0.67 & 0.44 & 0.76 \\
0.69 & 0.92 & 0.92 & 0.6 \\
0.0 & 0.0 & 0.53 & 0.0 \\
0.97 & 0.51 & 0.77 & 0.65 \\
0.58 & 0.64 & 0.24 & 0.0
\end{bmatrix}
$$

(b)

$$
\begin{bmatrix}
1 & 0 & 0 & 0 \\
0 & 0 & 0 & 1 \\
0 & 0 & 0 & 1 \\
0 & 0 & 1 & 0 \\
1 & 0 & 0 & 0 \\
0 & 1 & 0 & 0
\end{bmatrix}
$$

(c)

$$
\begin{bmatrix}
1 & 0 & 0 & 0 \\
0 & 0 & 0 & 1 \\
0 & 1 & 0 & 0 \\
0 & 0 & 1 & 0 \\
0 & 0 & 0 & 1 \\
1 & 0 & 0 & 0
\end{bmatrix}
$$

Figure 5.5 A rated qualification matrix, its rated solution, and weighted solution. (a) The qualification matrix. (b) The assignment matrix. (c) The assignment matrix in consideration of weights.

5.4.3 Weighted Role Assignment

Rated role assignment can be further extended. If the roles in a group have different importance, i.e. different weights, a weighted role assignment problem occurs.

Definition 5.10   Let $Q[i,j] \in [0,1]$ ($0 \le i < m, 0 \le j < n$) represent how well a given agent i plays a given role j. The *Weighted Group Role Assignment Problem (WGRAP)* aims to find a matrix T such that group g has the highest *weighted group performance* σ_w, i.e. $\sum_{j=0}^{n-1} W[j] \times \sum_{i=0}^{m-1} Q[i,j] \times T[i,j]$ is maximized.

For example, suppose Figure 5.5a is Q, $L = [2\ 1\ 1\ 2]$, and $W = [0.19\ 0.62\ 0.42\ 0.66]$, then Figure 5.5c is the WGRAP solution. Evidently, this solution is different from the RGRAP solution in Figure 5.5b. We do not use the soccer team example because different weights of positions do not make much difference from the assignment with GRA for the Q and L presented in Section 5.2. We encourage interested readers to investigate more to explain this kind of phenomenon.

5.5 General Assignment Problem and the K-M Algorithm

GRAPs are complex. The algorithms based on the exhaustive search for the GRAP are of exponential complexity (Zhu and Alkins 2009b). Therefore, these algorithms must be improved. Fortunately, the well-known K-M algorithm for the General Assignment Problem (GAP) has a complexity of $O(m^3)$ (Bertsekas 1981; Kuhn 1955; Munkres 1957), it is widely applied in industries (Riesen and Bunke 2009), and it is readily available in Java code format (Baker 2008). By transforming a GRAP to a General Assignment Problem (GAP), GRAP can be efficiently solved using the K-M algorithm. Although there are other algorithms (Bertsekas 1981; Toroslu and Üçoluk 2007) that are more efficient than the K-M algorithm, the K-M algorithm is used in this work because it has previously solved problems similar to the GRAP.

In addition, the K-M algorithm has a higher reputation, popularity, understandability, reliability, and code availability compared to the other algorithms.

A GAP (Baker 2008; Bertsekas 1981; Bourgeois and Lassalle 1971; Kuhn 1955; Munkres 1957) can be described as following.

Definition 5.11   Given an $n \times n$ matrix A of real numbers, a *GAP* is to find a permutation p (p_i; $i = 0, 1, 2, ..., n - 1$) of integers that minimizes $\sum_{i=0}^{n-1} A[i, p_i]$.

For example, for a 3×3 matrix $A = \begin{bmatrix} 7 & 5 & 11.2 \\ 5 & 4 & 1 \\ 9.3 & 3 & 2 \end{bmatrix}$, there are six possible permutations for which the associated sums are shown in Table 5.2.

Permutation (2) gives the smallest sum of 11.0 and is the solution to the assignment problem for this matrix.

The RGRAP can be transferred to a GAP by adjusting the number of agents and roles. If $m = \sum_{j=0}^{n-1} L[j]$ in *Definition* 5.1, the problem becomes a GAP.

The K-M algorithm is a minimization algorithm for square matrices of GAPs invented by Kuhn (1955) and improved by Munkres (1957). To understand it, *Definition* 5.12 and the König theorem (Theorem 5.1) are required.

Definition 5.12   A set of elements of a matrix are *independent* if none of them occupies the same row or column.

Theorem 5.1   (König Theorem, Munkres 1957): If M is a matrix and z is the number of independent zero elements of M, then there are z lines that contain all the zero elements of M.

Table 5.2   The permutations and sums for matrix A.

Assignment scheme number	P	$\sum_{i=0}^{n-1} A[i, p_i]$
(1)	0, 1, 2	13.0
(2)	**0, 2, 1**	**11.0**
(3)	1, 0, 2	12.0
(4)	1, 2, 0	15.3
(5)	2, 0, 1	19.2
(6)	2, 1, 0	24.5

The bold line is the solution of the problem.

The K-M algorithm starts by finding the smallest element in each row. This element is then subtracted from each row. This process is repeated for each column. Then, zeros are tagged and categorized as either starred or primed zeros. The "starred zeros" form an independent set of zeros, while the "primed zeros" are possible candidates. The algorithm also distinguishes lines (rows or columns): covered and uncovered. A zero is called covered (or non-covered) if it is in a covered (or non-covered) row or column.

To utilize the algorithm, the following initial conditions should be established: (i) no lines are covered and no zeros are starred or primed; (ii) for each row of the matrix M, subtract the value of the smallest element from each element in the row; and (iii) for each column of the resulting matrix, subtract the value of the smallest element from each element. The K-M algorithm is as follows:

Algorithm 5.1 K-M Algorithm

```
Input:
O    An m×m matrix M with the above preliminaries.
Output:
O    All the subscripts of the selected elements of M,
i.e., the starred zeroes.
begin
L1    for (all zeros z in M)
L2        if (there are no starred zeros in the row or
column of z)
L3            star z;
L4        while (true) // the main loop
L5            cover all the columns containing a 0*;
L6            if (all the columns are covered)
L7                return; // the optimal solution is
starred;
L8            endif
L9            while (!(all zeros are covered))//the
inner loop
L10               for (all non-covered zeros z)
L11                   prime z;
L12                   if (there is a starred zero z* in
this row)
L13                       cover row of z*;
L14                       uncover column of z*;
```

(Continued)

Algorithm 5.1 (Continued)

```
L15                     else
L16                         highlight z;
L17                         while (there is a starred zero
z* in z's column)
L18                             z=z*;
L19                             highlight z;
L20                             z:=the primed zero in z's
row; // This always exists.
L21                             highlight z;
L22                         endwhile
L23                         unstar highlighted starred
zeros;
L24                         star highlighted primed zeros
and remove    highlights;
L25                         remove primes and uncover all
rows and     columns;
L26                             continue to while (the main
loop);
L27                     endif
L28                 endfor
L29                 h:=smallest uncovered element in M;
L30                 add h to each covered row;
L31                 subtract h from each uncovered column;
L32             endwhile //the inner loop
L33         end while //the main loop
L34     endif
L35   endfor
end
```

For example, we process the matrix in Figure 5.6a. By initialization, we obtain the matrix in Figure 5.6b. With Steps L1–L3, we star the zeros (Figure 5.6c). From L4 and L7, we get Figure 5.6d. From L8–L13, we get Figure 5.6e. By Step L29, we get $h = 1$. By L30–L31, followed by a return to L9–L11 and L16, we get Figure 5.6f. From L23–L24, we get Figure 5.6g. Continuing to L25, we get Figure 5.6h. Through L26 and returning to L4 and L5, we get Figure 5.6i. Then, with L6–L7, we complete the process. Finally, we obtain the assignment result as $(0, 0)$, $(1, 2)$, and $(2, 1)$. The minimum sum of the problem is $7 + 1 + 3 = 11$.

(a)
$$\begin{bmatrix} 7 & 5 & 11.2 \\ 5 & 4 & 1 \\ 9.3 & 3 & 2 \end{bmatrix}$$

(b)
$$\begin{bmatrix} 0 & 0 & 6.2 \\ 2 & 3 & 0 \\ 5.3 & 1 & 0 \end{bmatrix}$$

(c)
$$\begin{bmatrix} 0^* & 0 & 6.2 \\ 2 & 3 & 0^* \\ 5.3 & 1 & 0 \end{bmatrix}$$

(d)
$$\begin{bmatrix} 0^* & 0' & 6.2 \\ 2 & 3 & 0^* \\ 5.3 & 1 & 0 \end{bmatrix}$$

(e)
$$\begin{bmatrix} 0^* & 0' & 6.2 \\ 2 & 3 & 0^* \\ 5.3 & 1 & 0 \end{bmatrix}$$

(f)
$$\begin{bmatrix} 0^* & 0' & 7.2 \\ 1 & 2 & 0^* \\ 4.3 & 0' & 0 \end{bmatrix}$$

(g)
$$\begin{bmatrix} 0^* & 0' & 7.2 \\ 1 & 2 & 0^* \\ 4.3 & 0^* & 0 \end{bmatrix}$$

(h)
$$\begin{bmatrix} 0^* & 0' & 7.2 \\ 1 & 2 & 0^* \\ 4.3 & 0^* & 0 \end{bmatrix}$$

(i)
$$\begin{bmatrix} 0^* & 0' & 7.2 \\ 1 & 2 & 0^* \\ 4.3 & 0^* & 0 \end{bmatrix}$$

Figure 5.6 An example processed by the K-M algorithm. (a) The matrix is the input. (b) The matrix is pre-processed. (c) Initial 0s are starred. (d) One 0 is primed. (e) The primed 0 is disqualified. (f) The second 0 is primed. (g) The new 0 is starred. (h) All correct 0s are starred. (i) The solution is obtained.

5.6 Solutions to GRA Problems

To apply the K-M algorithm in solving GRAPs, several issues should be considered. For one, note that the algorithm is always successful because it finds the smallest sum or the largest sum. However, in GRAPs, the K-M algorithm result might not be a valid solution for the group even when the group performance is optimal. For instance, $Q[i, j] = 0$ typically means that agent i is not qualified for role j, but the K-M solution may not take this into account and may utilize an unqualified agent. The improved algorithm for the rated group assignment problem (Zhu and Alkins 2009a) follows the same consideration. To generalize the GRA problem, we need a new definition.

Definition 5.13 *Qualification threshold τ is a real number in $[0, 1]$ to state that an agent is called qualified in a group only if its qualification value is greater than τ.*

With the introduction of τ, an agent may be unqualified for a role if its qualification value is not greater than τ, i.e. the agent should not be assigned to the role. That is to say, with $\tau = 0.0$, by the K-M algorithm, we may find an optimal result but not a workable assignment matrix. For example, in Figure 5.7a, if $L = [1\ 1\ 1]$ and $\tau = 0.0$, the K-M algorithm obtains the result shown in Figure 5.7b that is not a successful assignment matrix, while Figure 5.7c is.

Figure 5.7 A matrix that the K-M algorithm may obtain a wrong result. (a) A qualification matrix. (b) The solution with the K-M algorithm. (c) The correct solution with GRA.

(a)
$$\begin{bmatrix} 0.9 & 0.8 & 0.2 \\ 0.6 & 0.8 & 0.3 \\ 0.3 & 0.3 & 0.0 \end{bmatrix}$$

(b)
$$\begin{bmatrix} 1 & 0 & 0 \\ 0 & 1 & 0 \\ 0 & 0 & 1 \end{bmatrix}$$

(c)
$$\begin{bmatrix} 1 & 0 & 0 \\ 0 & 0 & 1 \\ 0 & 1 & 0 \end{bmatrix}$$

To avoid the result of Figure 5.7b, whether the group is acceptable after obtaining an assignment from the K-M algorithm needs to be checked:

For each role, there should be sufficient assigned agents, i.e. $\sum_{i=0}^{m-1} T[i,j] \times (\lceil Q[i,j]$ $-\tau \rceil) \geq L[j]$, where $0 \leq j < n$, where "$\lceil x \rceil$" means the smallest integer larger than x.

To ensure that the GRAP solution is valid (e.g. Figure 5.7c), we need to adjust the initial qualification matrix Q to avoid invalid solutions. Because the elements of the qualification matrix Q are within the range of $[0, 1]$, the sum of all the elements can never exceed m^2. Therefore, $-m^2$ can be used to replace unqualified elements (smaller than or equal to τ) in matrix Q (Figure 5.8). This number protects the K-M algorithm from obtaining an invalid result because the group performance will be negative if such a value is selected. If the group performance from the optimal result is still negative after the aforementioned adjustment, the group has no successful assignment.

By the definition of τ, the elements with the value of $-m^2$ in Q should not be considered in the assignment. If there are insufficiently qualified agents to play a specific role in a group, there is no solution for the group. In Figure 5.8, if $\tau = 0.6$, and $L = [1\ 1\ 3\ 1]$, no solution can be found. Therefore, prior to calling the K-M algorithm, we should ensure that the assignment is feasible by checking the following conditions:

- **Condition 1:** The total agent number should be larger than the required numbers, i.e. $m \geq$ $\sum_{j=0}^{n-1} L[j]$.

- **Condition 2:** For each role, there should be sufficient qualified agents, i.e. $\sum_{i=0}^{m-1} \lceil Q[i,j] - \tau \rceil \geq L[j]$, where $0 \leq j < n$.

$$\begin{bmatrix} 0.71 & -36 & -36 & -36 \\ -36 & 0.67 & -36 & 0.76 \\ 0.69 & 0.92 & 0.92 & 0.62 \\ -36 & -36 & -36 & -36 \\ 0.97 & -36 & 0.77 & 0.65 \\ -36 & 0.64 & -36 & -36 \end{bmatrix}$$

Figure 5.8 Qualification matrix for a group with no solution when $L = [1\ 1\ 3\ 1]$.

To convert a GRAP to a GAP, a square matrix must be created from Q because the original K-M algorithm only works with a square matrix. To form the square matrix used in the K-M algorithm, duplicate columns should be added based on the role range vector L.

Definition 5.14   A *role index vector* L' is an m $(\geq \sum_{j=0}^{n-1} L[j])$ dimensional vector created from a role range vector L. $L'[k]$ is a role number related with column k $(0 \leq k < m)$ in the adjusted qualification matrix Q' defined in Definition 5.15, i.e.

$$L'[k] = \begin{cases} 0 & (k < L[0]) \\ x & \left(\sum_{p=0}^{x-1} L[p] \leq k < \sum_{p=0}^{x} L[p](0 < x < n) \right) \\ n & \left(k \geq \sum_{p=0}^{n-1} L[p](0 \leq k < m) \right) \end{cases}$$

Definition 5.15   *Adjusted qualification matrix* Q' is an $m \times m$ matrix, where $Q'[i, j] \in [0,1]$ expresses the qualification value of agent i for role $L'[j]$, where,

$$Q'[i,j] = \begin{cases} Q[i, L'[j]] & \left(0 \leq i < m; 0 \leq j < \sum_{p=0}^{n-1} L[p] \right) \\ 1 & \left(0 \leq i < m; \sum_{p=0}^{n-1} L[p] \leq j < m \right) \end{cases}.$$

From Definitions 5.13–5.14, Q and L can be combined to create a square matrix Q' used in the GAP. For example, Q shown in Figure 5.8 with $L = [2\ 1\ 1\ 2]$ can create the Q' in Figure 5.9. The column numbers are recorded in a new vector $L' = [0\ 0\ 1\ 2\ 3\ 3]$.

$$\begin{bmatrix} 0.71 & 0.71 & -36 & -36 & -36 & -36 \\ -36 & -36 & 0.67 & -36 & 0.76 & 0.76 \\ 0.69 & 0.69 & 0.92 & 0.92 & -36 & -36 \\ -36 & -36 & -36 & 0.53 & -36 & -36 \\ 0.97 & 0.97 & -36 & 0.77 & 0.65 & 0.65 \\ -36 & -36 & 0.64 & -36 & -36 & -36 \end{bmatrix}$$

Figure 5.9 The square matrix produced by Definition 5.15 from Figure 5.8.

If $m > \sum_{j=0}^{n-1} L[j]$, we need to add $m - \sum_{j=0}^{n-1} L[j]$ columns of 1's to make a square matrix, i.e. every agent is fully qualified for an empty role. This assumption follows the logic implication: $A \to B$ (A implies B) is true if A is false. Here, we can replace A with "*there is a role*" and B with "*one is qualified to play it*". That is why we define

$$Q'[i,j] = 1 \ (0 \le i < m, \ \sum_{j=0}^{n-1} L[j] < j < m).$$

For example, the Q matrix with $L = [1\ 2\ 1]$ and $\tau = 0.6$ in Figure 5.10 has a square matrix of Q' shown in Figure 5.11. The column number vector Q' is $L' = [0\ 1\ 1\ 2\ 3\ 3]$, where both 3s correspond to the empty roles.

Because the K-M algorithm solves minimization problems, the problem needs to be transformed into a maximization problem. This is simply a matter of subtracting the entries of Q' from the largest entry of Q (i.e. 1) to obtain a new matrix M. Minimization of this new matrix results in a maximization of Q'.

For example, for the maximization of Q' shown in Figure 5.11, a matrix M (Figure 5.12) can be obtained by the following assignment:

$$M[i,j] = 1 - Q'[i,j], (0 \le i,j < m).$$

Now, we are ready to call the K-M algorithm with matrix M in Figure 5.12. The solution is obtained and shown in Figure 5.13 with a group performance of 3.2.

In summary, the algorithm to transfer an $m \times n$ qualification matrix to an $m \times m$ square matrix is the following:

$$\begin{bmatrix} 0.36 & 0.76 & 0.72 \\ 0.93 & 0.59 & 0.24 \\ 0.06 & 0.46 & 0.69 \\ 0.40 & 0.10 & 0.74 \\ 0.23 & 0.75 & 0.24 \\ 0.21 & 0.77 & 0.24 \end{bmatrix}$$

Figure 5.10 A matrix with $m > \sum_{j=0}^{n-1} L[j]$.

$$\begin{bmatrix} -36 & 0.76 & 0.76 & 0.72 & 1.0 & 1.0 \\ 0.93 & -36 & -36 & -36 & 1.0 & 1.0 \\ -36 & -36 & -36 & 0.69 & 1.0 & 1.0 \\ -36 & -36 & -36 & 0.74 & 1.0 & 1.0 \\ -36 & 0.75 & 0.75 & -36 & 1.0 & 1.0 \\ -36 & 0.77 & 0.77 & -36 & 1.0 & 1.0 \end{bmatrix}$$

Figure 5.11 A square matrix for the Q matrix in Figure 5.10.

$$\begin{bmatrix} 37 & 0.24 & 0.24 & 0.28 & 0.0 & 0.0 \\ 0.07 & 37 & 37 & 37 & 0.0 & 0.0 \\ 37 & 37 & 37 & 0.31 & 0.0 & 0.0 \\ 37 & 37 & 37 & 0.26 & 0.0 & 0.0 \\ 37 & 0.25 & 0.25 & 37 & 0.0 & 0.0 \\ 37 & 0.23 & 0.23 & 37 & 0.0 & 0.0 \end{bmatrix}$$

Figure 5.12 The square matrix transformed from the qualification matrix in Figure 5.11.

$$\begin{bmatrix} 0 & 1 & 0 \\ 1 & 0 & 0 \\ 0 & 0 & 0 \\ 0 & 0 & 1 \\ 0 & 0 & 0 \\ 0 & 1 & 0 \end{bmatrix}$$

Figure 5.13 The assignment matrix T for the group in Figure 5.10.

Algorithm 5.2 TransferQtoM

```
----------------------------
Input: m, n, an mxn matrix Q, an n vector L and a
qualification threshold τ;
Output: an mxm square M.
begin
for (0≤ i< m, 0≤ j< m)
     if (Q[i,j] > τ)
         Q′[i,j]:=Q[i,j];
     else
         Q′[i,j]:=- mxn;
     endif
endfor
i:= 0;
while (i<m)
    j:= 0; k:= 0;
    while (j< n)
         if (L[j]==1)
             M′[i, k]:= Q′[i, k]; L′[k]:= j; k:= k+1;
         else
           if (L[j]>1)
               x:= L[j];
               for (0≤g<x)
                   M′[i, g]:= Q′[i,j]; L′[k+g]:=j;
                   k:=k +x;
               endfor
             endif
           endif
       j:=  j+1;
    endwhile
    if (k<m)
           for (0≤h< m-k)
               M′[i, k+h]:= 1;
           endfor
    endif;
    i:= i+1;
 endlwhile
 for (0≤i< m, 0≤j< m)
    M[i,j]:=1- M′[i,j];
 endfor;
 end
----------------------------------------------
```

Theorem 5.2 An RGRAP can be transferred to a GAP.

[Proof]:

To prove the theorem, we only need to prove that the resulting subscript vector V from the K-M algorithm can be used to create the matrix T required by the RGRAP.

From the $m \times m$ matrix M, we obtain an n-dimensional subscript vector V by the K-M algorithm to make $\sum_{i=0}^{m-1} M[i, V[i]]$ the minimum, i.e. $\sum_{i=0}^{m-1} M[i, V[i]] \leq (\sum_{i=0}^{m-1} M[i,j] \ (j \neq [i]))$.

$$\because \sum_{i=0}^{m-1} M[i, V[i]] = \sum_{i=0}^{m-1}(1 - M'[i, V[i]]) = m - \sum_{i=0}^{m-1} M'[i, V[i]],$$

$$\therefore \sum_{i=0}^{m-1} M'[i, V[i]] = m - \sum_{i=0}^{m-1} M[i, V[i]] = \sum_{i=0}^{m-1}(1 - M[i, V[i]]).$$

$$\because \sum_{i=0}^{m-1} M[i, V[i]] \leq \sum_{i=0}^{m-1} M[i,j] \ (\ j \neq V[i]),$$

$$\therefore - \sum_{i=0}^{m-1} M[i, V[i]] \geq - \sum_{i=0}^{m-1} M[i,j] \ (\ j \neq V[i]),$$

$$\therefore m - \sum_{i=0}^{m-1} M[i, V[i]] \geq m - \sum_{i=0}^{m-1} M[i,j] \ (\ j \neq V[i]).$$

$$\because m - \sum_{i=0}^{m-1} M[i, V[i]] = \sum_{i=0}^{m-1}(1 - M[i, V[i]]) = \sum_{i=0}^{m-1} M'[i, V[i]] \text{ and}$$

$$m - \sum_{i=0}^{m-1} M[i,j] = \sum_{i=0}^{m-1}(1 - M[i,j]) = \sum_{i=0}^{m-1} M'[i,j],$$

$$\therefore \sum_{i=0}^{m-1} M'[i, V[i]] \geq \sum_{i=0}^{m-1} M'[i,j] (\ j \neq V[i]), \text{ i.e.,}$$

$\sum_{i=0}^{m-1} M'[i, V[i]]$ is the maximum sum for M'.

Corresponding to the RGRAP and the transfer algorithm, V can be used to create T, i.e.

for $(0 \leq i < m, 0 \leq j < n) \ T[i,j] := 0$;

for $(0 \leq i < m, 0 \leq j < n) \ T[i, L'[V[i]]] := 1$;

Therefore, if $T[i, j] = 1$, then $Q'[i, j] \times T[i, j] = M'[i, V[i]]$, i.e.

$$\sum_{i=0}^{m-1} \sum_{j=0}^{n-1} T[i,j] \times Q'[i,j] = \sum_{i=0}^{m-1} M'[i, V[i]].$$

Therefore, $\sum_{i=0}^{m-1} \sum_{j=0}^{n-1} T[i,j] \times Q'[i,j]$ is the maximum sum for Q'. Furthermore,

$\because Q[i,j] \geq Q'[i,j] \quad (0 \leq i < m, \; 0 \leq j < n),$

$\therefore \sum_{i=0}^{m-1} \sum_{j=0}^{n-1} T[i,j] \times Q[i,j]$ is the maximum sum for Q.

As a result, T is the matrix required by the RGRAP.■

Algorithm 5.3 Solve RGRAP

```
Input:
    O An n dimensional role's lower range vector L; and
    O An mxn rated qualification matrix Q.
Output:
    O Success: An mxn assignment matrix T in which, for
```
all columns j ($j = 0, \ldots, n-1$), $\sum_{i=0}^{m-1} T[i, j] \geq L[j]$, and v is
the maximum group performance.
```
    O Failure: An mxn assignment matrix T in which
```
there is at least one column j ($j = 0, \ldots, n-1$), $\sum_{i=0}^{m-1} T[i, j]$
$<L[j]$, and v is 0.
```
begin
  Step 1: If Conditions 1 and 2 are not satisfied, return
Failure;
  Step 2: TransferQtoM(Q, L, m, n, τ, M);//Call the
Transfer algorithm;
  Step 3: K-M(M); //Call the K-M algorithm.
  Step 4: Form the assignment matrix T based on the
result of K-M (M);
  Step 5: If T is a successful assignment return
Success, else Failure;
end
```

Steps 1, 2, 4, and 5 of Algorithm 5.3 have the complexity of $O(m^2)$. This is less than the complexity of the K-M algorithm (Step 3), i.e. $O(m^3)$ (Kuhn 1955; Munkres 1957; Riesen and Bunke 2009; Toroslu and Üçoluk 2007). Therefore, the overall complexity of the above algorithm is $O(m^3)$.

A WGRAP can also be solved by Algorithm 5.3.

Theorem 5.3 A WGRAP can be solved by Algorithm 5.3.

[Proof]:

To prove this theorem, we need to construct Q_1 from Q, and use Q_1 to obtain a T that maximizes $\sum\limits_{i=0}^{m-1} \sum\limits_{j=0}^{n-1} Q_1[i,j] \times T[i,j]$ using Algorithm 5.3, then we must prove that T also maximizes $\sum\limits_{j=0}^{n-1} \left\{ W[j] \times \sum\limits_{i=0}^{m-1} (Q[i,j] \times T[i,j]) \right\}$.

Suppose we have a qualification matrix Q, and a weight vector W for the WGRAP, we obtain a new matrix Q_1 by $Q_1[i,j] := W[j] \times Q[i,j]$.

Suppose that T is the assignment matrix obtained by Algorithm 5.3 based on Q_1, then $\sum\limits_{i=0}^{m-1} \sum\limits_{j=0}^{n-1} Q_1[i,j] \times T[i,j]$ is the maximum, i.e.

$$\sum_{i=0}^{m-1} \sum_{j=0}^{n-1} Q_1[i,j] \times T[i,j] \geq \sum_{i=0}^{m-1} \sum_{j=0}^{n-1} Q_1[i,j] \times T_1[i,j] (T \neq T_1);$$

Replace Q_1 with $W[j] \times Q[i,j]$, we have

$$\sum_{i=0}^{m-1} \sum_{j=0}^{n-1} W[j] \times Q[i,j] \times T[i,j] \geq \sum_{i=0}^{m-1} \sum_{j=0}^{n-1} W[j] \times Q[i,j] \times T_1[i,j] (T \neq T_1);$$

$$\because \sum_{i=0}^{m-1} \sum_{j=0}^{n-1} W[j] \times Q[i,j] \times T[i,j] = \sum_{j=0}^{n-1} W[j] \times \sum_{i=0}^{m-1} Q[i,j] \times T[i,j];$$

$$\therefore \sum_{j=0}^{n-1} W[j] \times \sum_{i=0}^{m-1} Q[i,j] \times T[i,j] \geq \sum_{j=0}^{n-1} W[j] \times \sum_{i=0}^{m-1} Q[i,j] \times T_1[i,j] (T \neq T_1);$$

$\therefore T$ makes $\sum\limits_{j=0}^{n-1} W[j] \times \sum\limits_{i=0}^{m-1} Q[i,j] \times T[i,j]$ the maximum.

From Definition 5.10, we have solved the WGRAP and the solution is T. ∎

5.7 Implementation and Performance Analysis

The K-M algorithm, when properly implemented, can operate with the computational complexity of $O(m^3)$ (Kuhn 1955; Munkres 1957; Riesen and Bunke 2009; Toroslu and Üçoluk 2007). This is significantly better than the exhaustive search-based algorithm (Zhu and Alkins 2009a).

Table 5.3 The time for the rated assignment algorithm ($\tau = 0.6$).

m / Time	Largest (ms)	Smallest (ms)	Average (ms)	Workable group rate
10	72.211 692	0.003 771	0.466 843	0.88
20	14.090 758	0.100 641	0.564 071	1
30	32.893 077	0.337 752	1.457 006	1
40	34.022 271	0.532 12	2.563 242	1
50	34.213 217	1.573 385	3.959 24	1
60	48.693 48	2.001 511	6.833 295	1
70	50.828 737	3.455 048	9.816 073	1
80	54.279 105	6.857 017	16.924 448	1
90	74.972 518	9.521 182	22.492 996	1
100	88.991 827	11.355 284	29.055 012	1

To verify the performance of the algorithm for RGRAPs, a program is implemented based on the K-M algorithm (Zhu and Alkins 2009a; Zhu et al. 2012). The hardware platform is a laptop with a CPU of 2.10 GHz and the main memory of 4 GB. The development environment is Microsoft Windows Vista (Home Edition) and Eclipse 3.2.

Ten experiments are designed for RGRAPs with different group sizes (Table 5.3), where the workable group rate is the percentage of workable groups among 300 random ones in one experiment. Almost all the cases have 100% workable groups. That is, it is easy to obtain an assignment solution with m agents for $m/2$ roles when $m \geq 20$, where one role may require 1 to 2 agents and $\tau = 0.6$.

In the experiments, a group is formed by randomly creating *an agent qualification matrix* where each agent is randomly assigned values of the range (0, 1] for n roles, i.e. each agent is somewhat qualified to play a role. The number of roles n is set as $m/2$ without loss of generality because the complexity of the K-M algorithm is mainly dependent on m. The lower ranges L of roles are randomly set as 1 or 2 to allow the corresponding group to have enough agents ($\sum_{j=0}^{n-1} L[j] \leq m$).

Each experiment repeats for 300 randomly created agent qualification matrices to show its generality. The results of the experiments for $\tau = 0.6$ are shown in Table 5.3 and the assignment time for $m = 10$ and $n = 5$ is between 0 and 72 ms.

Table 5.3 shows the typical data collected from the experiments stated above. The largest time required by random groups does not increase linearly because each experiment independently tests 300 random groups. It is possible for a smaller group to take more time than a larger group because the value distributions significantly affect the required time for the relevant algorithm. The fact that the largest time is oscillating also demonstrates the importance of members' qualification distributions in a group and where heuristic algorithms may exist.

The smallest time and average time used by the rated assignment algorithm increase slowly (slower than $O(m^3)$). This situation occurs because the square matrix used by the K-M algorithm is created from random groups. The random groups may create many columns full of 1s. Therefore, the time used by the algorithm is normally less than $O(m^3)$.

From Tables 5.3, it is clear that the proposed algorithm for the GRAPs is much better than the algorithm in Zhu and Alkins (2009a) and can be applied to large groups.

Please note that it is impractical to directly compare the data of the proposed algorithm with that of the exhaustive search algorithm. The latter has an exponential complexity and takes hours and even days if the agent number m exceeds 15. The estimation is as follows: The time used when $m = 10$ is 1.5s, then the time required to solve a problem with $m = 30$ is roughly $2^{30}/2^{10} \approx 2^{20} \times 1.5s \approx 18$ days. Hence, it is difficult and impractical to show the data of the exhaustive search algorithm.

As for the SimGRAPs, we do not have to make experiments because it is a programming problem that does not need optimization. The algorithm is straightforward with a complexity of $O(m^2)$ and it is taken as an exercise.

5.8 Case Study by Simulation

A training school hopes to hire one instructor for each section in its software engineering summer school that offers multicourses and multisections. The administrators hope to optimize the total performance of the summer school. They publish the call for instructors on a website and ask applicants to fill in an application form to indicate their requested salary (1000–5500), years of teaching experience (1–20), and the desired time segment for each course (1-morning, 2-afternoon, and 3-evening). They want to use the least amount of money to hire the most experienced instructors (one instructor can teach only one course to avoid overload), and at the same time, choose the instructors who would like to teach at the latest time segment in a day. It is a complex three-objective optimization problem with m applicants and n courses. Note that if an instructor is allowed to teach multisections, the problem becomes more complex, such that the proposed algorithm could not be applied directly.

Suppose that

- L is a vector that specifies the number of sections for each course, i.e. $L[j] \in \mathcal{N}$ $(0 \le j < n)$;

- $\vec{x} = [\vec{x}_0\ \vec{x}_1\ \vec{x}_2]$ is an array of matrices where $\vec{x}_0 =$ an $m \times n$ salary request matrix (applicants vs. courses), $\vec{x}_1 =$ an $m \times n$ years of experience matrix (applicants vs. courses), and $\vec{x}_2 =$ an m teaching time segment vector (corresponding to applicants);

- $f_0(\vec{x}) = \sum\limits_{i=0}^{m-1} \sum\limits_{j=0}^{n-1} \left(5500 - \vec{x}_0[i,j]\right) \times T[i,j];$

- $f_1(\vec{x}) = \sum\limits_{i=0}^{m-1} \sum\limits_{j=0}^{n-1} \vec{x}_1[i,j] \times T[i,j];$ and

- $f_2(\vec{x}) = \sum\limits_{i=0}^{m-1} \sum\limits_{j=0}^{n-1} \vec{x}_2[i] \times T[i,j].$

We need to find an $m \times n$ assignment matrix T where $T[i,j] \in \{0,1\}(0 \le i < m, 0 \le j < n)$, for the problem:

$$max\left\{ f_0\left(\vec{x}\right), f_1\left(\vec{x}\right), f_2\left(\vec{x}\right) \right\}$$

subject to

- $1000 \le \vec{x}_0[i,j] \le 5500,$

- $1 \le \vec{x}_1[i,j] \le 20,$

- $1 \le \vec{x}_2[i] \le 3,$

- $\sum\limits_{i=0}^{m-1} T[i,j] = L[j],$ and

- $\sum\limits_{i=0}^{m-1} T[i,j] \le 1(0 \le i < m,\ 0 \le j < n).$

Such multiobjective optimization problems are nondeterministic polynomial-time hard (NP-hard) (Gholamian et al. 2007) and may not have a feasible solution in most of the cases. Therefore, we need to use a heuristic algorithm to obtain a near-optimal solution. Without loss of generality, a simulation is conducted by randomly creating 20 applicants for 5 courses (total 10 sections, $L = [2\ 1\ 2\ 3\ 2]$). Then, we use a typical weighted sum method (Marler and Arora 2010) by normalizing the three factors. Each instructor is evaluated for each course with:

$$f(i,j) = 0.5 \times 1000/s(i,j) + 0.3 \times y(i,j)/20 + 0.2 \times t(i)/3.$$

where $f(i,j)$, $s(i,j)$, $y(i,j)$, and $t(i)$ are the total score, the salary requested, the years of teaching, and the time requested by instructor i, respectively, and

$$t(i) = \begin{cases} 1 & \text{when time segment is in the morning;} \\ 2 & \text{when time segment is in the afternoon; and} \\ 3 & \text{when time segment is in the evening.} \end{cases}$$

For example, the score of Instructor 5 for Course 0 is:

$$f(5,0) = 0.5 \times 1000/4500 + 0.3 \times 3/20 + 0.2 \times 1/3 = 0.223.$$

The evaluation results are shown in Table 5.4. The time used by the proposed algorithm is 24 ms. It is noted that another simulation for $m = 30$ ($n = 5$) took only 40 ms.

With the proposed algorithm, the best assignment with the weighted sum ($f(i, j)$) is shown in Table 5.5 with a total weighted sum of 7.6.

Now, let us analyze if the acquired assignment is acceptable with respect to the original optimization goals. The total salary is calculated as $1500 + 1500 + 1000 + 1000 + 1000 + 1500 + 1000 + 1500 + 1000 + 2500 = \$13\,500$. The ideal maximum salary is \$11 500 if we do not consider other factors. The closeness to the objectives

Table 5.4 Instructor evaluation.

Instructors\Courses	0	1	2	3	4
0	0.282	0.732	0.347	0.527	0.227
1	0.538	0.375	0.580	0.382	0.910
2	0.297	0.572	0.473	0.443	0.387
3	0.272	0.417	0.392	0.415	0.388
4	0.357	0.342	0.627	0.387	0.467
5	0.223	0.657	0.445	0.283	0.312
6	0.767	0.419	0.465	0.333	0.366
7	0.396	0.553	1.000	0.386	0.818
8	0.330	0.338	0.332	0.422	0.463
9	0.767	0.668	0.722	0.693	0.497
10	0.453	0.484	0.693	0.692	0.563
11	0.678	0.528	0.344	0.527	0.494
12	0.237	0.377	0.285	0.263	0.602
13	0.278	0.608	0.693	0.404	0.465
14	0.413	0.543	0.378	0.419	0.633
15	0.514	0.497	0.540	0.585	0.378
16	0.520	0.446	0.517	0.595	0.472
17	0.758	0.685	0.698	0.715	0.430
18	0.404	0.348	0.488	0.692	0.677
19	0.510	0.578	0.536	0.487	0.476

Table 5.5 Optimal instructor/course assignment.

Course instructors	0	1	2	3	4
0	0	1	0	0	0
1	0	0	0	0	1
2	0	0	0	0	0
3	0	0	0	0	0
4	0	0	0	0	0
5	0	0	0	0	0
6	1	0	0	0	0
7	0	0	1	0	0
8	0	0	0	0	0
9	1	0	0	0	0
10	0	0	0	1	0
11	0	0	0	0	0
12	0	0	0	0	0
13	0	0	1	0	0
14	0	0	0	0	1
15	0	0	0	0	0
16	0	0	0	0	0
17	0	0	0	1	0
18	0	0	0	1	0
19	0	0	0	0	0

can be computed as $(1 - (13\,500 - 11\,500)/11\,500) \times 100\% = 82.61\%$. The total years of experience are calculated as $20 + 20 + 11 + 20 + 4 + 15 + 1 + 15 + 14 + 20 = 140$ years. The ideal maximum years of experience are 186 years if other factors are not considered. The indication of closeness is $(1 - (186 - 140)/186) \times 100\% = 75.27\%$. The total scores for time segments $= 2 + 2 + 1 + 3 + 2 + 2 + 3 + 2 + 3 + 2 = 22$. The ideal maximum score is 25 if no other factors are considered. The indication of closeness is $(1 - (25.22)/25) \times 100\% = 88\%$.

Please note that 100% of closeness is impossible because different functions for optimizations may conflict. There must be trade-offs in the final assignment. Therefore, the proposed approach is practical.

5.9 Related Work

Although role assignment is important in fields such as management (Bostrom 1980), organizational behavior and performance (Ashforth 2001; Black 1988; Bostrom 1980), scheduling, training, and commanding (Turoff et al. 2004), there has been no systematic research. Some related researches that concern agent systems (Dastani et al. 2003; Odell et al. 2003; Stone and Veloso 1999; Vail and Veloso 2003; Wang et al. 2004; Xu et al. 2007), web services (Ng et al. 2008), access control (Ahn and Hu 2007; Al-Kahtani and Sandhu 2007), and sensor networks (Bhardwaj and Chandrakasan 2002) only concentrate on aspects of role assignment related to their field.

Al-Kahtani and Sandhu (2007) propose implicit user-role assignment by introducing role hierarchies in Role-Based Access Control (RBAC), i.e. Rule-Based (RB)-RBAC model. With rules, implicit user-role assignment is supported, i.e. no human intervention is needed and RB-RBAC automatically triggers authorization rules to assign users to roles.

Dastani et al. (2003) determine the conditions under which an agent can enact a role and what it means for the agent to enact a role. This research concerns the norms of how individual agents play individual roles. Their work, in fact, proposes a framework to solve part of *agent evaluation* as discussed in Section 5.1.

Ng et al. (2008) discuss the service assignment problems in web service applications. They assume that different service providers have different profits and commission rates that affect the service assignment. Their work demonstrates that service assignment is a potential application area for the algorithm discussed in this chapter.

Odell et al. (2003) point out that the roles played by an agent may change over time. They apply dynamic classification to deal with adding additional roles or removing roles beyond the minimum. They consider the robots' bids for a role and conduct role assignment in a fixed total order. Their contribution focuses on the third step of role assignment, i.e. role transfer.

Stone and Veloso (1999) introduce periodic team synchronization (PTS) domains as time-critical environments in which agents act autonomously. They state that dynamic role adjustment (the third step, i.e. role transfer) (Zhu and Zhou 2008, 2009) allows needed changes for a group of agents to collaborate. They use dynamic role assignments to change the formation of a robot soccer team based on the game situation, such as winning or losing.

Vail and Veloso (2003) extend the work (Stone and Veloso 1999) in role assignment and coordination in multi-robot systems, especially in highly dynamic tasks. They develop an approach to sharing sensed information and effective coordination through the introduction of shared potential fields. The potential fields are based on the positions (roles) of the other robots in the team and the position of the ball. The robots position themselves in the field by following the gradient to a minimum of the potential field.

Wang et al. (2004) propose to apply Minority Game (MG) strategies to dynamic role assignment in agent teams. They use soccer robots as an experiment platform to validate their proposed method. They mainly consider the first step, i.e. agent evaluation and the third step, i.e. role transfer.

Xu et al. (2007) build a prototype Role-based Agent Development Environment (RADE). In RADE, dynamic role assignment (the third step, i.e. role transfer) is emphasized. The dynamic process of role assignment is formalized as agent role mapping.

Zhang (2008) proposes a teamwork language RoB-MALLET (Role-Based Multi-Agent Logic Language for Encoding Teamwork) to support multi-agent collaboration. She designs rules and related algorithms to regulate the selection of roles and the assignment of roles to agents. Her major job is to evaluate agents' qualifications for roles (the first step).

The above research indicates a strong need to fundamentally investigate GRAPs and their solutions. These works also demonstrate the importance of the work discussed in this chapter. Al-Kahtani and Sandhu's work (2007) suggests the future work for GRA. Also, the algorithm proposed in this chapter can be applied to solve service assignment problems (Ng et al. 2008) and integrated with Xu et al.'s framework (2007) and Zhang's RoB-MALLET (2008). The comparison of some related work with our proposed algorithms based on the scope, goal, and efficiency is shown in Table 5.6.

5.10 Summary

Group role assignment is an important problem in role-based collaboration. Efficient algorithms are required for practical applications. This chapter discusses an efficient way to solve this problem. Because of the polynomial complexity, the newly proposed algorithm is highly practical for resolving the role assignment problems of large groups (Janiak et al. 2007; Leung and Hou 2005). Since this solution is established on the evaluation of agents, it emphasizes the importance of agent evaluation. Possibly, with this proposed solution, the complexity (being NP-hard) of constrained assignment problems (Lee 2010) can be improved by pertinent domain-oriented agent evaluation.

Further investigations can be conducted along the following directions:

1) GRAPs can be further generalized or specialized. For example, there might be a number of restrictions and constraints (Lee 2010) to assignment: some

Table 5.6 Comparisons between the related work and GRA.[a]

Methods	Scope	Goal	Algorithm	Efficiency
Al-Kahtani and Sandhu (2007)	RBAC	To control access to resources by assignment roles.	Not available	Not applicable
Bertsekas (1981)	Assignment	To improve the K-M algorithm.	Yes	$O(m^3)$
Dastani et al. (2003)	Agent collaboration	To verify if an agent is able to play a role.	Not available	Not applicable
K-M (Kuhn 1955; Munkres 1957)	Assignment	To minimize the sum of the selected values in a matrix.	Yes	$O(m^3)$
Ng et al. (2008)	Service composition	To obtain an optimized service composition within a special scenario based on some constraints.	Not available	Not applicable
Odell et al. (2003)	Agent collaboration	To demonstrate the requirement of dynamic role assignment.	Not available	Not applicable
Stone and Veloso (1999); Vail and Veloso (2003); Wang et al. (2004)	Robot collaboration	To make a robot soccer team to win a game. To concentrate on policies for robot collaborations.	Not available	Not applicable
Toroslu and Üçoluk (2007)	Incremental assignment	To solve a special assignment problem, i.e. incremental assignment.	Yes	$O(m^2)$
Xu et al. (2007), Zhang (2008)	Agent collaboration	To describe agent collaboration.	Not available	Not applicable
Zhu et al. (2012)	General collaboration	To optimize group performance by applying the K-M algorithm.	Yes	$O(m^3)$

[a] Note: m is the number of agents in a group.

conflicting agents cannot be assigned to a certain role at the same time (Ahn and Hu 2007); certain roles must be filled while other roles are optional; role hierarchies may need to be incorporated (Al-Kahtani and Sandhu 2007); and temporal requirements for roles may need to be considered.

2) Agent evaluation should be investigated based on domain requirements. The constraints mentioned in 1) may be transferred into a combined evaluation value when applying the proposed methodology.
3) Role assignment algorithms must be able to adapt to the changing requirements and minimize the load of task switching among agents.
4) The optimal scheduling of agents who perform more than one task can be considered. The maximization of group utility (other than the discussed performance) must be subject to availability and maximum workload restrictions.
5) It is valuable to investigate GRA with the application of the Bertsekas algorithm (1981) and the Toroslu and Üçoluk algorithm (2007), which have claimed to achieve better performance than the K-M algorithm.

References

Ahn, G.J. and Hu, H. (2007). Towards realizing a formal RBAC model in real systems. *Proceedings of ACM Symposium on Access Control Models and Technologies* (20–22 June 2007). Sophia Antipolis, France, pp. 215–224.

Al-Kahtani, M.A. and Sandhu, R. (2007). Induced role hierarchies with attribute-based RBAC. *Proceedings of ACM Symposium on Access Control Models and Technologies* (2–3 June 2003). Como, Italy, pp. 142–148.

Ashforth, B.E. (2001). *Role Transitions in Organizational Life: An Identity-based Perspective*. Mahwah, NJ: Lawrence Erlbaum Associates, Inc.

Baker, G. (2008). Java implementation of the classic Hungarian algorithm for the assignment problem. http://sites.google.com/site/garybaker/hungarian-algorithm/assignment (accessed 10 August 2020).

Bertsekas, D.P. (1981). A new algorithm for the assignment problem. *Mathematical Programming* 21: 152–171.

Bhardwaj, M. and Chandrakasan, A.P. (2002). Bounding the lifetime of sensor networks via optimal role assignments. *Proceedings of 21st Annual Joint Conference of the IEEE Computer and Communications Societies*, New York, USA (June 2002), vol. 3, pp. 1587–1596.

Black, J.S. (1988). Work role transitions: a study of American expatriate managers in Japan. *Journal of International Business Studies* 19 (2): 277–294.

Bostrom, R.P. (1980). Role conflict and ambiguity: critical variables in the MIS user-designer relationship. *Proceedings of Annual Computer Personnel Research Conference*, Miami, Florida, United States (June 1980). New York, NY, USA, pp. 88–115.

Bourgeois, F. and Lassalle, J.C. (1971). An extension of the Munkres algorithm for the assignment problem to rectangular matrices. *Communications of the ACM* 14 (12): 802–804.

Dastani, M., Dignum, V., and Dignum, F. (2003). Role-assignment in open agent societies. *Proceedings of the Second International Joint Conference on Autonomous Agents and Multiagent Systems* (14–18 July 2003). Melbourne, Australia, pp. 489–496.

Gholamian, M.R., Fatemi Ghomi, S.M.T., and Ghazanfari, M. (2007). A hybrid system for multiobjective problems – a case study in NP-hard problems. *Knowledge-Based Systems* 20 (4): 426–436.

Janiak, A., Kovalyov, M.Y., and Marek, M. (2007). Soft due window assignment and scheduling on parallel machines. *IEEE Transactions on Systems, Man and Cybernetics, Part A: Systems and Humans* 37 (5): 614–620.

Kuhn, H.W. (1955). The Hungarian method for the assignment problem. *Naval Research Logistic Quarterly* 2: 83–97. (Reprinted in vol. 52, no. 1, 2005, pp. 7 – 21).

Lee, M.-Z. (2010). Constrained weapon–target assignment: enhanced very large scale neighborhood search algorithm. *IEEE Transactions on Systems, Man and Cybernetics, Part A: Systems and Humans* 40 (1): 198–204.

Leung, Y.-W. and Hou, Y.-T. (2005). Assignment of movies to heterogeneous video servers. *IEEE Transactions on Systems, Man and Cybernetics, Part A: Systems and Humans* 35 (5): 665–681.

Marler, R.T. and Arora, J.S. (2010). The weighted sum method for multi-objective optimization: new insights. *Structural and Multidisciplinary Optimization* 41 (6): 853–862.

Munkres, J. (1957). Algorithms for the assignment and transportation problems. *Journal of the Society for Industrial and Applied Mathematics* 5 (1): 32–38.

Ng, V.T.Y., Chan, B., Shun, L.L.Y., and Tsang, R. (2008). Quality service assignments for role-based web services. *Proceedings of the IEEE Int'l Conference on Systems, Man and Cybernetics (SMC)*, Singapore (October 2008), pp. 2219–2224.

Odell, J.J., Van Dyke Parunak, H., Brueckner, S., and Sauter, J. (2003). Changing roles: dynamic role assignment. *Journal of Object Technology* 2 (5): 77–86.

Riesen, K. and Bunke, H. (2009). Approximate graph edit distance computation by means of bipartite graph matching. *Image and Vision Computing* 27 (7): 950–959.

Stone, P. and Veloso, M. (1999). Task decomposition, dynamic role assignment, and low-bandwidth communication for real-time strategic teamwork. *Artificial Intelligence* 110: 241–273.

Toroslu, I.H. and Üçoluk, G. (2007). Incremental assignment problem. *Information Sciences* 177 (6): 1523–1529.

Turoff, M., Chumer, M., Van de Walle, B., and Yao, X. (2004). The design of a Dynamic Emergency Response Management Information System (DERMIS). *Journal of Information Technology Theory and Application* 5 (4): 1–35.

Vail, D. and Veloso, M. (2003). Multi-robot dynamic role assignment and coordination through shared potential fields. In: *Multi-Robot Systems* (eds. A. Schultz, L. Parkera and F. Schneider), 87–98. Kluwer.

Wang, T., Liu, J., and Jin, X. (2004). Minority game strategies in dynamic multi-agent role assignment. *Proceedings of Int'l Conference on Intelligent Agent Technology* (20–24 September 2004). Beijing, China, pp. 316–322.

Xu, H., Zhang, X., and Patel, R. (2007). Developing role-based open multi-agent software systems. *International Journal of Computational Intelligence Theory and Practice* 2 (1): 39–56.

Zhang, Y. (2008). A role-based approach in dynamic task delegation in agent teamwork. *Journal of Software* 3 (6): 9–20.

Zhu, H. (2008). Fundamental issues in the design of a role engine. *Proceedings of the 6th Int'l Symposium on Collaborative Technologies and Systems (CTS)*, Irvine, CA, USA (19–23 May 2008), pp. 399–407.

Zhu, H. (2016). Avoiding conflicts by group role assignment. *IEEE Transactions on Systems, Man, and Cybernetics: Systems* 46 (4): 535–547.

Zhu, H. and Alkins, R. (2009a). Improvement to rated group role assignment algorithms. *Proceedings of the IEEE Int'l Conference On SMC*, San Antonio (October 2009), pp. 4861–4866.

Zhu, H. and Alkins, R. (2009b). Group role assignment. *Proceedings of Int'l Symposium On CTS*, Baltimore, MA (May 2009), pp. 431–439.

Zhu, H. and Zhou, M.C. (2006). Role-based collaboration and its kernel mechanisms. *IEEE Transactions on Systems, Man and Cybernetics, Part C* 36 (4): 578–589.

Zhu, H. and Zhou, M.C. (2008). Role transfer problems and algorithms. *IEEE Transactions on Systems, Man, and Cybernetics, Part A: Systems and Humans* 38 (6): 1442–1450.

Zhu, H. and Zhou, M. (2009). M–M role-Transfer Problems and Their Solutions. *IEEE Transactions on Systems, Man, and Cybernetics, Part A: Systems and Humans* 39 (2): 448–459.

Zhu, H., Zhou, M.C., and Alkins, R. (2012). Group role assignment via a Kuhn-Munkres algorithm-based solution. *IEEE Transactions on Systems, Man and Cybernetics, Part A* 42 (3): 739–750.

Exercises

1 Describe the symbols used in GRA, i.e. m, n, L, Q, T, and σ.

2 Define a workable role and workable group.

3 Describe and specify the GRA problem and its variations, i.e. simple GRA, rated GRA, and weighted GRA.

4 What is the general assignment problem? What is the K-M algorithm?

5 Describe the basic process to transfer a GRA problem to a general assignment problem.

6 Describe the process to transfer the result of the K-M algorithm to the result matrix T of GRA.

7 Why is L important when discovering and specifying the GRA problem?

8 In GRA, we assume that L and Q have been obtained. Based on the process of RBC, discuss the possible ways to obtain L and Q.

9 Discuss the applications of the GRA problem and describe a general scenario to model GRA.

10 Design an algorithm to solve the SimGRAP.

11 Investigate the possibility of applying the incremental assignment algorithm (Toroslu and Üçoluk 2007) to solve a GRA problem at the scale of $(m + 1) \times n$, and $(m + 1) \times (n + 1)$ based on the T obtained from GRA at the scale of $m \times n$.

6

Group Role Assignment with Constraints (GRA⁺)

6.1 Group Multi-Role Assignment (GMRA)

The Group Multi-Role Assignment (GMRA) problem is a variation of the Group Role Assignment (GRA) problem and a typical GRA with Constraints (GRA⁺) problem. By GMRA, we mean assigning a limited number of different roles to each agent within the scenario of GRA (Zhu and Alkins 2009; Zhu and Zhou 2006, 2009, 2012; Zhu et al. 2012a, b). GMRA is, in truth, a GRA problem that considers both the current and potential roles an the agent.

GMRA is common in the real world. For example, in a team, different people (agents) may be assigned several jobs (roles). In a university, a professor (agent) can be assigned with at most three different courses (roles) to teach and a student (agent) can register for at most six (of course, different) courses (roles) in a semester. In transportation, a group of cargo (agents) may be moved by several different containers (roles). In services computing, a server (agent) may provide several different services (roles).

Similar to the GRA problem, the GMRA problem is complex. The primary objective of this section is to provide a practical solution to the GMRA problem. Although GMRA problems originate from RBC research, they are important and applicable in domains such as administration, production, and engineering.

6.1.1 A Real-World Scenario

In company X, Ann, the Chief Executive Officer (CEO), recently signed a half-million-dollar contract. She asks Bob, the Human Resources (HR) officer, to organize a project team composed of company employees. Bob drafts a position list shown in Table 6.1 and a candidate staff shortlist shown in Table 6.2 for the team. Then, Bob initiates an evaluation process by asking the branch officers to evaluate

E-CARGO and Role-Based Collaboration: Modeling and Solving Problems in the Complex World, First Edition. Haibin Zhu.
© 2022 by The Institute of Electrical and Electronics Engineers, Inc.
Published 2022 by John Wiley & Sons, Inc.

Table 6.1 The required positions.

Position	Project manager	System analyst	Software developer	Tester
Required number	1	2	4	2

Table 6.2 The candidates and their position evaluations.

Positions Candidates	Project manager	System analyst	Software developer	Tester
Adam	0.18	**0.82**	0.29	0.10
Brian	0.35	0.80	**0.58**	0.35
Chris	0.84	**0.85**	**0.86**	0.36
Doug	0.96	0.51	**0.45**	**0.64**
Edward	0.22	0.33	**0.68**	0.33
Fred	**0.96**	0.50	0.12	**0.73**

The bold numbers and underlined numbers show different solutions to the problem discussed in the text.

Table 6.3 The maximum numbers of jobs to be assigned to each employee.

Names	Adam	Brian	Chris	Doug	Edward	Fred
Maximum number of jobs	1	2	3	2	2	2

candidates for each position (Table 6.2). After that, Bob informs Ann that he must assign some people to more than one position because the total number of staff members is less than the required number of positions. Ann agrees with Bob. However, she requests that if someone must be assigned to more than one job, then the jobs must be different and each employee should be limited to at most three jobs to avoid overload. Ann's demand is reasonable because an employee can alternate between different jobs but not among the same ones. Furthermore, an employee should be limited to a certain number of jobs in order to guarantee the quality of their work. Based on Ann's request, Bob composes a table (Table 6.3) to express the maximum number of jobs to be assigned to each employee. Then, Bob needs to conduct an optimized assignment. After some consideration, Bob suggests that the formulation of a satisfactory solution, in light of the situation and constraints, may require a significant amount of time. Fortunately, Ann, an experienced administrator, understands the complexity of the problem and does not demand an unreasonable response timeframe.

In the above scenario, Ann and Bob followed the initial steps of RBC but encountered a variation of the GRA problem that considers both current and potential roles collectively (Zhu and Zhou 2006, 2009, 2012; Zhu et al. 2012a, b). The final optimized solution is shown as **bold** and underlined text in Table 6.2, i.e. a tuple set as: {<Adam, {System Analyst}>, <Brian, {Software Developer}>, <Chris, {System Analyst, Software Developer}>, <Doug, {Software Developer, Tester}>, <Edward, {Software Developer}>, and <Fred, {Project Manager, Tester>}. The sum of all the assigned evaluation values is 6.57. This scenario clearly demonstrates the significance of GMRA problems. Doug's assignment seems not the best use of his talents. However, it can be supported if we consider the overall team performance.

6.1.2 Problem Formalization

In the previous chapters, we assume that an agent can only be assigned to one role at a time. In reality, when we consider a long-term project, there arise situations where one agent is assigned more than one role.

Definition 6.1 An *ability limit vector* is m-vector L^a, where $L^a[i]$ represents the maximum number of roles that can be assigned to agent i, $(0 \leq i < m)$.

Definition 6.2 A *Group Multi-Role Assignment (GMRA)* problem is to find a workable T to

$$max \quad \sigma = \sum_{i=0}^{m-1} \sum_{j=0}^{n-1} Q[i,j] \times T[i,j]$$

subject to

$$T[i,j] \in \{0,1\} \quad (0 \leq i < m, 0 \leq j < n), \tag{6.1}$$

$$\sum_{i=0}^{m-1} T[i,j] = L[j] \quad (0 \leq j < n), \tag{6.2}$$

$$\sum_{j=0}^{n-1} T[i,j] \leq L^a[i] \quad (0 \leq i < m), \tag{6.3}$$

where expression (6.1) is a 0-1 constraint; (6.2) makes the group workable, (6.3) allows agents to be assigned with a limited number of roles, i.e. $\mid |a_i.\mathcal{R}_e| \mid \leq L^a[i]$, or $|a_i.\mathcal{R}_p| + |a_i.\mathcal{R}_e| \leq L^a[i]$ $(0 \leq i < m)$.

The only difference between GMRA (Definition 6.2) and GRA (Definition 5.9) is that constraint (6.3) replaces "1" in constraint (5.3) with $L^a[i]$.

(a)

$$\begin{bmatrix} 0.18 & 0.82 & 0.29 & 0.01 \\ 0.35 & 0.80 & 0.58 & 0.35 \\ 0.84 & 0.85 & 0.86 & 0.36 \\ 0.96 & 0.51 & 0.45 & 0.64 \\ 0.22 & 0.33 & 0.68 & 0.33 \\ 0.96 & 0.50 & 0.10 & 0.73 \end{bmatrix}$$

(b)

$$\begin{bmatrix} 0 & 1 & 0 & 0 \\ 0 & 0 & 1 & 0 \\ 0 & 1 & 1 & 0 \\ 0 & 0 & 1 & 1 \\ 0 & 0 & 1 & 0 \\ 1 & 0 & 0 & 1 \end{bmatrix}$$

Figure 6.1 (a) A qualification matrix Q and (b) an assignment matrix T.

Figure 6.1b is a T obtained by Definition 6.2 with the Q in Figure 6.1a by following the constraints $L = [1, 2, 4, 2]$ and $L^a = [1, 2, 3, 2, 2, 2]$. The sum of the assigned evaluation values is 6.57.

6.1.3 The CPLEX Solution and Its Performance Experiments

The GMRA problem is solved using the IBM ILOG CPLEX optimization package (CPLEX), which is different from using the Optimization Programming Language (OPL) of the IBM ILOG CPLEX optimization studio (IBM 2019). Designing a Java program with CPLEX would result in better performance by bypassing the compiler of OPL. On the other hand, such a Java program can be used to solve a list of problems automatically. This feature is often a requirement in some applications, such as processor scheduling in supercomputing, and service allocations in a cloud environment.

To provide a solution with CPLEX, the major job is to provide the four required elements and transform the objective and the constraints into the required forms.

Step 1: determine the four elements (i.e. objective function coefficients, constraint coefficients, right-hand-side constraint values; and upper and lower bounds) required by the CPLEX package. Specifically, Q, L, L^a, and T are used to define a Linear Programmin (LP) problem in CPLEX. Matrix Q expresses the objective function coefficients and T expresses the variables. The upper and lower bounds of T are 1 and 0.

Step 2: add the objective and constraint expressions.

The objective of the GMRA problem can be expressed by a formula of the one-dimensional array forms of matrices Q and T. In the CPLEX package, we can maximize this formula based on the objective.

To add the optimization objective, we invoke the following methods in Java:

IloIntVar[]X = cplex.intVarArray(m∗n, 0, 1);

for T constraint (6.1),

cplex.addMaximize(cplex.scalProd(X, V));

for σ,

where $X[i \times n + j] = T[i, j]$ and $V[i \times n + j] = Q[i, j]$ $(0 \leq i < m, 0 \leq j < n)$.

Table 6.4 Test platform configuration.

	Hardware
CPU	Intel Core i7-4650U CPU @1.7 GHz 2.3 GHz
MM	8 GB
	Software
OS	Windows 7 Enterprise
Eclipse	Version: Luna Release (4.4.0)
JDK	Java 8 Update (45)

To add the constraints to CPLEX, we follow three steps:

1) Declare expression objects by calling:
   IloLinearNumExpr expr1 = cplex.linearNumExpr(); and
   IloLinearNumExpr expr2 = cplex.linearNumExpr();
2) Add all the terms by invoking the methods:
   "**for** $(0 \leq j < n)$ *expr1.addTerm*$(1, X[j + i*n])$," which corresponds to
   $\sum_{i=0}^{m-1} T[i,j]$ in constraint (6.2);
   "**for** $(0 \leq i < m)$ *expr2.addTerm*$(1, X[j + i*n])$," which corresponds to
   $\sum_{j=0}^{n-1} T[i,j]$ constraint (6.3); and
3) Add the constraint expressions to CPLEX by invoking: "*cplex.addEq(expr1, L)*"
   for $L[j]$ $(0 \leq j < n)$ constraint (6.2); and "*cplex.addLe(expr2, L^a)*" for $L^a[i]$ $(0 \leq i < m)$ constraint (6.3).

To check the applicability of the CPLEX solution, we conducted experiments with random groups of different sizes, where m is from 20 to 600 with an increment of 20, $n = m/2$, $L[j]$ is between 1 and 10, and $L^a[i]$ is between 1 and 6. For each m (n), we produced 100 random groups and recorded the maximum, minimum, and average processing time. The performance experiments are simulated using the platform shown in Table 6.4. The results show that the CPLEX solution is practical, that is, a problem with 500 agents can be done in a second.

6.1.4 Improvement of the CPLEX Solution

By looking into the GMRA problems, we find that there are many cases where there are no feasible solutions. When we observed the performance results of the CPLEX solutions, we noticed that problems with a similar scale consumed a similar amount of time, including those without a feasible solution. We believe that we can improve the CPLEX solution by providing a simple way to check if a GMRA problem has a feasible solution. With this improvement, more problems can be solved using CPLEX in the same time frame compared with before.

Theorem 6.1 The necessary condition for a GMRA problem to have a feasible solution is that $\sum_{i=0}^{m-1} L^a[i] \geq \sum_{j=0}^{n-1} L[j]$.

[**Proof**]:
To have a feasible solution means that there is a T, such that $T[i,j] \in \{0,1\}(0 \leq i < m,$ $0 \leq j < n)$, $\sum_{i=0}^{m-1} T[i,j] = L[j](0 \leq j < n)$, and $\sum_{j=0}^{n-1} T[i,j] \leq L^a[i]$ $(0 \leq j < n)$.

If $\sum_{i=0}^{m-1} L^a[i] < \sum_{j=0}^{n-1} L[j]$, i.e. $\sum_{i=0}^{m-1} L^a[i] = \sum_{j=0}^{n-1} L[j] - k$, where k is a positive integer.

To satisfy $\sum_{j=0}^{n-1} T[i,j] \leq L^a[i]$, we need $\sum_{i=0}^{m-1} \sum_{j=0}^{n-1} T[i,j] \leq \sum_{i=0}^{m-1} L^a[i] = \sum_{j=0}^{n-1} L[j] - k$,

i.e. $\sum_{j=0}^{n-1} \sum_{i=0}^{m-1} T[i,j] \leq \sum_{j=0}^{n-1} L[j] - k$.

For one case, there are at least k roles, such that

$$\sum_{i=0}^{m-1} T[i,j] = L[j] - 1.$$

In fact, for all other different cases, it does not satisfy the constraint

$$\sum_{i=0}^{m-1} T[i,j] = L[j](0 \leq j < n).$$

Therefore, the assignment matrix T is not a feasible solution. ∎

To check if the necessary condition is satisfied, the complexity is $O(m)$ if m and n are on the same scale because this checking only needs to compute $\sum_{i=0}^{m-1} L^a[i]$ and $\sum_{j=0}^{n-1} L[j]$.

Theorem 6.1 can help discard problems that do not meet the necessary condition and save the time spent in useless searching. However, there are still some problems that have no feasible solutions and make the CPLEX system try in vain. A necessary and sufficient condition for feasible solutions is required.

To find the necessary and sufficient condition for the GMRA problem to have a feasible solution, we need to introduce some symbols and theorems. Given vectors V and W, $|V|$ is the cardinality of V; $max^k V$ is the set of k biggest elements in V ($k \geq 0$), where $|V| \geq k$; V^* is the first element in V; specifically, V^* of an empty set is defined as 0; $sub(max^k V)$ is the function such that each of the k biggest elements in V minus 1; W_0 is the initial W, W_i is $W_{i-1} - \{W_{i-1}^*\}$ $(0 < i \le |W|)$; and WV_0 is the initial V, WV_i is $sub(max^{W_{i-1}^*} V_{i-1})$ $(0 < i \le |V|)$.

Theorem 6.2 *(Decidable Theorem)*: GMRA (with Q, T, L, and L^a) has a feasible solution if and only if there exists an integer q, such that L_q is empty and for each i, $L_i^* \le |L_{i-1} L^a_i|$ $(0 < i \le q)$.

[**Proof**]:
Left to Right (The necessary condition):

Proof by contradiction: if there is not an integer q such that L_q is empty and for each of i, $L_i^* \le |L_{i-1} L^a_i|$ $(0 \le i \le q)$, then it is equivalent to that for any integer q, L_q is not empty or there exists an integer i, such that $L_i^* > |L_{i-1} L^a_i|$.

Case 1 (L_q is not empty): The construction of a sequence of L_i and $L_{i-1} L^a_i$ is equal to assigning roles to agents that can play the most roles every time. This construction guarantees that $|L_{q-1} L^a_q|$ is the biggest after the current assignment, which can save the most agents for the next assignment. For any q, that L_q is not empty means that there always exists at least one role that has not enough agents to play it.

Case 2 $\left(L_i^* > |L_{i-1} L^a_i|\right)$: If there exists an integer i, such that $L_i^* > |L_{i-1} L^a_i|$. It means that, in the ith time, there are not enough agents to play role L_i^*.

In summary, GMRA has no feasible solution.

Right to Left (The sufficient condition):

If there exists an integer q, such that L_q is empty and for each of i, $L_i^* \le |L_{i-1} L^a_i|$ $(0 < i \le q)$, then we can construct a solution by assigning roles to agents that can play the most roles, i.e. choosing agent i, where $L^a[i] = \max \{L^a\}$. Therefore, we can obtain a workable T. ∎

Now, based on Theorem 6.2, we can construct a Cardinality Constraint Detection (CCD) algorithm as follows:

Algorithm 6.1 Cardinality Constraint Detection (CCD).

```
Input:
   • An n-vector L with the above preliminaries; and
   • An m-vector L^a with the above preliminaries.
Output:
   • True or False.
```

(Continued)

Algorithm 6.1 (Continued)

```
begin
while (L is not empty)
    if (L* > |Lᵃ|)
        return false;
    else
        Lᵃ = sub(maxᴸ˙Lᵃ);
        L = L- {L*};
    endif
endwhile
return true;
end
```

Note that n is the cardinality of L and m is that of L^a, i.e. n and m represent the numbers of elements in L and L^a, respectively. Hence, the complexity of the outside loop (while loop) is $O(n)$. The complexity of $L^a = sub(max^{L^*}L^a)$ is $O(m\log m)$, assuming that a quick sort method can be used to sort L^a. Then, the complexity of Algorithm 6.1 is $O(n \times m\log m)$. Suppose that m and n are on the same scale, the complexity of Algorithm 6.1 can be simplified to $O(m^2\log m)$. The two theorems are implemented in the improved CPLEX solution with Java in Eclipse.

6.1.5 Comparisons

To verify the proposed approach, we conducted comparisons between the improved CPLEX solution and the original CPLEX one (IBM 2019). The platform is the same as the one in Table 6.4. In our experiments, we used the case presented in Section 6.1.1 and three other cases. Here, we collected the total time used by the improved solution and the original CPLEX solution to process the same 100 random problems at each scale of $m = 100, 200, ..., 600$. To demonstrate the improvement, we collected the percentage of saved time, i.e. $(t_1-t_2)/t_1$, where t_1 is the time used by the original CPLEX solution to process 100 problems and t_2 is the time used by the improved solution to perform the same task.

The percentage of saved time can be 2–40% when processing the 100 random problems of various scales (m) (Zhu 2016). The results reflect the fact that the improved solution takes advantage of Theorems 6.1 and 6.2.

We need to state that GMRA is a new abstract assignment problem under the category of LP. Based on the authors' knowledge, there has not been a specific solution to this problem other than the LP solutions presented in this paper.

6.2 Group Role Assignment with Conflicting Agents (GRACA)

Conflicts are a common phenomenon in the real world. For example, in transportation, different chemicals (agents) cannot be stored in the same container (a role); in the animal world, a predator and a prey (agents) cannot be placed in the same cage (a role); and in wireless communication, various devices (agents) cannot be put in the same area (a role) due to potential interference (Guo et al. 2012). There are also many types of conflict among people in society (Tomlin et al. 1998; Walker et al. 2013). For example, unresolved emotional conflicts may prevent John and Matt from working together. Furthermore, multi-agent systems rarely exist without conflicts (Kilgour and Hipel 2005; Malsch and Weiss 2002; Rapoport and Chammah 1965; Tessier et al. 2002).

In the real world, frustrations arise in collaboration because pre-existing agent conflicts may not be considered during the role assignment process. Much effort can be expended in resolving or handling such conflicts (Bristow et al. 2014; Canbaz et al. 2014; Kilgour and Hipel 2005; Kinsara et al. 2015; Klein 1991) during collaboration. A primary objective of this section is to establish a mechanism whereby conflicts are dealt with before the cooperation process (i.e. role-playing) begins.

Successful role assignment while avoiding conflicts is an important step in collaboration. Conflicting agents should be prevented from playing the same roles (task, position, or area) through an assignment. This idea is called *conflict avoidance* (Tomlin et al. 1998).

In this section, we mainly present the formalization of the GRACA problem, verification of the benefits in solving GRACA problems, theoretical proof that the problem is a subproblem of an NP-Complete problem, and the provision of a practical solution.

6.2.1 A Real-World Scenario

In company X, after creating the position list in Table 6.1, Bob composes a candidate staff shortlist shown in Table 6.5. Then, Bob initiates an evaluation process by asking the branch officers to evaluate employees for each position (Table 6.5). Bob then meets with the branch officers to discuss possible conflicts between various employees based on factors such as past experiences, personality characteristics, working styles, emotional issues, and political beliefs (Table 6.6). Bob needs to

Table 6.5 The candidates and position evaluations.

Positions Candidates	Project manager	Senior programmer	Programmer	Tester
Adam	0.18	**0.82**	0.29	0.01
Brian	0.35	0.80	**0.58**	0.35
Chris	0.84	**0.85**	0.86	0.36
Doug	0.96	0.51	0.45	**0.64**
Edward	0.22	0.33	**0.68**	0.33
Fred	0.96	0.50	0.10	**0.73**
George	0.25	0.18	0.23	0.39
Harry	0.56	0.35	**0.80**	0.62
Ice	0.49	0.09	0.33	0.58
Joe	0.38	0.54	**0.72**	0.20
Kris	**0.91**	0.31	0.34	0.15
Larry	0.85	0.34	0.43	0.18
Matt	0.44	0.06	0.66	0.37

The bold numbers and underlined numbers show different solutions to the problem discussed in the text.

Table 6.6 Conflicts between candidates.[a]

Conflicts Candidates	1	2	3
Adam	Brian		
Brian	Adam		
Chris			
Doug			
Edward	Fred	Larry	Matt
Fred	Edward		
George			
Harry			
Ice			
Joe			
Kris			
Larry	Edward		
Matt	Edward		

[a] Note: Adam conflicts with Brian, and Edward conflicts with Fred, Larry, and Matt.

assign the most qualified candidates to positions while avoiding employee conflicts. That is to say, when constructing an agile group, the group performance should not be compromised due to employee conflicts. Evidently, this is not a trivial problem.

In the above scenario, Bob follows the initial steps of RBC but encounters a GRA problem with the additional constraint of conflicting agents. The final optimized assignment, after avoiding conflicting agents, is displayed in Table 6.5 (**bold**), i.e. a tuple set as: {<Adam, Senior Programmer>, <Brian, Programmer>, <Chris, Senior Programmer>, <Doug, Tester>, <Edward, Programmer>, <Fred, Tester>, <Harry, Programmer>, <Joe, Programmer>, and <Kris, Project Manager>}. The sum of all the assigned evaluation values is 6.73. This scenario clearly demonstrates the significance of GRACA problems. Though it appears that Doug's assignment is not the best use of his abilities, the assignment can be supported if we consider the improved overall team performance.

6.2.2 Problem Formalization

The state of a group after role assignment is highly dependent on various factors. Role assignment should consider these factors and endeavor to avoid conflicting agents in its assignment. Conflicting agents can be formally defined as follows (Zhu 2011, 2012, 2016):

Definition 6.3 *Two* agents i_1 and i_2 $(0 \le i_1, i_2 < m, i_1 \ne i_2)$ are *in conflict* on roles if i_1 and i_2 cannot be assigned to the same role. We say that i_1 is a *conflicting agent* of i_2 *on roles* and vice versa.

Definition 6.4 *Two* agents i_1 and i_2 $(0 \le i_1, i_2 < m, i_1 \ne i_2)$ are *in conflict* in groups if i_1 and i_2 cannot be assigned to the same group. We say that i_1 is a *conflicting agent* of i_2 *in groups* and vice versa.

Definition 6.5 A *conflicting agent matrix* is defined as an $m \times m$ matrix $A^c (A^c[i_1, i_2] \in \{0,1\}, 0 \le i_1, i_2 < m)$, where $A^c[i_1, i_2] = 1$ expresses that agent i_l is in conflict with agent i_2 and $A^c[i_1, i_2] = 0$ means no conflict. We define $A^c[i, i] = 0$ $(0 \le i < m)$ for simplicity.

Note: A^c is a symmetric matrix along the diagonal from $A^c[0, 0]$ to $A^c[m - 1, m - 1]$, i.e. $(A^c[i_1, i_2] = A^c[i_2, i_1])$ $(0 \le i_1, i_2 < m, i_1 \ne i_2)$.

For Table 6.5 in Section 6.2.1, we have the corresponding A^c shown in Figure 6.2a.

Definition 6.6 The *Group Role Assignment with Conflicting Agents on Roles (GRACAR)* problem is to find a workable T to

$$\max \sigma = \sum_{i=0}^{m-1} \sum_{j=0}^{n-1} Q[i,j] \times T[i,j]$$

(a)

$$
\begin{bmatrix}
0 & 1 & 0 & 0 & 0 & 0 & 0 & 0 & 0 & 0 & 0 & 0 & 0 & 0 \\
1 & 0 & 0 & 0 & 0 & 0 & 0 & 0 & 0 & 0 & 0 & 0 & 0 & 0 \\
0 & 0 & 0 & 0 & 0 & 0 & 0 & 0 & 0 & 0 & 0 & 0 & 0 & 0 \\
0 & 0 & 0 & 0 & 0 & 0 & 0 & 0 & 0 & 0 & 0 & 0 & 0 & 0 \\
0 & 0 & 0 & 0 & 0 & 1 & 0 & 0 & 0 & 0 & 0 & 1 & 1 \\
0 & 0 & 0 & 0 & 1 & 0 & 0 & 0 & 0 & 0 & 0 & 0 & 0 & 0 \\
0 & 0 & 0 & 0 & 0 & 0 & 0 & 0 & 0 & 0 & 0 & 0 & 0 & 0 \\
0 & 0 & 0 & 0 & 0 & 0 & 0 & 0 & 0 & 0 & 0 & 0 & 0 & 0 \\
0 & 0 & 0 & 0 & 0 & 0 & 0 & 0 & 0 & 0 & 0 & 0 & 0 & 0 \\
0 & 0 & 0 & 0 & 0 & 0 & 0 & 0 & 0 & 0 & 0 & 0 & 0 & 0 \\
0 & 0 & 0 & 0 & 0 & 0 & 0 & 0 & 0 & 0 & 0 & 0 & 0 & 0 \\
0 & 0 & 0 & 0 & 1 & 0 & 0 & 0 & 0 & 0 & 0 & 0 & 0 & 0 \\
0 & 0 & 0 & 0 & 1 & 0 & 0 & 0 & 0 & 0 & 0 & 0 & 0 & 0
\end{bmatrix}
$$

(b)

$$
\begin{bmatrix}
0 & 1 & 0 & 0 \\
0 & 0 & 1 & 0 \\
0 & 1 & 0 & 0 \\
0 & 0 & 0 & 1 \\
0 & 0 & 1 & 0 \\
0 & 0 & 0 & 1 \\
0 & 0 & 0 & 0 \\
0 & 0 & 1 & 0 \\
0 & 0 & 0 & 0 \\
0 & 0 & 1 & 0 \\
1 & 0 & 0 & 0 \\
0 & 0 & 0 & 0 \\
0 & 0 & 0 & 0
\end{bmatrix}
$$

Figure 6.2 (a) The conflict matrix A^c. (b) The assignment matrix T.

subject to 6.1, 6.2, and

$$
\sum_{j=0}^{n-1} T[i,j] \le 1 \quad (0 \le i < m) \tag{6.4}
$$

$$
A^c[i_1, i_2] \times (T[i_1, j] + T[i_2, j]) \le 1 \quad (0 \le i_1, i_2 < m, i_1 \ne i_2, 0 \le j < n) \tag{6.5}
$$

where constraint (6.4) means that each agent is assigned to at most one role and (6.5) mandates that no conflicting agents are assigned to the same role.

For example, Figure 6.2b is a T following *Definition 6.6*, where Q is shown in Table 6.5, A^c is in Figure 6.2a, and L is in Table 6.1. The sum of the assigned evaluation values is 6.73. We can compare it with the sum of 6.96 (underlined numbers in Table 6.2) from the GRA solution, which has conflicts, i.e. Adam and Brian are in conflict. This comparison indicates that the avoidance of conflicts does not significantly degrade the GRA performance. On the other hand, if conflicts arise during the role-playing phase, considerable effort may be required to resolve the issue.

Definition 6.7 The *Group Role Assignment with Conflicting Agents in a Group (GRACAG)* problem is to find a workable T to

$$
max \ \sigma = \sum_{i=0}^{m-1} \sum_{j=0}^{n-1} Q[i,j] \times T[i,j]
$$

subject to (6.1), (6.2), (6.4) and

$$
A^c[i_1, i_2] \times (T[i_1, j_1] + T[i_2, j_2]) \le 1 \quad (0 \le i_1, i_2 < m, i_1 \ne i_2, 0 \le j_1, j_2 < n) \tag{6.6}
$$

(a)

$$
\begin{bmatrix}
0.71 & 0.6 & 0.0 & 0.22 \\
0.29 & 0.67 & 0.44 & 0.76 \\
0.69 & 0.92 & 0.92 & 0.6 \\
0.0 & 0.0 & 0.53 & 0.0 \\
0.97 & 0.51 & 0.77 & 0.65 \\
0.58 & 0.64 & 0.24 & 0.0
\end{bmatrix}
$$

(b)

$$
\begin{bmatrix}
1 & 0 & 0 & 0 \\
0 & 0 & 0 & 1 \\
0 & 0 & 0 & 1 \\
0 & 0 & 1 & 0 \\
1 & 0 & 0 & 0 \\
0 & 1 & 0 & 0
\end{bmatrix}
$$

(c)

$$
\begin{bmatrix}
0 & 0 & 0 & 0 & 1 & 0 \\
0 & 0 & 1 & 0 & 0 & 0 \\
0 & 1 & 0 & 0 & 1 & 0 \\
0 & 0 & 0 & 0 & 0 & 0 \\
1 & 0 & 1 & 0 & 0 & 0 \\
0 & 0 & 0 & 0 & 0 & 0
\end{bmatrix}
$$

Figure 6.3 An example of GRACAG. (a) A Q matrix. (b) A T matrix, (c) A conflict matrix A^c.

Figure 6.4 Two assignment matrices for two different problems. (a) The T for $L = [2\ 1\ 1\ 2]$. (b) The T for $L = [1\ 1\ 1\ 1]$.

(a)

$$
\begin{bmatrix}
1 & 0 & 0 & 0 \\
0 & 0 & 0 & 1 \\
0 & 1 & 0 & 0 \\
0 & 0 & 1 & 0 \\
0 & 0 & 0 & 1 \\
1 & 0 & 0 & 0
\end{bmatrix}
$$

(b)

$$
\begin{bmatrix}
0 & 0 & 0 & 0 \\
0 & 0 & 0 & 1 \\
0 & 0 & 0 & 0 \\
0 & 0 & 1 & 0 \\
1 & 0 & 0 & 0 \\
0 & 1 & 0 & 0
\end{bmatrix}
$$

where constraint (6.6) means that no conflicting agents can be assigned to the same group.

From the above definitions, GRACAG has more restrictions in obtaining T than GRACAR does. For example, suppose that Q is shown as Figure 6.3a, $L = [2\ 1\ 1\ 2]$, then T in Figure 6.3b is a solution when no conflict is considered. However, if the conflicting agent matrix in Figure 6.3c is considered in a GRACAR problem, a solution is shown in Figure 6.4a. If Figure 6.3c is considered in a GRACAG problem, then there is no solution for $L = [2\ 1\ 1\ 2]$. Figure 6.4b shows a solution for $L = [1\ 1\ 1\ 1]$. From these examples, it is understandable that there might be no solution for a group expressed by Q, L, and A^c. Also, we know that if $L[j] = 1$ for all $(0 \leq j < n)$, no conflict occurs in GRACAR. This fact indicates the importance of L in the E-CARGO model.

To help solve GRACAR/G problems, we need to know if a problem is solvable.

Definition 6.8 The *number of required agents* n_a is defined as $\sum_{j=0}^{n-1} L[j]$.

Definition 6.9 The *number of conflicts* n_c is defined as $\sum_{i_1=0}^{m-2} \sum_{i_2=i_1+1}^{m-1} A^c[i_1, i_2]$.

Theorem 6.3 The GRACAR/G problems are solvable if $m - n_c \geq n_a$.

[Proof]: To solve the problems of GRACAR/G, we need to find a subset $\mathcal{A}'$ from the original agent set $\mathcal{A}$, so that there are no conflicts in $\mathcal{A}'$ and the number of agents must meet the requirement of L, i.e. $|\mathcal{A}'| \geq n_a$.

$\because m - n_c \geq n_a,$

$\therefore$ We may delete one agent from each conflicting pair of agents, i.e. let set $\mathcal{X} = \{i|$ $A^c[i, i'] = 1 (0 \leq i < m, i + 1 \leq i' < m)\}$ and $\mathcal{A}' = \mathcal{A} - \mathcal{X}$. Now, $\mathcal{A}'$ is free of conflicts.

$\because$ One agent may be in conflict with more than one agent,

$\therefore |\mathcal{X}| \leq n_c.$

$\therefore |\mathcal{A}'| = |\mathcal{A}| - |\mathcal{X}| \geq n_a.$

Theorem 6.3 is proved. ∎

Note that Theorem 6.3 is a sufficient condition but not necessary. For the example discussed in Section 6.2.1, we do not have to remove any agents for GRACAR. For GRACAG, we only need to remove two (Adam and Edward), but not four to acquire the solution. Although Theorem 6.3 only states the sufficient condition, it is useful for HR officers when shortlisting candidates.

6.2.3 The Benefits of Avoiding Conflicts

To quantitatively assess the benefits of avoiding conflicts, we conducted simulations with randomly generated groups. In the simulation, a specific group size of $m = 100$ and $n = 16$ was chosen. Note that the generality of the simulation was not lost from choosing a specific group size since we are concentrating only on the benefits of conflict avoidance, not the time efficiency. Furthermore, this group size is typical of a mid-sized company or factory. We define two new concepts for this analysis.

Definition 6.10 The *number of assigned conflicts* n_{ac} is defined as $\sum_{j=0}^{n-1} \sum_{i_1=0}^{m-2} \sum_{i_2=i_1+1}^{m-1} A^c[i_1, i_2] \times T[i_1, j] \times T[i_2, j]$, where T is obtained without considering A^c, i.e. the result of GRA.

Definition 6.11 The *conflict rate* p_c (named after **p**ercentage of **c**onflicts) is defined as $n_c/[m \times (m - 1)/2]$, i.e. the number of conflicts divided by the total number of pairs of agents.

For example, the p_c in Table 6.5 is $4/(13 \times 12) \approx 2.56\%$.

To estimate the benefits of conflict avoidance, we need to make some assumptions regarding the nature of conflicts. Conflicts are diverse and affect collaborations in varying degrees (Tessier et al. 2002, Tomlin et al. 1998). Some conflicts can be ignored, some must be resolved, and some may even lead to loss of life, e.g. on the battlefield, soldiers accomplishing the same task must not be hampered by conflicts. By drawing an analogy between bugs in software (Pressman 2004) and conflicts in collaboration, we can categorize conflicts by their severity, the percentage loss in productivity: ignorable (0% loss), mild (20% loss), annoying (40%

loss), disturbing (60% loss), serious (80% loss) and extreme (100% loss). Some may produce even larger losses (>100%). Because we are discussing collaboration, conflicts that are categorized as worse than extreme are ignored. In fact, if productivity loss is more than 40%, conflict avoidance will generally provide a higher benefit than otherwise. The justification for this claim will be discussed later in the chapter. Therefore, we will consider conflicting agents to be pairs of agents who have annoying conflicts with one another; as such, the agents in conflict lose 40% of their original qualifications and the conflict affects both parties.

Each group was randomly generated according to the following parameters, range, and definitions:

- $Q[i,j]$ ($0 \leq i < m, 0 \leq j < n$), i.e. the qualification value of each agent for each role;
- $A^c[i_1, i_2]$($0 \leq i_1, i_2 < m$), i.e. the conflicts between agents; and
- $L[j]$ ($0 \leq j < n$), $1 \leq L[j] \leq 6$ to make $m \geq n_a$.

For each group, we also collected several statistics:

1) σ_1, the maximum group performance without considering conflicts;
2) σ_2, the maximum group performance after avoiding conflicts;
3) σ_2/σ_1, the comparison between σ_2 and σ_1 in percentage;
4) n_a, the number of required agents;
5) n_c, the number of conflicts; and
6) n_{ac}, the number of assigned conflicts in the T corresponding to σ_1.

The benefit, denoted as λ, is calculated as follows:

1) We obtain each agent's average qualification, i.e.

σ_1/n_a.

2) Then, the lost group performance is

$\sigma_1 \times 2 \times 0.4 \times n_c/n_a$.

3) The adjusted group performance is

$(1 - 0.8 \times n_c/n_a) \times \sigma_1$.

4) Then, we have the benefit

$$\lambda = [\sigma_2 - (1 - 0.8 \times n_c/n_a) \times \sigma_1]/[(1 - 0.8 \times n_c/n_a) \times \sigma_1].$$

To observe the differences between the configurations, we change the conflict rate p_c from 1/3 to 1/18. To make the data more convincing, 300 groups for each conflict rate were generated. Finally, we computed the averages of the 300 trials for each of the seven statistics. Table 6.7 shows the simulation results.

By observing the simulation results, it is evident that conflict avoidance does not significantly degrade group performance. The average group performance ranges

Table 6.7 Simulation averages ($m = 100$, $n = 16$).

p_c	σ_1	σ_2	σ_2/σ_1 (%)	n_a	n_c	n_{ac}	λ (%)
1/3	54.05	53.30	98.61	56.14	859.47	14.58	24.37
1/4	53.80	53.35	99.16	56.85	618.94	10.74	17.12
1/5	53.54	53.21	99.38	56.57	492.52	08.18	12.60
1/6	54.01	53.73	99.48	56.10	397.67	06.86	10.25
1/7	54.83	54.59	99.56	56.95	343.22	06.41	09.36
1/8	54.29	54.06	99.58	56.37	321.61	06.01	08.84
1/9	53.86	53.69	99.68	56.90	267.02	04.72	06.88
1/10	53.43	53.26	99.68	56.47	247.63	04.50	06.59
1/11	53.70	53.55	99.72	56.78	227.13	04.14	06.99
1/12	53.94	53.80	99.74	56.01	203.97	03.71	06.32
1/13	54.04	53.92	99.78	56.14	186.10	03.31	04.70
1/14	53.96	53.84	99.78	56.01	182.60	03.28	04.65
1/15	53.38	53.27	99.79	56.45	169.86	03.04	04.36
1/16	53.73	53.63	99.81	56.79	151.13	02.95	04.21
1/17	54.19	54.09	99.82	56.25	146.60	02.78	03.90
1/18	54.05	53.97	99.85	56.13	132.33	02.40	03.36

from 98.61% to 99.85% depending on different conflict rates. Conflicts in these groups are introduced based on the conflicting rate p_c.

Based on the above simulation data and estimations, the right-most column λ of Table 6.7 confirms the benefit of avoiding conflicts. However, it is not a trivial matter to solve GRACAR/G problems. We tried several initial algorithms (Zhu 2011, 2012), the complexity of which is exponential. Previous experiments also demonstrated the complexity of the solutions of GRACAR/G problems (Zhu 2011, 2012, 2016; Zhu and Feng 2013; Zhu et al. 2015). Therefore, we need to analyze the complexity of GRACAR/G problem and develop a practical solution.

6.2.4 GRACAR/G Problems Are Subproblems of an NP-Complete Problem

In this section, we provide proof that GRACAR/G problems are subproblems of the extended Integer Linear Programming (x-ILP) that can be restricted to the Integer Linear Programming (ILP) problem that is NP-Complete (Lyamin and Sloan 2002; Papadimitriou 1981; Rardin 1997; Schrijver 2000; Wolsey and Nemhauser 1999). This proof suggests that GRACAR/G problems can be solved using tools that solve ILP problems.

Theorem 6.4 The GRACAR/G problems are of the Non-Polynomial (NP) class.

[Proof]: According to Garey and Johnson (1979), a problem is in the NP class if there is a nondeterministic algorithm used to guess a solution and check in polynomial time whether the solution adheres to the constraints.

For GRACAR, we may discuss a problem Π derived from GRACAR to answer whether $\sigma \geq b$ (where $\sigma = \sum_{i=0}^{m-1} \sum_{j=0}^{n-1} Q[i,j] \times T[i,j]$ and b is a specific number) but not to determine the maximum value of σ. If a guessed T', which makes the algorithm nondeterministic, for Π is provided, we can use the following algorithm to check if $\sigma \geq b$ and T' follows constraints (6.1), (6.2), (6.4), and (6.5), where $\sigma \geq b$, (6.1), (6.2), and (6.4) are straightforward.

For (6.5), the checking stage is (in Java-like format):

for j **from** 0 **to** n-1 **do**
  for i_1 **from** 0 **to** m-1 **do**
    for i_2 **from** i_1+1 **to** m-1 **do**
      if $(A^c[i_1, i_2] \times (T'[i_1, j] + T'[i_2, j]) > 1)$ **return** *false*;
return *true*;

The time used by this algorithm is polynomial, i.e. $O(n \times m^2) \leq O(m^3)$.

By considering other constraints, the total time is still $O(m^3)$, i.e. polynomial.

Now, Π is of the NP class.

Therefore, GRACAR is of the NP class.

In the same way, we determine that the checking time for GRACAG is also $O(m^3)$.

Theorem 6.4 is proved. ∎

Definition 6.12 The extended ILP (x-ILP) problem is a class of problems to find an $X = [x_0\ x_1\ ...\ x_{n'-1}]$ to obtain

$$\max \sum_{j=0}^{n'-1} c_j x_j$$

subject to

$$x_j \in \{0, 1\} \quad (0 \leq j < n') \tag{6.7}$$

$$\sum_{j=0}^{n'-1} a_{ij} x_j < b_i \quad (0 \leq i < m') \tag{6.8}$$

where "$<$" expresses "$<$", "$>$", "$=$", "$\neq$", "$\leq$" or "$\geq$".

Theorem 6.5 The x-ILP problem is NP-Complete.

[Proof]: Based on the method provided in (Garey and Johnson 1979, pp. 63–65), we prove by restriction. The x-ILP problem can be restricted to the ILP problem (Papadimitriou 1981; Rardin 1997; Schrijver 2000) by replacing "$<$" with "$\leq$", i.e.

$$\sum_{j=0}^{n'-1} a_{ij}x_j \leq b_i \ (0 \leq i < m') \tag{6.9}$$

Because the ILP problem is NP-Complete (Burkard et al. 2012), the x-ILP problem is NP-Complete.

Theorem 6.5 is proved.∎

Note that even though the expression of the ILP problem here is a little different from that in (Garey and Johnson 1979), it is common to express the ILP problem in the form of

$$max \ \sum_{j=0}^{n'-1} c_j x_j$$

subject to constraints (6.7) and (6.9) in other contributions (Papadimitriou 1981; Rardin 1997; Schrijver 2000).

To simplify descriptions, we let $C = [c_0, c_1, ..., c_{n'-1}]$, $B = [b_0, b_1, ..., b_{m'-1}]$, and $A = [a_{ij}](0 \leq i < m', 0 \leq j < n')$.

Theorem 6.6 The GRACAR problem is a subproblem of the x-ILP problem.

[**Proof**]: According to Garey and Johnson (1979), a problem Π can be expressed as $<D, Y>$, where D is the set of all the problem instances of Π and Y is the set of all the problem instances that can be successfully solved. Another problem Π' expressed as $<D', Y'>$ is a subproblem of Π if $D' \subseteq D$, and $Y' = Y \cap D'$.

Now, let x-ILP = Π and GRACAR = Π'.

We prove $\forall d' \in D' \rightarrow d' \in D$.

From a GRACAR problem d', we have L, Q, A^c, and T.

Let

$$n' = m \times n;$$
$$c_j = Q[\,j/m, j\%m]\ \ (0 \leq j < n');$$
$$x_j = T[\,j/m, j\%m]\ \ (0 \leq j < n');$$

We have $\sum_{j=0}^{n'-1} c_j x_j = \sum_{i=0}^{m-1} \sum_{j=0}^{n-1} Q[i,j] \times T[i,j]$.

That is, the objective of the ILP and that of the GRACAR problem is the same.

At the same time, constraint (6.1), $T[i,j] \in \{0, 1\}(0 \leq i < m, 0 \leq j < n)$ is the same as constraint (6.7), i.e. $x_j \in \{0,1\}, (0 \leq i < n')$.

Let

$$m' = n + m + n \times n_c.$$
$$b_j = L[j]\ (0 \leq j < n);$$
$$b_j = 1\ (n \leq j < m');$$

$a_{ij} = 0 \; (0 \leq i < m', 0 \leq j < n')$;

$a_{i(i+k \times n)} = 1 \; (0 \leq i < n, 0 \leq k < m)$;

$a_{i(k \times n + j\%n)} = 1 \; (n \leq i < m+n, 0 \leq j < n', 0 \leq k < n, k+n = i)$;

$a_{(m+n+i \times n + k)(i \times n + k)} = 1$

$(0 \leq i < m, \exists j (0 \leq j < m) \ni A^c[i,j] = 1, \; 0 \leq k < n)$; and

$a_{(m+n+i \times n + k)(j \times n + k)} = 1 \; (0 \leq i, j < m, A^c[i,j] = 1, 0 \leq k < n)$.

Then, constraints (6.2), (6.4), and (6.5) are the same as a constraint (6.8), i.e.

$$\sum_{j=0}^{n'-1} a_{ij} x_j = b_i \; (0 \leq i < n), \text{ and } \sum_{j=0}^{n'-1} a_{ij} x_j \leq b_i \; (n \leq i < m').$$

Now, a GRACAR problem is expressed as an x-ILP problem, i.e. $\forall d' \in D' \rightarrow d' \in D$.

$\because \forall d' \in D' \rightarrow d' \in D$.

$\therefore D' \subseteq D$.

At the same time, in the above transformation, no new constraints or conditions are introduced. That is to say, if problem d' is solvable in D', then it is solvable in D.

Now, let us prove $Y' = Y \cap D'$.

We define $S(d)$ as a predicate that d is solvable and use "$\leftrightarrow$" to express "is equivalent to."

$\because d \in Y \leftrightarrow d \in D \wedge S(d)$,

$d' \in Y' \leftrightarrow d' \in D' \wedge S(d')$, and

$d' \in D' \rightarrow d' \in D$,

$\therefore d' \in D' \wedge S(d') \rightarrow d' \in D \wedge S(d')$, i.e.

$\forall d' \in Y' \rightarrow d' \in D \wedge S(d') \wedge d' \in D'$.

$\because d' \in D \wedge S(d') \leftrightarrow d' \in Y$,

$\therefore \forall d' \in Y' \rightarrow d' \in Y \wedge d' \in D'$.

$\therefore Y' \subseteq Y \cap D'$.

$\because d' \in D \wedge S(d') \wedge d' \in D' \rightarrow d' \in D' \wedge S(d') \rightarrow d' \in Y'$,

$\therefore \forall d' \in D \wedge S(d') \wedge d' \in D' \rightarrow d' \in Y'$.

$\because d' \in D \wedge S(d') \leftrightarrow d' \in Y$,

$\therefore \forall d' \in Y \wedge d' \in D' \rightarrow d' \in Y'$,

i.e. $Y \cap D' \subseteq Y'$.

$\therefore Y' \subseteq Y \cap D'$.

Therefore, Theorem 6.6 is proved.∎

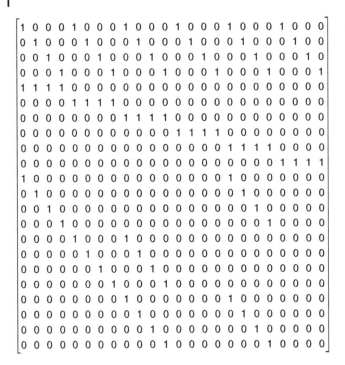

Figure 6.5 Matrix A for the GRACAR problem.

To better understand the proof, consider the example shown in Figure 6.3.

From Figures 6.3 and 6.4, $m = 6$, $n = 4$, $L = [2\ 1\ 1\ 2]$. Q and A^c are as shown in these figures.

Let $n' = m \times n = 24$;

$m' = n + m + n \times n_c = 4 + 6 + 4 \times 3 = 22$;

$$C = [0.71\ \ 0.6\ \ 0.0\ \ 0.22\ \ 0.29\ \ 0.67\ \ 0.44\ \ 0.76\ \ 0.69\ \ 0.92\ \ 0.92\ \ 0.6\ \ 0.0$$
$$0.0\ \ 0.53\ \ 0.0\ \ 0.97\ \ 0.51\ \ 0.77\ \ 0.65\ \ 0.58\ \ 0.64\ \ 0.24\ \ 0.0];$$

$$B = [2\ \ \ 1\ \ \ 1\ \ \ 2$$
$$1\ \ \ 1\ \ \ 1\ \ \ 1\ \ \ 1\ \ \ 1$$
$$1\ \ \ 1\ \ \ 1\ \ \ 1\ \ \ 1\ \ \ 1\ \ \ 1\ \ \ 1\ \ \ 1\ \ \ 1\ \ \ 1\ \ \ 1];\ \text{and}$$

A is shown in Figure 6.5.

Theorem 6.7 The GRACAG problem is a subproblem of the x-ILP problem.

[Proof]: We can use the technique as shown in the proof of Theorem 6.6 to prove that $D' \subseteq D$ and $Y' = Y \cap D'$.

The only additional part we need to prove is the process to determine A. The process is as follows:

Let

$$a_{ij} = 0 \ (0 \le i < m', 0 \le j < n');$$

$$a_{i(i + k \times n)} = 1 \ (0 \le i < n, 0 \le k < m);$$

$$a_{i(k \times n + j\%n)} = 1 \ (n' \le i < n + n', 0 \le j < n', 0 \le k < n, k + n' = i);$$

$$a_{(2n' + i \times n + k)(i \times n + k)} = 1$$
$$(0 \le i < m, \exists j(0 \le j < m) \ni A^c[i,j] = 1, \ 0 \le k < n); \text{ and}$$

$$a_{(2n' + i \times n + k)(i \times n + k + k')} = 1 \ (0 \le i, j < m, A^c[i,j] = 1, 0 \le k, k' < n).$$

Then, constraints 6.2, 6.4, and 6.5 are the same as constraint (6.8), i.e. $\sum_{j=0}^{n'-1} a_{ij} x_j = b_i \ (0 \le i < n)$, and

$$\sum_{j=0}^{n'-1} a_{ij} x_j \le b_i \ (n \le i < m').$$

Now, the GRACAG problem is expressed as an x-ILP problem.

Therefore, Theorem 6.7 is proved. ∎

For example, for the GRACAG problem shown in Figure 6.3, m', n', B, and C are the same as those for the GRACAR problem, and A is shown in Figure 6.6.

Figure 6.6 Matrix A for the GRACAG problem.

$$
\begin{bmatrix}
1&0&0&0&1&0&0&0&1&0&0&0&1&0&0&0&1&0&0&0&1&0&0&0\\
0&1&0&0&0&1&0&0&0&1&0&0&0&1&0&0&0&1&0&0&0&1&0&0\\
0&0&1&0&0&0&1&0&0&0&1&0&0&0&1&0&0&0&1&0&0&0&1&0\\
0&0&0&1&0&0&0&1&0&0&0&1&0&0&0&1&0&0&0&1&0&0&0&1\\
1&1&1&1&0&0&0&0&0&0&0&0&0&0&0&0&0&0&0&0&0&0&0&0\\
0&0&0&0&1&1&1&1&0&0&0&0&0&0&0&0&0&0&0&0&0&0&0&0\\
0&0&0&0&0&0&0&0&1&1&1&1&0&0&0&0&0&0&0&0&0&0&0&0\\
0&0&0&0&0&0&0&0&0&0&0&0&1&1&1&1&0&0&0&0&0&0&0&0\\
0&0&0&0&0&0&0&0&0&0&0&0&0&0&0&0&1&1&1&1&0&0&0&0\\
0&0&0&0&0&0&0&0&0&0&0&0&0&0&0&0&0&0&0&0&1&1&1&1\\
1&0&0&0&0&0&0&0&0&0&0&0&1&1&1&1&0&0&0&0&0&0&0&0\\
0&1&0&0&0&0&0&0&0&0&0&0&1&1&1&1&0&0&0&0&0&0&0&0\\
0&0&1&0&0&0&0&0&0&0&0&0&1&1&1&1&0&0&0&0&0&0&0&0\\
0&0&0&1&0&0&0&0&0&0&0&0&1&1&1&1&0&0&0&0&0&0&0&0\\
0&0&0&0&1&0&0&0&1&1&1&1&0&0&0&0&0&0&0&0&0&0&0&0\\
0&0&0&0&0&1&0&0&1&1&1&1&0&0&0&0&0&0&0&0&0&0&0&0\\
0&0&0&0&0&0&1&0&1&1&1&1&0&0&0&0&0&0&0&0&0&0&0&0\\
0&0&0&0&0&0&0&1&1&1&1&1&0&0&0&0&0&0&0&0&0&0&0&0\\
0&0&0&0&0&0&0&1&0&0&0&0&0&0&0&0&1&1&1&1&0&0&0&0\\
0&0&0&0&0&0&0&0&1&0&0&0&0&0&0&0&1&1&1&1&0&0&0&0\\
0&0&0&0&0&0&0&0&0&1&0&0&0&0&0&0&1&1&1&1&0&0&0&0\\
0&0&0&0&0&0&0&0&0&0&1&0&0&0&0&0&1&1&1&1&0&0&0&0
\end{bmatrix}
$$

Theorem 6.8 GRACAR/G problems can be transformed into an x-ILP problem in polynomial time.

[Proof]: We may estimate the time of the transformation from GRACAR to x-ILP using the following steps:

1) To form C, we use $O(m \times n) < O(m^2)$ due to $0 \le n < m$;
2) To form B:

$\because 0 \le n < m$ and $0 \le n_c < m \times (m - 1)/2,$

$\therefore O(m')$

$= O(n + m + n \times n_c)$

$\le O(m + m + m \times m \times (m - 1)/2)$

$= O(2m + m^3/2 - m^2/2)$

$< O(m^3/2 + m^3/2)$

$= O(m^3)$ if $m \ge 2$;

3) To form A:

$\because m' = n + m + n \times n_c,\ n' = m \times n,\ 0 \le n < m$ and $0 \le n_c < m \times (m - 1)/2,$

$\therefore O(m' \times n' + m \times n' + m^2 + m^2)$

$<\ O[(m + n + n \times m \times (m-1)/2)\ \times m \times n\ + m \times m \times n + 2m^2] < O(m^5)$ if $m \ge 3.$

In summary, the time used to transform GRACAR to x-ILP is $O(m^5)$.

We can use the same technique to estimate that the time used to transform GRA-CAR to x-ILP is also $O(m^5)$.

Therefore, Theorem 6.8 is proved. ∎

Note: Based on Theorems 6.4–6.8, we may not state that GRACAR/G are NP-Complete. However, based on our initial experiments (Zhu 2016), GRACAR/G are much more difficult than the GRA problem, of which an efficient solution is outlined in Chapter 5. We can state only that GRACAR/G are still *OPEN* problems based on the theory of Garey and Johnson (1979), i.e. they *may* or *may not* have efficient (in polynomial-time) solutions.

6.2.5 Solutions with CPLEX

From the above theoretical analysis, we know that the GRACAR/G problems are subproblems of the x-ILP problem. If we find an approach that solves the x-ILP problem in an acceptable amount of time, the GRACAR and GRACAG problems will have practical solutions. Fortunately, CPLEX is available to solve x-ILP problems.

Our experiments suggest that the CPLEX solution is useful and a problem with $m = 400$ can be solved in a second (Zhu 2016). Note that the 10 000 variables in the current ILOG package indicate a similar scale limitation.

6.3 Group Role Assignment with Cooperation and Conflict Factors

This section discusses a new GRA problem that considers both Cooperation and Conflict Factors (CCFs) (GRACCF) between agents. Intuitively, when an administrator assigns tasks to people, they may need to consider many factors, e.g. individual/collective performance, preferences, future conflict, and cooperation potentials. Task assignment can be conducted without considering such factors. However, such assignments may lead to difficulties or performance losses when people execute tasks. Our proposed method is to avoid future problems by careful assignments, i.e. to assign tasks to people with consideration of possible influence from people's conflict or cooperation factors.

To solve a GRACCF problem is to maximize group performance by assigning roles in consideration of cooperation and conflict factors. In fact, this problem is a significant extension of GRA with Conflicting Agents on Roles/Groups (GRACAR/G)) (Zhu 2016) (Section 6.2) that avoids conflicting agents being assigned to the same role/group. GRACCF considers both conflicts and cooperation, i.e. cooperating agents should be assigned together to play the same roles, but conflicting agents should not.

6.3.1 A Real-World Scenario

In company X, after creating Table 6.1 for the positions, Table 6.8 is created to include the candidate staff shortlist and the evaluation values of each candidate for each position using an evaluation method, e.g. Multi-Criteria Decision Making (MCDM) (Schöbel 2011, Shen et al. 2003). Then, Ann and Bob arranged a meeting with the branch officers to discuss the positions for candidate employees. The officers reported that based on the previous history, some employees work well together, while others conflict. This can be due to a number of reasons, such as personalities, working styles, emotional issues, and political beliefs. Subsequently, Ann asks the branch officers to evaluate the degrees of cooperation and conflict between the employees. The evaluation results are shown in Table 6.9.

In Table 6.9, the 0s mean no effect exists; the numbers less than 0 mean that there is conflict, but those larger than 0 indicate cooperation. The nonzero numbers mean the changing percentage of employees' qualification values, i.e. the cross point of "Edward (SP)(4, 1)" and "Bret (T)(1, 3)" is 0.3, meaning that the qualification value of Edward as a Senior Programmer (SP) increases by 30% if Bret takes the position of a Tester (T). In other words, employee pairs with good relationships will increase their performance while working on related jobs, whereas those pairs in conflict will decrease their performance.

Finally, Ann asks Bob to find the best assignment in consideration of the influences of cooperative and conflicting relations among employees. Bob informs Ann

Table 6.8 The candidates and evaluations on positions.[a]

Positions / Candidates	Project manager (0)	Senior programmer (1)	Programmer (2)	Tester (3)
Adam(0)	0.18	**0.82**	0.29	0.01
Bret(1)	0.35	**0.80**	0.58	0.35
Chris(2)	0.84	0.85	**0.86**	0.36
Doug(3)	0.96	0.51	0.45	**0.64**
Edward(4)	0.22	0.33	**0.68**	0.33
Fred(5)	0.96	0.50	0.10	**0.73**
George(6)	0.25	0.18	0.23	0.39
Harry(7)	0.56	0.35	**0.80**	0.62
Ice(8)	0.49	0.09	0.33	0.58
Joe(9)	0.38	0.54	**0.72**	0.20
Kris(10)	**0.91**	0.31	0.34	0.15
Larry(11)	0.85	0.34	0.43	0.18
Matt(12)	0.44	0.06	0.66	0.37

[a] Note: The numbers in parentheses are the indices of roles and agents, e.g. Fred is agent 5 and Tester is role 3.

that a significant amount of time may be required to formulate a satisfactory solution. Ann understands the situation and allows a reasonable response time. From the above scenario, Ann and Bob follow the initial steps of RBC but Bob encounters a GRA problem with the constraints of cooperation and conflict factors.

6.3.2 Problem Formalization

Some agents in a group may conflict when they play related roles. Conversely, other agents' performance may improve due to cooperation between agents.

Definition 6.13 A *cooperation and conflict Factor (CCF) matrix* C^f is an $(m \times n) \times (m \times n)$ matrix: $(A \times R) \times (A \times R) \rightarrow [-1, +1]$, where $C^f[i, j, i', j'] \in [-1, +1]$ expresses the changing degree of agent i's qualification value when agent i plays role j and agent i' plays role j' ($0 \leq i, i' \leq m - 1, 0 \leq j, j' \leq n - 1$), and $C^f[i, j, i', j'] > 0$ (<0) means cooperation (conflict), i.e. an increase (decrease) of the group qualification.

By this definition, the group qualification increases/decreases by a percentage of agent i's qualification when agent i plays role j and agent i' plays role j'. In other

Table 6.9 The cooperation or conflict factors[a].

Person (position) Person (position)	Adam (SP) (0,1)	Adam (P) (0,2)	Bret (SP) (1,1)	Bret (P) (1,2)	Bret (T) (1,3)	Edward (SP) (4,1)	Edward (P) (4,2)	Edward (T) (4,3)	Fred (PM) (5,0)	Fred (SP) (5,1)	Fred (T) (5,3)	Larry (PM) (11,0)	Larry (P) (11,2)	Matt (P) (12,2)	Matt (T) (12,3)
Adam (SP)(0,1)	0	0	-0.3	0.35	0.35	0	0	0	-0.4	0	0	0	0	0.8	0.9
Adam (P)(0,2)	0	0	-0.2	-0.2	0.2	0	0	0	-0.5	0	0	0	0	0.5	0.6
Bret (SP)(1,1)	-0.2	0.2	0	0	0	0.2	0.2	0.3	0	0	0	-0.3	0.35	0.7	0.6
Bret (P)(1,2)	-0.35	-0.2	0	0	0	0.2	0.2	0.3	0	0	0	-0.2	-0.2	0	0
Bret (T)(1,3)	0.35	0.4	0	0	0	0.2	0.2	0.2	0	0	0	-0.3	0.35	0	0
Edward (SP)(4,1)	0	0	0.2	0.2	0.3	0	0	0	-0.5	-0.4	-0.3	0	0	0.6	0.7
Edward (P)(4,2)	0	0	0.2	0.2	0.3	0	0	0	-0.4	-0.45	-0.3	0	0	0	0
Edward (T)(4,3)	0	0	0.2	0.2	0.2	0	0	0	-0.2	-0.2	-0.3	0	0	0	0
Fred (PM)(5,0)	0	0	0	0	0	0.3	0.2	0.3	0	0	0	0	0	0.8	0.7
Fred (SP)(5,1)	0	0	0	0	0	-0.4	-0.45	-0.2	0	0	0	0	0	0.6	0.5
Fred (T)(5,3)	0	0	0	0	0	-0.2	-0.2	-0.3	0	0	0	0	0	0	0
Larry (PM)(11,0)	0.3	0.2	-0.3	0.35	-0.3	0.3	0.2	0.1	-0.5	-0.4	-0.3	0	0	0.7	0.7
Larry (P)(11,2)	0	0	-0.2	-0.2	0.2	0	0	0	0	0	0	0	0	0.6	0.6
Matt (P)(12,2)	0	0	0	0	0	0	0	0	-0.4	0	0	0.8	0.6	0	0
Matt (T)(12,3)	0	0	0	0	0	0	0	0	-0.3	0	0	0.8	0.6	0	0

[a] Note: 1) The other person (position) pairs not in the table are all 0s; 2) PM: Project Manager, SP: Senior Programmer, P: Programmer, and T: Tester; 3) (i, j) means an assignment, e.g. (4, 2) means person 4 is assigned with position 2; and 4) The values are produced according to the same evaluation criteria.

words, agents i and i' affect each other when they play roles j and j'. The range of $[-1, +1]$ assumes that an agent can obtain up to double its original productivity from cooperation, or may be entirely unproductive due to conflicts.

The above assumption is reasonable and there are many such cases in reality. For example, in a software development team shown in Table 6.9, if Edward takes the position of *senior programmer* (row), he would be happy and increasing his performance by 20% when Bret takes the position of *programmer* (column). Edward would be even happier and increasing his performance by 30% if Bret plays the role of tester. However, if Bret takes the position of *tester*, he would be happy and increasing his performance by 20% if Edward plays a position of *senior programmer*, *programmer*, or *tester*. This case demonstrates that matrix C^f is asymmetrical.

Definition 6.14   Given Q, L, and C^f, the *GRA with Cooperation or Conflict Factors (GRACCF)* problem is to find a T to

$$
max \left\{ \sum_{i=0}^{m-1} \sum_{j=0}^{n-1} Q[i,j] \times T[i,j] + \sum_{i=0}^{m-1} \sum_{j=0}^{n-1} \sum_{i'=0}^{m-1} \sum_{j'=0}^{n-1} Q[i,j] \times \right.
$$

$$
\left. C^f[i,j,i',j'] \times T[i,j] \times T[i',j'] \right\}
$$

(6.10)

subject to (6.1), (6.2), and (6.4).

The above problem GRACCF presents a real-world problem that expresses the important aspect of cooperation/conflict. With the CCFs introduced, an individual may contribute to a team more or less with one assignment than with another. For example, individually, the performances of agents i_1 and i_2 playing role j are 0.7 and 0.8, respectively, i.e. $Q[i_1, j] = 0.7$, $Q[i_2, j] = 0.8$, and their simple sum is 1.5. If the cooperation factors of i_1 with i_2 on role j are 0.3 and 0.4, respectively, i.e. $C^f[i_1, j, i_2, j] = 0.3$ and $C^f[i_2, j, i_1, j] = 0.4$, then the sum of the real performances of the two individual agents is $0.7 + 0.8 + 0.7 \times 0.3 + 0.8 \times 0.4 = 2.03$. Definition 6.14 reveals the benefits of cooperation, i.e. "$1 + 1 > 2$"; Conversely, it also shows the drawbacks of conflict, i.e. "$1 + 1 < 2$."

The GRACCF problem is a nonlinear 0-1 programming problem (Balas and Mazzola 1984; Bradley et al. 1977; Dastani et al. 2003; Gonzaga 1995; Hansen et al. 1993; Kuhn 2014; Munkres 1957; Papadimitriou 1981; Rardin 1997; Wolsey and Nemhauser 1999). It is also a special case of the quadratic assignment problem (QAP) (Burkard 2013; Pitsoulis and Pardalos 2009). The Sahni and Gonzalez Theorem (Burkard 2013) states that the quadratic assignment problem is strongly NP-hard. Therefore, there is no general way to solve the GRACCF problem.

Definition 6.15 (Burkard 2013): The QAP is to

$$min\left\{\sigma = \sum_{\substack{i,k=0 \\ i\neq k}}^{n-1} \sum_{\substack{j,l=0 \\ j\neq l}}^{n-1} C[i,j,k,l] \times x[i,k] \times x[j,l] + \sum_{i,j=0}^{n-1} B[i,j] \times x[i,j]\right\}$$

(6.11)

subject to

$$x[i,j] \in \{0,1\} \quad (0 \leq i,j < n),$$ (6.12)

$$\sum_{i=0}^{n-1} x[i,j] = 1 \quad (0 \leq j < n),$$ (6.13)

$$\sum_{j=0}^{n-1} x[i,j] = 1 \quad (0 \leq i < n),$$ (6.14)

where C is a $(n \times n) \times (n \times n)$ matrix of cost values, $C[i,j,k,l]$ and $B[i,j]$ $(0 \leq i,j,k,l < n)$ are constants, and x is a $n \times n$ variable matrix.

Theorem 6.9 The GRACCF problem is strongly NP-hard.

[**Proof**]: Theorem 6.9 is proved if we can transform the GRACCF problem to the QAP form through restrictions and equivalent transformations (Garey and Johnson 1979).

Restriction 1: let $m = n$.

Restriction 2: let $L[j] = 1$ $(0 \leq j < n)$.

Restriction 3: let "$\leq$" be restricted to "=" in (6.14).

Let $C[i,j,k,l] = -C^f[i,j,k,l] \times Q[i,j]$ $(0 \leq i,j,k,l < n)$, then the time complexity is about $O(n^4)$.

Let $B[i,j] = -Q[i,j]$ $(0 \leq i,j < n)$, then the time complexity is about $O(n^2)$.

That is, the total time to transform the GRACCF problem to the QAP is within polynomial time, $O(n^4)$.

Because min (x) can be transferred to max$(-x)$, the objective of the QAP, i.e. (6.11), is equivalent to the objective of the GRACCF problem, i.e. (6.10).

Now, the GRACCF problem becomes a QAP. Because the GRACCF problem has become a QAP by restriction, the original GRACCF problem is more complex than the QAP. Note that the range of the coefficients in C does not affect the complexity of the problem.

Therefore, the GRACCF problem is strongly NP-hard.

Theorem 6.9 is proved. ∎

As a matter of fact, we tried the IBM ILOG CPLEX optimization studio (CPLEX) with Definition 6.14 but failed because CPLEX does not support such a nonlinear expression (6.10). It is worth noting that in most situations, a nonlinear objective function cannot be transformed to a linear one, and therefore no efficient method to solve the general GRACCF problem exists. However, we may find a special way to transform the objective of the GRACCF problem into a linear one.

Because the variables in the objective of the GRACCF problem only take 0 or 1 as their values, i.e. $T[i,j] \in \{0, 1\}$, we can transfer the objective into a linear one by logical expression transformations.

$T[i,j] \times T[i',j'] = 1$ if both $T[i,j]$ and $T[i',j']$ are 1, otherwise, $T[i,j] \times T[i',j'] = 0$. We may introduce additional variables $\overline{T}[i,j,i',j'] = T[i,j] \times T[i',j']$ and the constraint can be expressed as a linear expression. The constraints are as follows:

$$\overline{T}[i,j,i',j'] \in \{0,1\} \quad (0 \le i, i' < m, 0 \le j, j' < n) \tag{6.15}$$

$$2\overline{T}[i,j,i',j'] \le T[i,j] + T[i',j'] \le \overline{T}[i,j,i',j'] + 1 \quad (0 \le i, i' < m, 0 \le j, j' < n) \tag{6.16}$$

Now, we have a new expression of GRACCF.

Definition 6.16   the new expression of GRACCF is to

$$max \left\{ \sum_{i=0}^{m-1} \sum_{j=0}^{n-1} Q[i,j] \times T[i,j] + \sum_{i=0}^{m-1} \sum_{j=0}^{n-1} \sum_{i'=0}^{m-1} \sum_{j'=0}^{n-1} C^f[i,j,i',j'] \times \right.$$

$$\left. Q[i,j] \times \overline{T}[i,j,i',j'] \right\} \tag{6.17}$$

subject to (6.1), (6.2), (6.4), (6.15), and (6.16).

Therefore, we make the objective function linear by adding some variables that are related to C^f. The number of variables is determined by the number of nonzero values in C^f plus $m \times n$, that is, the number of variables in T.

Theoretically, C^f is a matrix that has a large dimension, i.e. $(m \times n) \times (m \times n)$. This dimension will increase the complexity of any solution. Therefore, any solution to the GRACCF problem with an LP solver is still very complex.

6.3.3   A Practical Solution

From a practical point of view, C^f is a sparse matrix. That is to say, only a limited number of agents may have significant cooperation or conflict factors. We use n_{cf} to express the number of elements in C^f that are not zeros. Now we define a new matrix to replace C^f.

Definition 6.17 A *Compact Cooperation and Conflict Factor matrix* C^{cf} is an $n_{cf} \times$ 5 matrix, where $C^{cf}[k,4] \in [-1, 0) \cup (0, 0]$ $(0 \le k < n_{cf})$ expresses that the changing degree of agent $C^{cf}[k, 0]$'s qualification on role $C^{cf}[k, 1]$ while agent $C^{cf}[k, 2]$ plays role $C^{cf}[k, 3]$, where n_{cf} is called *the significance number*, i.e. the total number of nonzero elements in C^f, i.e. $n_{cf} = \sum\limits_{i=0}^{m-1} \sum\limits_{j=0}^{n-1} \sum\limits_{i'=0}^{m-1} \sum\limits_{j'=0}^{n-1} \lceil |C^f[i,j,i',j']| \rceil$).

From Definition 6.17, we may simplify the $(m \times n) \times (m \times n)$ matrix $\overline{T}$ into an n_{cf} vector $\overline{T'}$.

Now, the GRACCF becomes:

$$max \left\{ \sum_{i=0}^{m-1} \sum_{j=0}^{n-1} Q[i,j] \times T[i,j] + \sum_{k=0}^{n^{cf}-1} C^{cf}[k,4] \times Q[C^{cf}[k,0], C^{cf}[k,1]] \times \overline{T'}[k] \right\}$$

$$(6.18)$$

subject to (6.1), (6.2), (6.4), and

$$\overline{T'}[k] \in \{0,1\} \quad (0 \le k < n_{cf}) \tag{6.19}$$

$$2\overline{T'}[k] \le T[C^{cf}[k,0], C^{cf}[k,1]] + T[C^{cf}[k,2], C^{cf}[k,3]] \quad (0 \le k < n_{cf}) \tag{6.20}$$

$$T[C^{cf}[k,0], C^{cf}[k,1]] + T[C^{cf}[k,2], C^{cf}[k,3]] \le \overline{T'}[k] + 1 \quad (0 \le k < n_{cf}) \tag{6.21}$$

Similar to the solution to GMRA, we use the CPLEX package to solve the GRACCF problem. The two major tasks are acquiring the four required elements and transforming the objective and the constraints into the forms required by CPLEX.

Step 1: determine the four elements (objective function coefficients, constraint coefficients, right-hand-side constraint values; and upper and lower bounds) required by the CPLEX package. Specifically, we use Q, L, C^{cf}, T, and $\overline{T'}$ to define an LP problem in CPLEX. In this case, matrices Q and C^{cf} express the objective function coefficients, and T and $\overline{T'}$ express the variables. The upper and lower bounds of both T and $\overline{T'}$ are 1 and 0, respectively.

Step 2: add the objective and constraint expressions.

The objective of the GRACCF problem can be expressed by a formula formed by the one-dimensional array forms of matrices Q and T and a linear expression of C^{cf} and $\overline{T'}$. In the CPLEX package, we can maximize this formula based on the objective.

The final optimized assignment calculated with CPLEX for the GRACCF problem mentioned in Section 6.3.1 is shown in Table 6.8 as the <u>underlined numbers</u>, i.e. a tuple set as: {<Adam, Senior Programmer>, <Bret, Tester>, <Chris, Programmer>, <Doug, Tester>, <Edward, Senior Programmer>, <Harry, Programmer>, <Joe, Programmer>, <Larry, Project Manager>, and <Matt, Programmer>}. From Table 6.8, the total sum of the assigned evaluation values plus the effect of CCFs is 9.46. However, by GRA (**bold**), the total sum (6.96) plus the CCFs (negative, $C^f[0,1,1,1] \times Q[0,1] + C^f[1,1,0,1] \times Q[1,1] = -0.3 \times 0.82 - 0.2 \times 0.80 = -0.406$) is 6.554. This scenario clearly demonstrates the significance of GRACCF problems. Furthermore, the cooperation and conflict factor is shown to considerably affect group performance, i.e. $(9.45 - 6.554)/6.554 = 44.19\%$.

6.3.4 Performance Experiments

To verify the practicality of the proposed approach, we conduct performance experiments using the platform shown in Table 6.4. The first experiment is conducted to present the trend of the consumed time. In each step, we repeated the test for 100 rounds. In each round, Q, L, and C^{cf} are randomly generated (uniform distributions), where m changes from 10 to 200 with an increment of 10 per step. To compare the impact of the ratio of n/m on performance, we form two groups of tests with n/m ratios of 1/3 and 1/4, respectively. To determine the influence of the number of nonzero CCFs, we set n_{cf} as $5m$ and $7m$, respectively.

Based on the results, we confirm that the proposed solution based on CPLEX is practical, that is, a problem with $m = 200$ can be solved in two seconds (Zhu et al. 2018). From the results, it is clear that n_{cf} affects the performance significantly, i.e. a larger n_{cf} results in a longer computation time. Note that the ranges for n, n_{cf}, and $L[j](0 \leq j < n)$ are set to guarantee a feasible solution for each random group.

From the results, we verify that the most important parameters affecting the performance of the proposed solution are m and n_{cf}, i.e. m and n_{cf} determine the complexity scale of the GRACCF problem.

In the experiments, fluctuations in the maximum curve mean that sometimes, solving a problem with a larger m may require less time than that of a problem with a smaller m. These fluctuations are present because the corresponding groups have different parameter configurations in m, n, L, Q, and C^{cf} (Zhu et al. 2018).

The processing time is between 2.5 and 5.2 seconds in an experiment with 100 random groups with $Q[i,j] \in [0, 1]$ $(0 \leq i < m, 0 \leq j < n)$ and $C^{cf}[k,4] \in [-1, -0.01] \cup [0.01, 1]$ $(0 \leq k < n_{cf})$ (Zhu et al. 2018).

The second experiment (Figure 6.7) studies the processing time for a group with $m = 50$, but the number of rows of the compact CCF matrix ranges from 500 to 2000 with an increment of 250 for each step. Based on all the above experiments, we conclude that the proposed approach is practical.

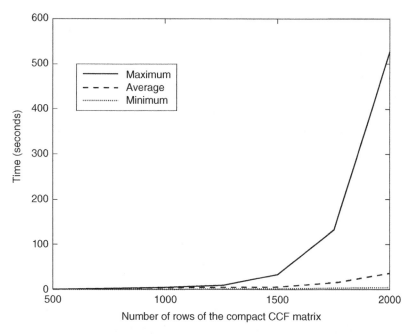

Figure 6.7 Processing time of groups with $m = 50$, $n = m/4$, $0 \leq L[j] \leq 12$, and $0 \leq j < n$.

6.3.5 The Benefits

To quantitatively assess the benefits of considering both cooperation and conflict factors in GRA, we conducted simulations on the same platform shown in Table 6.4 using random groups. Two typical sizes, i.e. ($m = 30$, $n = 5$) and ($m = 100$, $n = 16$), were chosen for the simulation without a loss of generality. These two group sizes are very common in small- and mid-sized companies or factories. In this study, we are concentrating only on the benefits of GRACCF, not the efficiency. As in Definition 6.8, $n_a = \sum_{j=0}^{n-1} L[j]$.

For each group, we randomly generated data with the following ranges:

- $Q[i,j] \in [0, 1]$ ($0 \leq i < m$, $0 \leq j < n$), i.e. the qualification value of each agent for each role;
- $C^{cf}[k,4] \in [-1, -0.01] \cup [0.01, 1]$ ($0 \leq k < n_{cf}$), i.e. the cooperation and conflict factors; and
- $L[j]$ ($0 \leq j < n$), $1 \leq L[j] \leq m/n$ to make $m \geq n_a$.

For each group, we collected several statistics:

- σ_1, the maximum group performance with GRA;

- σ_2, the maximum group performance with GRA after considering cooperation and conflict factors, i.e. the real group performance with GRA; and
- σ_3 the maximum group performance with GRACCF.

The benefit λ is calculated as $(\sigma_3 - \sigma_2)/\sigma_2$.

To observe the differences among different configurations, we also changed *the significance number $n_{cf} = m \times x$*, where x is from 1 to 10 with an increment of 1 for each step. To make the data more realistic, 100 groups were generated and used for each n_{cf}. Every assignment considers a random CCF corresponding to the significance number n_c. Finally, we computed the averages of the 100 groups for each set of statistics. By observing the simulation results, CCFs significantly affect group performance. Group performance increases between 2% and 38% based on the configurations of the groups (Zhu et al. 2018).

To check the effect of the range of the CCFs, we made a new experiment with $C^{cf}[k,4] \in [-0.9, -0.2] \cup [0.2, 0.9]$ $(0 \leq k < n_c)$ using the same configurations as the above case. The results of the new experiment are almost identical to the above. This follows common sense that the random creation of C^{cf} would make the elements of cooperation and conflict factors evenly distributed. If only cooperation factors are considered, higher cooperation factors mean better performance. Parameter n_{cf} affects the benefits significantly, i.e. a larger n_{cf} means more benefits regardless of the configuration. In other words, if you allow more people to present their opinions in role assignment, the group performance increases more. However, it is not a trivial matter to collect CCFs based on a large n_c. A larger n_{cf} increases the complexity of the GRACCF problem and the burden of collecting the CCFs (Zhu et al. 2018).

6.3.6 Cooperation and Conflict Factor Collection

GRACCF pursues optimized group performance by considering cooperation and conflict among agents. The solution to the GRACCF problem has been verified to have significant benefits based on the above simulations. However, the solution to a GRACCF problem is based on the establishment of matrix C^{cf} and we must provide a feasible way to collect CCFs in order to apply the proposed GRACCF solution.

We propose using a questionnaire to collect the opinions of all the group members to form the C^{cf} matrix. An example of the questionnaire for an enterprise is shown in Figure 6.8. These collected data are then transformed into the compact matrix C^{cf}. In designing the questionnaire, we recommend an n_{cf} of 5 to 10 times of m, i.e. employees should be asked to fill in the form for 5 to 10 <employee, position> tuples that significantly affect their work in the team. Although a GRACCF problem with a larger n_{cf} may provide more benefits, the time, complexity, and effort of processing the survey must be considered.

Name: ___(#3)___ .

If I work in __(#1)__ , I __(#2)__ to cooperate with __(#3)__ in __(#1)__ .
If I work in __(#1)__ , I __(#2)__ to cooperate with __(#3)__ in __(#1)__ .
If I work in __(#1)__ , I __(#2)__ to cooperate with __(#3)__ in __(#1)__ .
If I work in __(#1)__ , I __(#2)__ to cooperate with __(#3)__ in __(#1)__ .
If I work in __(#1)__ , I __(#2)__ to cooperate with __(#3)__ in __(#1)__ .
If I work in __(#1)__ , I __(#2)__ to cooperate with __(#3)__ in __(#1)__ .
If I work in __(#1)__ , I __(#2)__ to cooperate with __(#3)__ in __(#1)__ .
If I work in __(#1)__ , I __(#2)__ to cooperate with __(#3)__ in __(#1)__ .

#1: a) Position A b) Position B c) Position C d) Position D
#2: a) Strongly like b) like c) weakly like d) weakly dislike e) dislike f) strongly dislike
#3: a) Adam b) Brett c) Chris d) Doug e) Edward f) Fred g) George h) Harry i) Ice j) Joe

Figure 6.8 An empty questionnaire.

This proposed method of CCF collection is verified by a case study. The survey was sent to all the graduate students in the Department of Automatic Control and System Engineering at Nanjing University, China. Ten replies were received. Intuitively, this is a reasonable and practical number of replies for a team with up to 50, or even 100 people, because most people only care what positions they take but not the positions others may take. The rate of replies does not affect the GRACCF result very much because it is believed that the qualifications (performances) of those people who do not reply are not affected by others' role assignments.

In the questionnaire, each employee should select an option from the numbered list and fill in the corresponding blanks. In most cases, it is difficult for employees to express their cooperation and conflict factors mathematically. As a result, we use some fuzzy descriptions, e.g. strongly like, like, weakly like, etc. In fact, the questionnaire borrows the ideas from the Likert-type scale. However, in our questionnaire, we hope to create values that reflect the factors of cooperation or conflict. In our proposed method, we assign "*weakly like*" with a value of 0.1 to express cooperation but "*weakly dislike*" with a value of −0.1 to express conflict. The neutral Likert-type scale can be reflected with "no answer" in our questionnaire.

In the questionnaire, we use $n_c = 8m$. Figure 6.9 is an example of a returned survey. Note that the surveys in Figures 6.8 and 6.9 were translated from Chinese into English. Specifically, the Chinese names in the survey were replaced with English names for comprehension and anonymity.

In the survey, positions mean roles, and employees mean agents. In order to generate a C^{cf}, fuzzy descriptions such as "strongly like" must be transformed into numerical CCFs. There are a variety of transformation rules. We are using the rules shown in Table 6.10.

Name: #3 a .

If I work in #1 c , I #2 b to cooperate with #3 e in #1 c .
If I work in #1 c , I #2 b to cooperate with #3 c in #1 c .
If I work in #1 c , I #2 e to cooperate with #3 d in #1 b .
If I work in #1 a , I #2 b to cooperate with #3 j in #1 c .
If I work in #1 d , I #2 e to cooperate with #3 j in #1 a .
If I work in #1 b , I #2 e to cooperate with #3 e in #1 a .
If I work in #1 d , I #2 b to cooperate with #3 b in #1 a .
If I work in #1 c , I #2 e to cooperate with #3 g in #1 d .

#1: a) Position A b) Position B c) Position C d) Position D
#2: a) strongly like b) like c) weakly like d) weakly dislike e) dislike f) strongly dislike
#3: a) Adam b) Brett c) Chris d) Doug e) Edward f) Fred g) George h) Harry i) Ice j) Jeo

Figure 6.9 A returned questionnaire.

Table 6.10 Rules of transferring fuzzy descriptions to numbers.

Intentions	Number
Strongly like	0.9
Like	0.5
Weakly like	0.1
Weakly dislike	−0.1
Dislike	−0.5
Strongly dislike	−0.9

Table 6.11 shows a part of the final CCFs collected and transformed from the surveys in our case study. Using the CCFs, we are able to solve the GRACCF problem for our team.

6.4 Related Work

Assignment problems have been investigated extensively in the past decades (Bhardwaj and Chandrakasan 2002; Burkard 2013; Burkard et al. 2012; Kuhn 2005; Kumar et al. 2013; Munkres 1957; Pitsoulis and Pardalos 2009; Shen et al. 2003). However, the problems in this chapter have not been investigated due to different research and modeling methodologies in the literature. This situation

Table 6.11 Part of the collected C^{cf}.

$C^{cf}[k,0]$	$C^{ce}[k,1]$	$C^{cf}[k,2]$	$C^{cf}[k,3]$	$C^{cf}[k,4]$
0	2	4	2	0.5
0	2	2	2	0.5
0	2	3	1	−0.5
0	0	2	2	0.5
0	3	9	0	−0.5
0	1	5	0	−0.5
0	3	1	0	0.5
0	3	6	3	−0.5

demonstrates that RBC and E-CARGO (Sheng et al. 2016; Zhu 2011 2012, 2016; Zhu and Zhou 2006; Zhu et al. 2012a, b) are promising to discover and extend the research of assignment problems.

Past research on role assignment focuses on organizational performance (Ashforth 2001), system modeling (Odell et al. 2003), system design (Canbaz et al. 2014; Hou et al. 2014; Hou et al. 2011), system management (Bradley et al. 1977; Shen et al. 2003), robot task allocations (Choi et al. 2010a, b; Stone and Veloso 1999; Vail and Veloso 2003), multi-agent systems (Dastani et al. 2003; Ferber et al. 2004; Jumadinova et al. 2014), and nodes in networked systems (Bhardwaj and Chandrakasan 2002; Hui and Zhang 2015). Some others are concerned with conflict management in multi-agent systems (Bristow et al. 2014; Canbaz et al. 2014; Kilgour and Hipel 2005; Klein 1991; Malsch and Weiss 2002; Nyanchama and Osborn 1999; Tessier 2002; Walker et al. 2013).

It is noteworthy to mention that Burkard et al. published a textbook on assignment problems in 2012. The book presents and discusses the assignment problems known in research before 2012. Our proposed GRA, GMRA, and GRACCF problems were not covered in this book. This result unveils the fact that RBC provides an innovative way to investigate abstract assignment problems in the sense of collaboration based on the E-CARGO model. This fact indicates a strong need to investigate the GRA$^+$ problems discussed in this chapter.

6.5 Summary

In this chapter, we discussed several types of GRA problems called GRA$^+$. We are using GRA$^+$ to emphasize that there are many more constraints in real-world GRA problems than those discussed in Chapter 5.

The GMRA problem adds a new data structure L^a and replaces the constraint of GRA (5.3) with (6.3). The GRACAR adds a new data structure A^c and a new constraint (6.5). The GRACCF is much more complex than GRA and is one of the instances of GRA⁺. The major extension of GRACCF is that we replace A^c with a new data structure C^f, which is more complex than A^c. Simply, A^c is an $m \times m$ matrix, while C^f is an $(m \times n) \times (m \times n)$ matrix. Fortunately, C^f is a sparse matrix and we can simplify it into a compact data structure to improve its problem-solving efficiency.

We may also note that GRA⁺ significantly extends the representability of general assignment (Burkard et al. 2012) problems. This facilitates the investigation of more related problems and the application of GRA in the industry.

GRA⁺ covers a wide range of assignment problems that can be specified by the E-CARGO model using specific data structures to include extra information to express a variety of different constraints.

References

Ashforth, B.E. (2001). *Role Transitions in Organizational Life: An Identity-Based Perspective*. Mahwah, NJ: Lawrence Erlbaum Associates, Inc.

Balas, E. and Mazzola, J.B. (1984). Nonlinear 0-1 programming: I. Linearization techniques. *Mathematical Programming* 30: 1–21.

Bhardwaj, M. and Chandrakasan, A.P. (2002). Bounding the lifetime of sensor networks via optimal role assignments. *Proceedings of Twenty-First Annual Joint Conference of the IEEE Computer and Communications Societies (INFOCOM 2002)*, New York, USA (June 2002), vol. 3, pp. 1587–1596.

Bradley, S.P., Hax, A., and Magnanti, T. (1977). *Applied Mathematical Programming*. Boston, MA: Addison-Wesley.

Bristow, M., Fang, L., and Hipel, K.W. (2014). Agent-based modeling of competitive and cooperative behavior under conflict. *IEEE Transactions on Systems, Man, and Cybernetics: Systems* 44 (7): 834–850.

Burkard, R.E. (2013). Quadratic assignment problems. In: *Handbook of Combinatorial Optimizations* (eds. P.M. Pardalos, D.-Z. Du and R.L. Graham), 2741–2814. New York: Springer.

Burkard, R.E., Dell'Amico, M., and Martello, S. (2012). *Assignment Problems (Revised Reprint)*. Philadelphia, PA: SIAM.

Canbaz, B., Yannou, B., and Yvars, P.-A. (2014). Resolving design conflicts and evaluating solidarity in distributed design. *IEEE Transactions on Systems, Man, and Cybernetics: Systems* 44 (8): 1044–1055.

Choi, H.L., Brunet, L., and How, J.P. (2010a). Consensus-based decentralized auctions for robust task allocation. *IEEE Transactions on Robotics* 25 (4): 912–926.

Choi, H.L., Whitten, A.K., and How, J.P. (2010b). Decentralized task allocation for heterogeneous teams with cooperation constraints. *American Control Conference (ACC)*, Baltimore, MD (30 June 30–2 July 2010), pp. 3057–3062.

Dastani, M., Dignum, V., and Dignum, F. (2003). Role-assignment in open agent societies. *Proceedings of the Second Int'l Joint Conference on Autonomous Agents and Multiagent Systems*, Melbourne, Australia (14–18 July 2003), pp. 489–496.

Ferber, J., Gutknecht, O., and Michel, F. (2004). From agents to organizations: an organizational view of multi agent systems. In: *Agent-Oriented Software Engineering (AOSE) IV*, Melbourne, LNCS 2935 (eds. P. Giorgini, J. Müller and J. Odell), 214–230. Berlin, Heidelberg, Germany: Springer.

Garey, M.R. and Johnson, D.S. (1979). *Computers and Intractability: A Guide to the Theory of NP-Completeness*. New York, NY: W. H. Freeman and Comp.

Gonzaga, C.C. (1995). On the complexity of linear programming. *Rensenhas IME-USP Journal* 2 (2): 197–207.

Guo, W., Healy, W.M., and Zhou, M.C. (2012). Impacts of 2.4-GHz ISM band interference on IEEE 802.15.4 wireless sensor network reliability in buildings. *IEEE Transactions on Instrumentation and Measurement* 61 (9): 2533–2544.

Hansen, P., Jaumard, B., and Mathon, V. (1993). State-of-the-art survey—constrained nonlinear 0–1 programming. *ORSA Journal on Computing* 5 (2): 97–119.

Hou, M., Zhu, H., Zhou, M., and Arrabito, G.R. (2011). Optimizing operator-agent interaction in intelligent adaptive interface design: a conceptual framework. *IEEE Transactions on Systems, Man and Cybernetics, Part C: Applications and Reviews* 41 (2): 161–178.

Hou, M., Banbury, S., and Burns, C. (2014). *Intelligent Adaptive Systems: An Interaction-Centered Design Perspective*. Boca Raton, FL: CRC Press, Taylor and Francis Group.

Hui, Q. and Zhang, H. (2015). Optimal balanced coordinated network resource allocation using swarm optimization. *IEEE Transactions on Systems, Man, and Cybernetics: Systems* 45 (5): 770–787.

IBM (2019). ILOG CPLEX optimization studio. http://www-01.ibm.com/software/integration/optimization/cplex-optimization-studio/.

Jumadinova, J., Dasgupta, P., and Soh, L.-K. (2014). Strategic capability-learning for improved multi-agent collaboration in ad hoc environments. *IEEE Transactions on Systems, Man, and Cybernetics: Systems* 44 (8): 1003–1014.

Kilgour, D.M. and Hipel, K.W. (2005). The graph model for conflict resolution: past, present, and future. *Group Decision and Negotiation* 14 (6): 441–460.

Kinsara, R.A., Kilgour, D.M., and Hipel, K.W. (2015). Inverse approach to the graph model for conflict resolution. *IEEE Transactions on Systems, Man, and Cybernetics: Systems* 45 (5): 734–742.

Klein, M. (1991). Supporting conflict resolution in cooperative design systems. *IEEE Transactions on Systems, Man, and Cybernetics* 21 (6): 1379–1390.

Kuhn, H.W. (2005). The Hungarian method for the assignment problem. *Naval Research Logistic Quarterly* 2, 1955: 83–97. (Reprinted in vol. 52, no. 1, 2005, pp. 7 – 21.

Kuhn, H.W. (2014). Nonlinear programming: a historical view. In: *Traces and Emergence of Nonlinear Programming* (eds. G. Giorgi and T.H. Kjeldsen), 393–414. Basel: Springer.

Kumar, A., Dijkman, R., and Song, M. (2013). Optimal resource assignment in workflows for maximizing cooperation. *Lecture Notes in Computer Science* 8094: 235–250.

Lyamin, A.V. and Sloan, S.W. (2002). Lower bound limit analysis using non-linear programming. *International Journal for Numerical Methods in Engineering* 55 (5): 573–611.

Malsch, T. and Weiss, G. (2002). Conflicts in social theory and multi-agent systems. In: *Conflicting Agents* (eds. C. Tessier, L. Chaudron and H.-J. Müller), 111–149. Springer.

Munkres, J. (1957). Algorithms for the assignment and transportation problems. *Journal of the Society for Industrial and Applied Mathematics* 5 (1): 32–38.

Nyanchama, M. and Osborn, S. (1999). The role graph model and conflict of interest. *ACM Transactions on Information and System Security* 2 (1): 3–33.

Odell, J.J., Van Dyke Parunak, H., Brueckner, S., and Sauter, J. (2003). Changing roles: dynamic role assignment. *Journal of Object Technology* 2 (5): 77–86.

Papadimitriou, C.H. (1981). On the complexity of integer programming. *Journal of the ACM* 28 (4): 765–768.

Pitsoulis, L. and Pardalos, P.M. (2009). Quadratic assignment problem. In: *Encyclopedia of Optimization* (eds. L. Pitsoulis and P.M. Pardalos), 3119–3149. New York: Springer-Verlag.

Pressman, R.S. (2004). *Software Engineering: A Practitioner's Approach*, 6e. New York, NY: McGraw-Hill Higher Education.

Rapoport, A. and Chammah, A.M. (1965). *Prisoner's Dilemma: A Study in Conflict and Cooperation*, vol. 165. Ann Arbor, MI: University of Michigan Press.

Rardin, R.L. (1997). *Optimization in Operations Research*. Upper Saddle River, NJ: Prentice Hall.

Schöbel, A. (2011). On the similarities of some multi-criteria decision analysis methods. *Journal of Multi-Criteria Decision Analysis* 18 (3–4): 219–230.

Schrijver, A. (2000). *Theory of Linear and Integer Programming*. John Wiley & Sons.

Shen, M., Tzeng, G.-H., and Liu, D.-R. (2003). Multi-criteria task assignment in workflow management systems. *Proceedings of the 36th Annual Hawaii Int'l Conference on System Sciences*, Hawaii, USA (6–9 January 2003), pp. 1–9.

Sheng, Y., Zhu, H., Zhou, X., and Hu, W. (2016). Effective approaches to adaptive collaboration via dynamic role assignment. *IEEE Transaction on Systems, Man, and Cybernetics: Systems* 46 (1): 76–92.

Stone, P. and Veloso, M. (1999). Task decomposition, dynamic role assignment, and low-bandwidth communication for real-time strategic teamwork. *Artificial Intelligence* 110: 241–273.

Tessier, C., Müller, H.J., Fiorino, H., and Chaudron, L. (2002). Agents' conflicts: new issues. In: *Conflicting Agents* (eds. C. Tessier, L. Chaudron and H.-J. Müller), 1–30. Springer.

Tomlin, C., Pappas, G.J., and Sastry, S. (1998). Conflict resolution for air traffic management: a study in multiagent hybrid systems. *IEEE Transactions on Automatic Control* 43 (4): 509–521.

Vail, D. and Veloso, M. (2003). Multi-robot dynamic role assignment and coordination through shared potential fields. In: *Multi-Robot Systems* (eds. A. Schultz, L. Parkera and F. Schneider), 87–98. Kluwer.

Walker, S.B., Hipel, K.W., and Xu, H. (2013). A matrix representation of attitudes in conflicts. *IEEE Transactions on Systems, Man, and Cybernetics: Systems* 43 (6): 1328–1342.

Wolsey, L.A. and Nemhauser, G.L. (1999). *Integer and Combinatorial Optimization*. New York: Wiley-Interscience.

Zhu, H. (2011). Group role assignment with conflicting agent constraints. *Proceedings of the ACM/IEEE Int'l Symposium on Collaborative Technologies and Systems*, Philadelphia, PL, USA (23–27 May 2011), pp. 431–439.

Zhu, H. (2012). Role assignment for an agent group in consideration of conflicts among agents. In: *The Proceedings of Canadian Conference on Artificial Intelligence*, vol. 7310 (eds. L. Kosseim and D. Inkpen): Canadian AI 2012, 267–279. Toronto, Canada: LNAI.

Zhu, H. (2016). Avoiding conflicts by group role assignment. *IEEE Transactions on Systems, Man, and Cybernetics: Systems* 46 (4): 535–547.

Zhu, H. and Alkins, R. (2009). Group role assignment. *Proceedings of the ACM/IEEE Int'l Symposium on Collaborative Technologies and Systems*, Baltimore, MA (May 2009), pp. 431–439.

Zhu, H. and Feng, L. (2013). An efficient approach to group role assignment with conflicting agents. *The IEEE Int'l Conference on Computer-Supported Cooperative Work in Design (CSCWD)*, Whistler, Canada (27–29 June 2013), pp. 145–152.

Zhu, H. and Zhou, M.C. (2006). Role-based collaboration and its Kernel mechanisms. *IEEE Transactions on Systems, Man and Cybernetics, Part C* 36 (4): 578–589.

Zhu, H. and Zhou, M. (2009). M–M role-transfer problems and their solutions. *IEEE Transactions on Systems, Man and Cybernetics, Part A: Systems and Humans* 39 (2): 448–459.

Zhu, H. and Zhou, M.C. (2012). Efficient role transfer based on Kuhn–Munkres algorithm. *IEEE Transactions on Systems, Man and Cybernetics, Part A: Systems and Humans* 42 (2): 491–496.

Zhu, H., Hou, M., and Zhou, M.C. (2012a). Adaptive collaboration based on the E-CARGO model. *International Journal of Agent Technologies and Systems* 4 (1): 59–76.

Zhu, H., Zhou, M.C., and Alkins, R. (2012b). Group role assignment via a Kuhn-Munkres algorithm-based solution. *IEEE Transactions on Systems, Man and Cybernetics, Part A* 42 (3): 739–750.

Zhu, H., Sheng, Y., and Zhou, X. (2015). The benefits of conflict avoidance in collaboration. *IEEE 14th Int'l Conference on Cognitive Informatics & Cognitive Computing (ICCI∗CC)*, Beijing, China (6–8 July 2015), pp. 221–229.

Zhu, H., Sheng, Y., Zhou, X.-Z., and Zhu, Y. (2018). Group role assignment with cooperation and conflict factors. *IEEE Transactions on Systems, Man, and Cybernetics: Systems* 48 (6): 851–863.

Exercises

1 What do we mean by L^a, A^c, C^f, and C^{cf}?

2 What do we mean by n_c, n_a, n_{ac}, n_{cf}, and p_c?

3 Why is $m - n_c \geqslant n_a$ a sufficient condition for GRACAR to have a feasible solution?

4 Why is solving GRACAR more beneficial than solving GRA?

5 Could you write the formalization of GMRA, GRACAR, GRACAG, and GRACCF?

6 Why are the problems in this chapter categorized as GRA⁺?

7 What is a QAP? Why does GRACCF belong to QAP?

8 What is the major difference between GRACAR and GRACCF?

9 Why can GRACCF be transferred to a linear programming problem?

10 If we extend the conflict matrix A^c, where $A^c[i_1, i_2] \in [0, 1](0 \leq i_1, i_2 < m)$, of GRACAR to a new matrix $A^{c'}$ where $A^{c'}[i_1, i_2] \in [-1, 1]$ $(0 \leq i_1, i_2 < m)$, may we define a new GRA⁺? Justify your answer.

7

Group Role Assignment with Multiple Objectives (GRA^{++})

7.1 Group Role Assignment with Budget Constraints (GRABC)

In general, group performance is an indicator of excellence for the group to accomplish a task. In Group Role Assignment (GRA), group performance is a simple sum of all the performances of individual agents on their assigned roles in the group. An agent's performance on a role is expressed by its qualification value. Many factors are found to have an impact on the group's performance. Budget limits are such a factor (Andelman and Mansour 2004; Bhattacharya and Dupas 2012; Cao et al. 2014; Karabakal et al. 2000; Sakellariou et al. 2007; Yu and Buyya 2006). This section discusses a new GRA problem that considers the team budget. To solve the Group Role Assignment with Budget Constraints (GRABC) problem is to maximize the group performance while minimizing the budget.

GRABC (Zhu 2020b) is different from GRA (Zhu et al. 2012a, b, Chapter 5), GRA with Constraints (GRA^{+}) (Zhu 2016; Zhu et al. 2016, 2017, 2018; Chapter 6) due to the introduction of the multiple objectives. The existing solutions to the GRA^{+} problems cannot be used to solve the GRABC problem without a significant extra effort.

7.1.1 A Real-World Scenario

In a training school, Ann, the Principal, has recently developed a summer curriculum in Advanced Computing. She asks Bob, the Human Resources (HR) officer, to recruit a team of instructors for the school. Bob drafts a position list shown in Table 7.1 for the team and posts a call for applications on the school website. After a month, a candidate shortlist is gathered and shown in Table 7.2, where the qualification values are obtained by matching the required skills of each position with the curriculum vitae of the candidates. In Table 7.2, the candidate evaluation value

E-CARGO and Role-Based Collaboration: Modeling and Solving Problems in the Complex World, First Edition. Haibin Zhu.

Table 7.1 The required positions.

Position	Big data	Artificial intelligence	Cloud computing	Software engineering
Required number	1	2	4	2

Table 7.2 The candidates and evaluations on positions.

	Positions			
Candidates	Big data	Artificial intelligence	Cloud computing	Software engineering
Adam	0.18	**(0.82)**	0.29	0.01
Bret	0.35	0.80	[0.58]	0.35
Chris	0.84	**(0.85)**	**(0.86)**	0.36
Doug	**(0.96)**	0.51	0.45	**(0.64)**
Edward	0.22	0.33	**(0.68)**	0.33
Fred	0.96	0.50	0.10	**(0.73)**
George	0.25	0.18	0.23	0.39
Harry	0.56	[0.35]	**(0.80)**	0.62
Ice	0.49	0.09	[0.33]	[0.58]
Joe	[0.38]	0.54	**(0.72)**	0.20
Kris	0.91	0.31	[0.34]	0.15
Larry	0.85	[0.34]	[0.43]	0.18
Matt	0.44	0.06	0.66	[0.37]

is a dimensionless number. It expresses the degree of excellence for an agent to work on a role. It can be understood as the grade of a student in a course.

Using the prediction of the tuitions and the budget distribution shown in Table 7.3, where the stipend is counted in terms of $10 000, or $10K, Ann requests Bob to (i) maximize the group performance while minimizing the total requested stipends; (ii) assign each instructor at most two courses; and (iii) assign at most one section of the same course due to parallel sections. Then, Bob sends a notification email to all the shortlisted candidates and asks for their desired stipends for each position. The collected requests are shown in Table 7.4, where the stipend is counted in terms of $10K. For example, Adam requests $10 000 to teach the course of Big Data and Chris requests $40 000 to teach the course of Artificial Intelligence.

Bob notices that in most cases, the candidates with higher qualifications request higher stipends. He tells Ann that a significant processing time may be required to

Table 7.3 The budget distribution (in ten thousand dollars, i.e. $10K).

Position	Big data	Artificial intelligence	Cloud computing	Software engineering
Budget limits	5	8	16	6

Table 7.4 The candidates' stipend requests on positions ($10K).

	Positions			
Candidates	Big data	Artificial intelligence	Cloud computing	Software engineering
Adam	1	(4)	2	1
Bret	1	4	[2]	2
Chris	4	(4)	(4)	2
Doug	(5)	3	2	(3)
Edward	1	1	(3)	1
Fred	5	2	1	(3)
George	2	2	2	2
Harry	2	[1]	(4)	3
Ice	2	1	[1]	[2]
Joe	[1]	2	(3)	1
Kris	5	1	[1]	1
Larry	4	[1]	[1]	1
Matt	2	1	3	[1]

formulate a solution to such an assignment problem. Fortunately, Ann, a highly experienced administrator, understands the situation and allows for a reasonable response time.

In the above scenario, Ann and Bob are working through the steps of Role-Based Collaboration (RBC) (Zhu and Zhou 2006) before role-playing and Bob encounters a GRABC problem.

7.1.2 Problem Formalization

As a tradition in dealing with GRA, agents and roles are the two major concerns. The related factors and constraints are formalized in this section. In the following discussions, our context is a group, and environments and groups are simplified into vectors and matrices, respectively.

Definition 7.1 Given Q, L, L^a, and τ, the *Group Multi-Role Assignment Extended (GMRA$^+$)* problem is to find a matrix T to obtain

$$\sigma'_1 = max\left\{\sum_{i=0}^{m-1}\sum_{j=0}^{n-1}Q[i,j] \times T[i,j]\right\}$$

subject to

$$T[i,j] \in \{0,1\} \quad (0 \le i < m, 0 \le j < n) \tag{7.1}$$

$$\sum_{i=0}^{m-1}T[i,j] = L[j] \quad (0 \le j < n) \tag{7.2}$$

$$\sum_{j=0}^{n-1}T[i,j] \le L^a[i] \quad (0 \le i < m) \tag{7.3}$$

$$Q[i,j] \times T[i,j] > \tau \times T[i,j] \quad (0 \le i < m, 0 \le j < n) \tag{7.4}$$

where constraint (7.1) indicates whether an agent is assigned; (7.2) makes the group workable; (7.3) means that each agent can only be assigned to a limited number of roles, and (7.4) informs that the assigned Q values should be greater than τ.

We use T^* to express the assignment matrix that produces the maximum σ. Compared with GMRA (Zhu et al. 2017), GMRA$^+$ has one more constraint, i.e. constraint (7.4). For example, Figure 7.1a is a qualification matrix for Table 7.2. Figure 7.1c is the T^* that makes the group (Table 7.2) work with vectors $L = [1\ 2\ 4\ 2]$ from Table 7.1, $L^a = [2\ 2\ 2\ 2\ 2\ 2\ 2\ 2\ 2\ 2\ 2\ 2\ 2]$, and $\tau = 0$ in GMRA$^+$. $\sigma'_1 = 7.06$. Note, $\tau = 0$ if it is not explicitly stated otherwise.

(a)

0.18	0.82	0.29	0.01
0.35	0.80	0.58	0.35
0.84	0.85	0.86	0.36
0.96	0.51	0.45	0.64
0.22	0.33	0.68	0.33
0.96	0.50	0.10	0.73
0.25	0.18	0.23	0.39
0.56	0.35	0.80	0.62
0.49	0.09	0.33	0.58
0.38	0.54	0.72	0.20
0.91	0.31	0.34	0.15
0.85	0.34	0.43	0.18
0.44	0.06	0.66	0.37

(b)

0	1	0	0
0	1	0	0
0	0	1	0
0	0	0	1
0	0	1	0
0	0	0	1
0	0	0	0
0	0	1	0
0	0	0	0
0	0	1	0
1	0	0	0
0	0	0	0
0	0	0	0

(c)

0	1	0	0
0	0	0	0
0	1	1	0
1	0	0	1
0	0	1	0
0	0	0	1
0	0	0	0
0	0	1	0
0	0	0	0
0	0	1	0
0	0	0	0
0	0	0	0
0	0	0	0

Figure 7.1 Matrix examples. (a) A qualification matrix. (b) A GRA solution. (c) A GMRA$^+$ solution.

Theorem 7.1   (Zhu et al. 2016): The necessary conditions for GMRA$^+$ to have a feasible solution are

1) $\sum_{i=0}^{m-1} L^a[i] \geq \sum_{j=0}^{n-1} L[j]$ $(0 \leq i < m, 0 \leq j < n)$, and

2) $\sum_{i=0}^{m-1} \lceil Q[i,j] - \tau \rceil \geq L[j]$ $(0 \leq j < n)$.

[**Proof**]: Condition 1 has been proved in Chapter 6.

For Condition 2, we prove $S_1 \rightarrow S_2$, where S_1 is the statement of "GMRA$^+$ has a feasible solution" and S_2 is the statement of "$\sum_{i=0}^{m-1} \lceil Q[i,j] - \tau \rceil \geq L[j]$ $(0 \leq j < n)$."

Suppose S_1 is true.

Then, $\sum_{i=0}^{m-1} \lceil Q[i,j] - \tau \rceil$ $(0 \leq j < n)$ is the number of agents that have the qualification values greater than τ on role j, and $L[j]$ is the required number of agents to be assigned on role j.

Constraint   (7.4),   i.e. $\forall (0 \leq i < m, \;\; 0 \leq j < n) Q[i, \;\; j] \times T[i, \;\; j] > \tau \times T[i, \;\; j]$ informs that:

1) for all $[i, j]$s that make $T[i, j] = 0$, $Q[i, j]$ is ignored, and

2) for all $[i, j]$s that make $T[i, j] = 1$, we have

$Q[i, j] > \tau$, i.e. $Q[i, j] - \tau > 0$, i.e. $\lceil Q[i, j] - \tau \rceil = 1$, because $Q[i, j] \in [0,1]$.

$\therefore \sum_{i=0}^{m-1} \lceil Q[i,j] - \tau \rceil \geq \sum_{i=0}^{m-1} T[i,j] = L[j]$, i.e.

$$\sum_{i=0}^{m-1} \lceil Q[i,j] - \tau \rceil \geq L[j],$$

$\therefore S_2$ is true. Theorem 7.1 is proved. ∎

Definition 7.2   A *budget request matrix* Q' is an $m \times n$ matrix, where $Q'[i, j] \in (0, +\infty)$ expresses the budget request value (in \$10K) of agent $i \in \mathcal{N}$ $(0 \leq i < m)$ for role $j \in \mathcal{N}$ $(0 \leq j < n)$.

Definition 7.3   Given Q', L, L^a, and τ, the *Group Budget Assignment (GBA)* problem is to find a matrix T to obtain

$$\sigma_b = min \left\{ \sum_{i=0}^{m-1} \sum_{j=0}^{n-1} Q'[i,j] \times T[i,j] \right\}$$

subject to (7.1)–(7.4).

GBA is an assignment that only takes care of the budget requests of agents on roles. GBA is an instance of GRA$^+$ and its solution may not be useful for role assignment in reality. However, it can be used as a benchmark to conduct comparisons with other problems to be discussed.

In this section, we use ω and ω_p to express the total budget and performance limit, respectively, and p and p_b to represent the rate factor of the original group performance and budget, respectively.

Definition 7.4 Given Q, Q', L, and L^a, the *Group Role Assignment with Budget Constraints (**GRABC**)* problem in a general form is to find a workable T to obtain

$$max \sum_{i=0}^{m-1} \sum_{j=0}^{n-1} Q[i,j] \times T[i,j] \quad \text{and}$$

$$min \sum_{i=0}^{m-1} \sum_{j=0}^{n-1} Q'[i,j] \times T[i,j]$$

subject to (7.1)–(7.4).

The GRABC problem is a typical Multi-Objective Optimization Problem (MOOP). The basic ways to solve such a problem include the weighted sum approach and the goal programming approach, which are extensions of linear programming (Rardin 1997).

Inspired by the goal programming approach of dealing with MOOPs (Rardin 1997), we replace one objective with a new constraint and make it an Integer Linear Programming (ILP) problem. Here, there are two kinds of goals, team performance and budget goals. In the following, we discuss the related problems in obtaining the maximum team performance under different budget goals and those in obtaining the least budget under different performance goals.

To make the names short and simple without confusions, we use GRABC-* to express different forms of the GRABC problem in this section, where * is a placeholder that expresses Weighted Sum (WS), Synthesis (Syn), Performances (P_1, P_2, and P_3), or Budgets (B_1, B_2, and B_3). We also simply call "Group Multi-Role Assignment Extended (GMRA⁺)" as "Group Role Assignment (GRA)" if there is no confusion.

7.1.2.1 Optimization of Group Performance

Definition 7.5 A *role budget limit vector* B is an n-vector, where $B[j] \in (0, +\infty)$ expresses the total budget for role $j \in \mathcal{N}$ $(0 \le j < n)$.

We use B' to express the assigned role budget vector, i.e. $B'[j] = \sum_{i=0}^{m-1} Q'[i,j] \times T^*[i,j]$ $(0 \le j < n)$.

Definition 7.6 Given Q, Q', L, L^a, τ, and B, the *Group Role Assignment with Budget Constraints (**GRABC**)* problem in Form 1 of Performance (**GRABC-P₁**) is to find a workable T to obtain

$$\sigma_{p1} = max \sum_{i=0}^{m-1} \sum_{j=0}^{n-1} Q[i,j] \times T[i,j]$$

subject to (7.1)–(7.4), and

$$\sum_{i=0}^{m-1} Q'[i,j] \times T[i,j] \le B[j] \quad (0 \le j < n) \tag{7.5}$$

where constraint (7.5) means that the total budget for each role is limited by the given number $B[j]$.

Back to Section 7.1.1, the scenario is an instance of the GRABC-P_1 problem, i.e. Q and Q' are shown in Tables 7.2 and 7.4, respectively, L is shown in Table 7.1, $L^a = [2\ 2\ 2\ 2\ 2\ 2\ 2\ 2\ 2\ 2\ 2\ 2\ 2]$, and B is shown in Table 7.3.

Here, we need to point out that some instances of the GRABC problem may not have feasible solutions. We should check the conditions necessary for a feasible solution first.

To discuss the following theorems, we need some special symbols. Given vector V, $|V|$ is the cardinality of V; $min^k V$ ($max^k V$) (Zhu et al. 2016, 2017) is the ordered vector that includes the k smallest (largest) elements in V (from small to large, $k \geq 0$), where $|V| \geq k$. Given matrix $\mathcal{X}$, $|\mathcal{X}|$ is the number of elements in $\mathcal{X}$; $min^k \mathcal{X}$ ($max^k \mathcal{X}$) is the ordered vector that includes the k smallest (largest) elements in $\mathcal{X}$ (from small to large, $k \geq 0$), where $|\mathcal{X}| \geq k$; and $\mathcal{X}[i]$ denotes the vector of the elements at row i in $\mathcal{X}$ and $\mathcal{X}[j]$ that at column j in $\mathcal{X}$.

Theorem 7.2 **(P_1):** The necessary condition for GRABC-P_1 to have a feasible solution is $\sum_{k=0}^{L[j]-1} min^{L[j]} Q'[j][k] \leq B[j]$ $(0 \leq j < n)$.

To assist the following proofs, we let $Q' \cdot T$ express the matrix computed from Q' and T, where $Q' \cdot T[i, j] = Q'[i, j] \times T[i, j] (0 \leq i < m, 0 \leq j < n)$.

[**Proof**]: We prove by contradiction. Suppose that $\sum_{k=0}^{L[j]-1} min^{L[j]} Q'[j][k] > B[j]$ $(0 \leq j < n)$.

For $(0 \leq j < n)$, then we have $\sum_{i=0}^{m-1} Q'[i,j] \times T[i,j] = \sum_{i=0}^{m-1} Q' \cdot T[j][i]$.

For each j, $\sum_{i=0}^{m-1} T[i,j] = L[j]$ from (7.2).

To remove those items $T[i,j] = 0$, we form Q^{Tj}, which is a $L[j]$ – vector formed by the jth column of $Q' \cdot T$, by calling Algorithm 7.1 with $SV(Q^{Tj}, Q' \cdot T[j])$.

The correctness of the algorithm SV (Theorem 7.3(P_{1s})) can be proved by induction on $L[j]$. Now, we have

$$\sum_{i=0}^{m-1} Q' \cdot T[j][i] = \sum_{k=0}^{L[j]-1} Q^{Tj}[k]$$

Replacing $Q^{Tj}[k]s$ $(0 \leq k < L[j])$ with the minimum $L[j]$ elements of $Q'[j]$, we have

$$\sum_{i=0}^{m-1} Q' \cdot T[j][i] \geq \sum_{k=0}^{L[j]-1} min^{L[j]} Q'[j][k], \text{ i.e.}$$

$$\sum_{i=0}^{m-1} Q'[i,j] \times T[i,j] \geq \sum_{k=0}^{L[j]-1} min^{L[j]} Q'[j][k] > B[j].$$

This is contradictory with constraint (7.5).

Theorem 7.2 is proved. ∎

Algorithm 7.1 SV

```
Input: an empty L[j] - vector V₁ and an m - vector V₂ (m≥L[j]);
                                                              (S₁)
Output: vector V₁                                             (S₂)
begin
  k₂ = 0;                                                     (S₃)
  for (0 ≤ k₁ < L[j])                                         (S₄)
        while (V₂[k₂] = 0)                                     (S₅)
              k₂ = k₂ + 1;                                     (S₆)
        endwhile;                                              (S₇)
        V₁[k₁] = V₂[k₂];                                       (S₈)
        k₂ = k₂ + 1;                                           (S₉)
  endfor                                                       (S₁₀)
end
```

Theorem 7.3 (P_{1S}): Algorithm **SV** (V_1, V_2) is correct.

[**Proof**]: We prove by induction on $L[j]$. For all the cases, V_2 is an m-vector, $L[j] \leq m$.

When $L[j] = 1$, V_1 is a 1-vector.

(S_4)–(S_{10}) work once at $k_1 = 0$. (S_5)–(S_7) are executed $k' \geq 0$ times until $V_2[k'] \neq 0$, i.e. $k_2 = k'$. (S_8) makes $V_1[0] = V_2[k']$; then Algorithm **SV** ends at $k_1 = 0$ and $k_2 = k' + 1 \leq m$.

Suppose $L[j] = k(k > 1)$. Algorithm **SV** ends correctly at $k_1 = k - 1$ and $k_2 = k'' + 1 \leq m$ ($k'' \geq 0$).

Now, let us check the situation when $L[j] = k + 1(k > 1)$.

(S_4)–(S_{10}) run one more time than when $L[j] = k(k > 1)$, i.e. $k_1 = k$.

(S_5)–(S_7) are executed $k'''(k''' \geq 0)$ times until $V_2[k'' + 1 + k'''] \neq 0$, i.e. $k_2 = k'''$. (S_8) makes $V_1[k] = V_2[k'' + 1 + k''']$, then Algorithm **SV** ends at $k_2 = k'' + k''' + 2$.

Now, Theorem 7.3 is proved. ∎

The meaning of Theorem 7.2 is that the sum of the $L[j]$ (the number of required agents for role j) least budget requests of role j should be less than the budget limit for that role. In other words, if role j requires $L[j]$ agents, the total budget request for $L[j]$ agents that request the lowest stipends should be less than the budget limit for role j.

The time complexity (Garey and Johnson 1979) of applying Theorem 7.2 can be computed as follows:

Because $(0 \leq j < n)$, for each j:

1) to get $\sum_{k=0}^{L[j]-1} min^{L[j]}Q'[j][k]$, the complexity is $O(m \times L[j])$; and
2) to check $\sum_{k=0}^{L[j]-1} min^{L[j]}Q'[j][k] \leq B[j]$, the complexity is $O(1)$.

That is, the total complexity is $O(n \times m \times L[j]) \leq O(m^3)$ due to $n \leq m$, and $L[j] \leq m$. Compared with the complexity of an ILP solver $\geq O(m^3)$ (Bradley et al. 1977; Burkard et al. 2009; Hansen et al. 1993; Kuhn 1955, 2014; Munkres 1957; Papadimitriou 1981; Wolsey and Nemhauser 1999), it is beneficial to apply Theorem 7.2.

Definition 7.7 Given Q, Q', L, L^a, τ, and ω, the *Group Role Assignment with Budget Constraints* problem in Form 2 of Performance (**GRABC-P$_2$**) is to find a workable T to obtain

$$\sigma_{p2} = max \sum_{i=0}^{m-1}\sum_{j=0}^{n-1} Q[i,j] \times T[i,j]$$

subject to (7.1)–(7.4), and

$$\sum_{i=0}^{m-1}\sum_{j=0}^{n-1} Q'[i,j] \times T[i,j] \leq \omega \qquad (7.6)$$

where constraint (7.6) means that the total budget for the group is limited by the given number ω.

Definition 7.8 We define $n_a = \sum_{j=0}^{n-1} L[j]$, i.e. the *total number of required agents in a group*.

Theorem 7.4 (**P$_2$**): The necessary condition for GRABC-P$_2$ to have a feasible solution is $\sum_{k=0}^{n_a-1}(min^{n_a}Q')[k] \leq \omega$.

[**Proof**]: We prove by contradiction. Suppose that $\sum_{k=0}^{n_a-1}(min^{n_a}Q')[k] > \omega$. For $(0 \leq i < m, 0 \leq j < n)$, with the expression of $Q' \cdot T$, we have

$$\sum_{i=0}^{m-1}\sum_{j=0}^{n-1} Q'[i,j] \times T[i,j] = \sum_{i=0}^{m-1}\sum_{j=0}^{n-1} Q' \cdot T[i,j].$$

Let Q^{TV} be an $(m \times n)$-vector, where $Q^{TV}[i_1] = Q' \cdot T[i_1/n, i_1 \% n]$ $(0 \leq i_1 < m \times n)$.

Using a similar algorithm to **SV** in proving **Theorem 7.2**, we form an n_a-vector Q^{TV1} from Q^{TV} by calling **SV** (Q^{TV1}, Q^{TV}), where $(0 \leq k_1 < n_a)$, then we have

$$\sum_{i=0}^{m-1}\sum_{j=0}^{n-1} Q'[i,j] \times T[i,j] = \sum_{i_1=0}^{m \times n-1} Q^{TV}[i_1]$$

$$= \sum_{k=0}^{n_a-1} Q^{TV1}[k].$$

Now we replace $Q^{TV1}[k]$s $(0 \leq k < n_a)$ with the minimum n_a elements of Q'.

$$\sum_{i=0}^{m-1} Q'[i,j] \times T[i,j] \geq \sum_{k=0}^{n_a-1}(min^{n_a}Q')[k] > \omega,$$

i.e. $\sum_{i=0}^{m-1} Q'[i,j] \times T[i,j] > \omega$. This inequality contradicts constraint (7.6). Theorem 7.4 is proved. ∎

The meaning of Theorem 7.4 is that the total sum of the least n_a (the total number of required agents) budget requests should be less than the budget limit, i.e. the n_a agents who request the lowest stipends should be covered by the budget limit. To apply Theorem 7.4, we need to:

1) compute $\sum_{i=0}^{m-1}\sum_{j=0}^{n-1}Q'[i,j] \times T[i,j]$ in $O(m \times n)$; and

2) check $\sum_{i=0}^{m-1}\sum_{j=0}^{n-1}Q'[i,j] \times T[i,j] \leq \omega$ in $O(1)$.

The total time complexity is $O(m \times n)$ or simply in $O(m^2)$. Therefore, Theorem 7.4 is also beneficial.

Definition 7.9 Given Q, Q', L, L^a, τ and p_b (≥ 1), the *Group Role Assignment with Budget Constraints* problem in Form 3 of Performance (**GRABC-P₃**) is to find a workable T to obtain

$$\sigma_{p3} = max \sum_{i=0}^{m-1}\sum_{j=0}^{n-1}Q[i,j] \times T[i,j]$$

subject to (7.1)–(7.4), and

$$\sum_{i=0}^{m-1}\sum_{j=0}^{n-1}Q'[i,j] \times T[i,j] \leq p_b \times \sigma_b (0 \leq i < m, 0 \leq j < n) \qquad (7.7)$$

where constraint (7.7) expresses that the total assigned budget should be less than $p_b \times \sigma_b$ (σ_b is the ideal budget obtained by GBA).

Theorem 7.5 (**P₃**): The condition necessary for GRABC-P₃ to have a feasible solution is $\sum_{k=0}^{n_a-1}(min^{\,n_a}Q')[k] \leq p_b \times \sigma_b$.
[**Proof**]: Let $\omega = p_b \times \sigma_b$, then GRABC-P₃ becomes GRABC-P₂. ∎

To apply Theorem 7.5 seems complex:

1) Call GBA to obtain σ_b in $O(m^3)$ (Zhu et al. 2016, 2017); and
2) Other computations are the same as those in Theorem 7.4.

Therefore, the time used to check Theorem 7.5 is $O(m^3)$ ($=O(m^3)$+ the time to check Theorem 7.4, i.e. $O(m^2)$). Note that the ILP solver also needs to compute σ_b. Hence, it is beneficial to apply Theorem 7.5.

7.1.2.2 Optimization of Budgets

Definition 7.10 A *role performance requirement vector P^r* is an n-vector, where $P^r[j]\epsilon[0, +\infty)$ expresses the required qualification value for role $j\epsilon N$ ($0 \leq j < n$).

We use $P^{r'}$ to express the assigned role performance vector, where $P^{r'}[j] = \sum_{i=0}^{m-1}Q[i,j] \times T^*[i,j](0 \leq j < n)$.

Definition 7.11 Given Q, Q', L, L^a, τ, and P^r, the *Group Role Assignment with Budget Constraints* problem in Form 1 of Budget (***GRABC-B_1***) is to find a workable T to obtain

$$\sigma_{b_1} = min \sum_{i=0}^{m-1} \sum_{j=0}^{n-1} Q'[i,j] \times T[i,j]$$

subject to (7.1)–(7.4), and

$$\sum_{i=0}^{m-1} Q[i,j] \times T[i,j] \geq P^r[j] \quad (0 \leq j < n) \tag{7.8}$$

where constraint (7.8) means that the sum of all the assigned agent qualification values on a role should not be less than the required value.

Theorem 7.6 (**B_1**): The necessary condition for GRABC-B_1 to have a feasible solution is $\sum_{k=0}^{L[j]-1} max^{L[j]} Q[j][k] \geq P^r[j] \quad (0 \leq j < n)$

The meaning of Theorem 7.6 is that the total sum of the $L[j]$ (the number of required agents for role j) best qualification values on role j should not be less than the required role performance, i.e. the top $L[j]$ agents' qualification values together should be larger than or equal to the required role performance.

[**Proof**]: It can be proved in the same way as that proves Theorem 7.2. ■

Definition 7.12 Given Q, Q', L, L^a, τ, and ω_p, the *Group Role Assignment with Budget Constraints* problem in Form 2 of Budget (***GRABC-B_2***) is to find a workable T to obtain

$$\sigma_{b2} = min \sum_{i=0}^{m-1} \sum_{j=0}^{n-1} Q'[i,j] \times T[i,j]$$

subject to (7.1)–(7.4), and

$$\sum_{i=0}^{m-1} \sum_{j=0}^{n-1} Q[i,j] \times T[i,j] \geq \omega_p \quad (0 \leq j < n) \tag{7.9}$$

where constraint (7.9) means that the assigned agents should have a qualification value not less than the required value, i.e. ω_p.

Theorem 7.7 (**B_2**): The necessary condition for GRABC-B_2 to have a feasible solution is $\sum_{k=0}^{n_a-1} (max^{n_a} Q)[k] \geq \omega_p$.

The meaning of Theorem 7.7 is that the sum of the best n_a (the number of required agents) qualification values should not be less than the required role performance, i.e. the n_a best qualification values together should be larger than or equal to the required group performance.

[**Proof**]: It can be proved in the same way as Theorem 7.4. ■

Definition 7.13 Given Q, Q', L, L^a, τ, and $p_p(\leq 1)$, the *Group Role Assignment with Budget Constraints* problem in Form 3 of Budget (**GRABC-B3**) is to find a workable T to obtain

$$\sigma_{b3} = min \sum_{i=0}^{m-1} \sum_{j=0}^{n-1} Q'[i,j] \times T[i,j]$$

subject to (7.1)–(7.4), and

$$\sum_{i=0}^{m-1} \sum_{j=0}^{n-1} Q[i,j] \times T[i,j] \geq p_p \times \sigma'_1 \quad (0 \leq i < m, \ 0 \leq j < n) \quad (7.10)$$

where constraint (7.10) means that the total qualification values of the assigned agents should not be less than the given percentage of the original GMRA+ group performance σ'_1.

Theorem 7.8 (**B3**): The necessary condition for GRABC-B3 to have a feasible solution is $\sum_{k=0}^{n_a-1} (max^{n_a} Q)[k] \geq p_p \times \sigma'_1$.

[**Proof**]: It can be proved in the same way as that proves Theorem 7.4. ■

Note that the gains of applying **Theorems 7.6–7.8** are the same as those of **Theorems 7.2**, **7.4**, and **7.5**, respectively.

Now since we have discussed the above different variations of the GRABC problem, the question is "are there more variations?" We may not answer "absolutely NOT." However, we believe the above variations plus GRABC-WS and GRABC-Syn (Section 7.1.3) can cover most of the real-world applications related to the GRABC problem.

7.1.3 Solutions with an ILP Solver

Section 7.1.2 has discussed different ways to solve the GRABC problem. Note that different solutions to the GRABC problem are all ILP problems. Therefore, we can use the IBM ILOG CPLEX (CPLEX) (IBM 2017) package. When using the CPLEX solution to GMRA (Chapter 6; Zhu et al. 2016), we need a new constraint to be added, i.e. constraint (7.4).

In most cases, the solution to GMRA+ does not differ from that of GMRA if the elements of the Q matrix are greater than zero and τ is not large. For example, the GMRA+ solution to the problem expressed with Tables 7.1–7.4, and $L^a = [2\ 2\ 2\ 2\ 2\ 2\ 2\ 2\ 2\ 2\ 2\ 2]$ does not change the result from the solution to GMRA (Zhu et al. 2017). However, if $\tau > 0.64$, the GMRA+ reports failure, meaning that no feasible solution exists. The new GMRA+ solution may seem insignificant, but it serves as a good foundation to help solve other forms of the GRABC problem.

The weighted sum method is a normal way to solve a MOOP. We at first compose the two objectives into one by providing weights for each. Before the composition, we need to normalize the data in Q' to be comparable with Q. We let Q'' represent the normalized Q'.

Definition 7.14 The *normalized budget matrix Q''* is defined as:

$$Q''[i,j] = Q'[i,j]/\max\{Q'[i,j]\} \quad (0 \le i < m, 0 \le j < n) \tag{7.11}$$

Now, GRABC can be transformed into an ILP problem.

Definition 7.15 Given Q, Q' (Q''), L, L^a, τ, w_1, and w_2, the *Group Role Assignment with Budget Constraints* problem in the Weighted Sum form (***GRABC-WS***) aims to find a workable T to obtain

$$\sigma_{(w1,w2)} = \max\left\{ w_1 \times \sum_{i=0}^{m-1}\sum_{j=0}^{n-1} Q[i,j] \times T[i,j] \right.$$
$$\left. - w_2 \times \sum_{i=0}^{m-1}\sum_{j=0}^{n-1} Q''[i,j] \times T[i,j] \right\}$$

subject to (7.1)–(7.4).

With CPLEX, the problem presented in Section 7.1.1 can be solved as a GRABC-WS problem with parameters $L = [1\,2\,4\,2]$ in Table 7.1, $L^a = [2\,2\,2\,2\,2\,2\,2\,2\,2\,2\,2\,2]$, $\tau = 0$, $w_1 = 0.5$, and $w_2 = 0.5$. The final result is shown in Table 7.2 (bracketed numbers with "[]"). From Table 7.2, the total sum of the assigned evaluation values (in "[]") is 3.7, the assigned total budget is \$110K, and $B' = [1\ 2\ 5\ 3]$. Here, the total team performance 3.7 is not acceptable compared with the original 7.06. We try another weight distribution ($w_1 = 0.8$ and $w_2 = 0.2$). This result is shown in Table 7.2 (parenthesized numbers with "()"). Using the new weight distribution, the total qualification value is 7.06, the total budget is \$330K, and $B' = [5\ 8\ 14\ 6]$. This result seems better than the previous one. The above situations demonstrate that it is crucial to choose an acceptable and reasonable weight distribution for the weighted sum method.

Definition 7.16 Given Q, Q', L, L^a, and τ, the *Group Role Assignment with Budget Constraints* problem in the synthesized form (***GRABC-Syn***) is to find a workable T to obtain

$$\sigma_{syn} = \max\left\{ \sum_{i=0}^{m-1}\sum_{j=0}^{n-1} Q[i,j]/Q'[i,j] \times T[i,j] \right\}$$

subject to (7.1)–(7.4).

GRABC-Syn aims to optimize the total performance per 10 000 dollars, i.e. performance-cost ratio. Using CPLEX, the problem presented in Section 7.1.1 can be solved as a GRABC-Syn problem with $L = [1\,2\,4\,2]$ and $L^a = [2\,2\,2\,2\,2\,2\,2\,2\,2\,2\,2\,2]$. The total sum of the assigned evaluation values is 3.45, the assigned total budget is

$100K, and $B' = [1\ 2\ 5\ 2]$. Here, the total team performance 3.45 is unacceptable compared with the previous value of 7.06. However, it is the assignment with the best performance-cost ratio.

Using CPLEX, we can solve GRABC-P_1 by introducing the B vector. With $B = [5\ 8\ 16\ 6]$, and $L^a = [2\ 2\ 2\ 2\ 2\ 2\ 2\ 2\ 2\ 2\ 2\ 2\ 2\ 2\ 2\ 2]$, the problem presented in Section 7.1.1 can be solved and the total qualification value is 7.06, the total budget is $330K, and the assigned budgets $B' = [5\ 8\ 14\ 6]$.

GRABC-P_2 can also be solved by using CPLEX. Let $\omega = 35$, then the total qualification value is 7.06, the total budget is $330K, and $B' = [5\ 8\ 14\ 6]$. The result of GRABC-P_2 is the same as that of GRABC-P_1 if $\omega = 35$. If $\omega = 30$, the total qualification value is 6.79, the total budget is $300K, and $B' = [4\ 8\ 13\ 5]$.

GRABC-P_3 requires that σ_b is initially obtained by calling GMRA⁺. After obtaining σ_b, GRABC-P_3 becomes GRABC-P_2, where $\omega = p_b \times \sigma_b$. For the problem in Section 7.1.1 with GBA, $\sigma_b = 9.0$. If $p_b = 2$, the total qualification value is 4.99, the total budget is $180 ($p_b \times \sigma_b = 2 \times 9.0 = 18.0$), and $B' = [2\ 3\ 10\ 3]$.

Other forms of GRABC problems can be solved in the same ways stated above.

7.1.4 Simulations of GRABC-WS and GRABC-Syn

To gain an insight into the solutions to the GRABC problem, we conduct simulations using random Qs and Q's. In the first simulation, we attempt to find the correlation between weight distributions and the two objectives of the GRABC problem. Based on the correlation, practitioners may be able to choose a weight distribution between the group performance and the budget to meet their needs. In the second simulation, we attempt to determine the relationship between the best performance-cost ratio and the two objectives. Our simulations are designed as three group sizes with the parameters $1 \le L[j] \le 3$ and $1 \le L^a[j] \le 5$ ($0 \le i < m$, $0 \le j < n$), i.e. $m = 20$, $n = 6$; $m = 50$, $n = 16$; and $m = 100$, $n = 33$.

For each group size, w_1 ranges from 0.1 to 1.0, $w_2 = 1.0 - w_1$. Under each weight distribution, we create 100 random Qs and Q's:$Q[i, j] \in [0, 1]$; and $Q'[i, j] \in [max(Q[i, j] \times 10 - 3, 0), Q[i, j] \times 10 + 3]$ to indicate that the requested salary matches the qualifications in different degrees.

We collect the maximum, minimum, and average values of group performance (i.e. $\sigma_{p(w1,\ w2)}$) and total budget (i.e. $\sigma_{b(w1,\ w2)}$s in $10K). Our experiments demonstrate that the group performance and the budget change linearly according to the weights. There are no exceptions to the theoretical equations.

GRABC-Syn has a definite result to obtain the maximum sum of performance-cost ratios. We use the settings of $m = 10$ to 100 with a step of 10, $n = m/3$, $1 \le L[j] \le 3$, and $1 \le L^a[j] \le 3$ ($0 \le i < m$, $0 \le j < n$).

For each group size, we create 100 random Qs and Q's with parameters $Q[i, j] \in [0, 1]$ and $Q'[i, j] \in [max(Q[i, j] - 0.3, 0), Q[i, j] + 0.3]$ to indicate that the requested salary (in $10K) follows the qualifications with different degrees.

Because GRABC-Syn aims to obtain the maximum sum of performance ratios, the σ_ps and σ_bs are based on specific Qs and Q's. The experiments show no significant connections among σ_ps, σ_bs, and σ_{syn}s. The only clear evidence is that σ_ps, σ_bs, and σ_{syn}s all increase linearly with the group size, i.e. m.

7.1.5 Performance Experiments and Improvements

We conduct an experiment to see if GRABC solutions are practical in dealing with real-world problems. The experimental platform is shown in Table 7.5.

The following settings are used to check the time used by each GRABC-P_1 solution ($0 \leq i < m$, $0 \leq j < n$):

1) $m \in \mathcal{N}$, $10 \leq m \leq 400$ with steps of 20;
2) $k \in \mathcal{N}$, $3 \leq k \leq 10$ randomly, $n = m/k$;
3) $\tau = 0.5$ as a passing qualification mark;
4) $1 \leq L^a[i] \leq 5$ so that each agent can be assigned with 1 to 5 roles;
5) $1 \leq L[j] \leq \sum_{i=0}^{m-1} L^a[i]/n$ so that enough agents are ready to play roles;
6) $Q[i, j] \in [0.3, 1]$ so that not all the agents have a passing mark at each position and the mark on each position is not very low, i.e.,<30%; and
7) $Q'[i, j] \in [max(Q[i, j] \times 10 - 2, 0), Q[i, j] \times 10 + 2]$ so that the requested salary follows the qualifications but with variations.

Using the above settings, we create 100 random groups with different group sizes, i.e. m. For each m, we collect the maximum, minimum, and average time used for the 100 groups. The results show that a problem with $m = 400$ can be solved in 4.5 seconds. In the experiments, there are a few exceptions when $m = 120, 260$, and 300. The maximum time does not follow the curve nicely. This is understandable because the processing time is determined by the ILP algorithms and the data distribution among several different data structures, such as Q, Q', L, and B. A few exceptions are reasonable (Zhu 2020a).

Table 7.5 Test platform configuration.

Hardware	
CPU	Intel core i7-2677U CPU @1.8 GHz 1.80 GHz
MM	6 GB
Software	
OS	Windows 7 Enterprise
Eclipse	Version: Version: Neon.1a Release (4.6.1)
JDK	Java 1.8.0_91

7.1.5.1 Improvements

The necessary conditions (theorems) we described in Section 7.1.4 can be used in programming. Theoretically, the theorems can help save time. Here, we make experiments to verify the savings by comparing the improved CPLEX solution by checking the conditions to have feasible solutions with the CPLEX ones without checking. The platform used is also shown in Table 7.5. To save space, only the experiments of Theorem 7.2 are presented.

In the experiments, we set $m \in \mathcal{N}$, $10 \leq m \leq 200$, $k \in \mathcal{N}$, $3 \leq k \leq 10$, $n = m/k$. Others are the same as those in 3)–7) discussed in Section 7.1.5.

Following the above premise, we conduct four different experiments. The first two experiments apply the necessary conditions proposed in this section, i.e. GMRA+ and Theorem 7.2.

1) Exp. 1: $B[j] = L[j] \times 4$ so that the average budget for each agent is \$400K.
2) Exp. 2: $B[j] = L[j] \times [3, 10]$ so that the average budget for each role is randomly from \$30K to \$100K.

  The second two experiments apply all the necessary conditions related to the problems, i.e. GMRA (Zhu et al. 2016, 2017), GMRA⁺, and Theorem 7.2.

3) Exp. 3: $B[j] = L[j] \times 4$ so that the average budget for each agent is \$400K.
4) Exp. 4: $B[j] = L[j] \times [3, 10]$ so that the average budget for each role is randomly from \$30K to \$100K.

With each experiment, we collect the total time used by the improved solution and the original CPLEX solution to process the same 200 random problems. To present the improvement, we collect the percentage of saved time, i.e. $(t_1 - t_2)/t_1$, where t_1 is the time used by the CPLEX solution to process 200 problems and t_2 is the time used by the improved solutions (Zhu 2020a).

The experiment result shows the time savings of 8–15% and indicates that the average process time of the improved solution to GRABC-P_1 is less than 200 ms for the group size between $m = 10$ and $m = 200$. The experiment shows that checking necessary conditions helps save total process time if we need to deal with numerous problems. The conditions used in solving GMRA (Zhu et al. 2017) problems indeed affect the consumed time, but not much. This result demonstrates that the contributions of the theorems in Sections 7.1.3 and 7.1.4 are significant, at least, in the scope expressed by the experiments.

A few worst cases of the 200 random groups are acceptable because the CPLEX solution may quickly find the solution but the improved CPLEX may check the necessary conditions. Fortunately, there are not many such cases, i.e. less than 5%. The four experiments verify that the discussed necessary conditions are useful in processing the proposed problems. Such improvements may also be useful in cloud computing for service providers to deal with multiple clients quickly (Zhu 2020a).

As for other improved forms of GRABC, we omit the descriptions of the experiments to save space. They have the same improvement scales as that of GRABC-P_1.

7.1.6 Synthesis

Through the discussion of the solutions to the GRABC problems in Sections 7.1.3–7.1.5, we notice that different constraints produce different assignments with variable group performances and budgets. In practice, one could choose a corresponding solution to meet the specific requirements. Figure 7.2 gives a flowchart to utilize the solutions of this section. Figure 7.2 also reveals the requirements and tasks necessary to organize a team by role assignment in consideration of budget, e.g. one may need to collect Q, Q', τ, B, P', and other parameters. Because the GRABC problem mainly considers two objectives, group performance and budgets, the conditions are σs and B's as shown in Figure 7.2.

The GRABC problems are abstract. Many real-world problems can be instances of GRABC problems, e.g. the recruiting problem in Section 7.1.1 is a typical example.

7.2 Good at Many Things and Expert in One (GMEO)

It is important to organize a team efficiently and maintain the team. Usually, administrators try to avoid critical members. Based on this requirement, administrators prefer team members that are "good at many things and expert in one (GMEO)." However, if too many people are "good at many things," it is undoubtedly a waste. RBC and E-CARGO are good means to provide modeling and solutions to the GMEO problem.

Note that, conventionally, GMEO (Zhu 2020a) has no exact solutions even for a small group (e.g. ten people) and is normally managed by highly qualified administrators. The proposed solutions provide numerical solutions with algorithms that form a solid foundation for decision making in dealing with issues related to GMEO.

Critical members are a headache for an administrator. In many cases, the unavailability of critical members may result in a team's failure or leave it in a very inefficient state (Schreiber and Carley 2004). Therefore, it is a big challenge for an administrator to organize a team and balance its robustness with the least extra resources. Conventional ways of dealing with such situations are based on experience, wisdom, logical analysis, or intuitive judgments of administrators. It is difficult for such a method to make a robust team with the least number of backups, especially if the team is large.

The GMEO problem is very complex because there are many situations in organizing a team to meet the requirement of GMEO. In most cases, the administrator

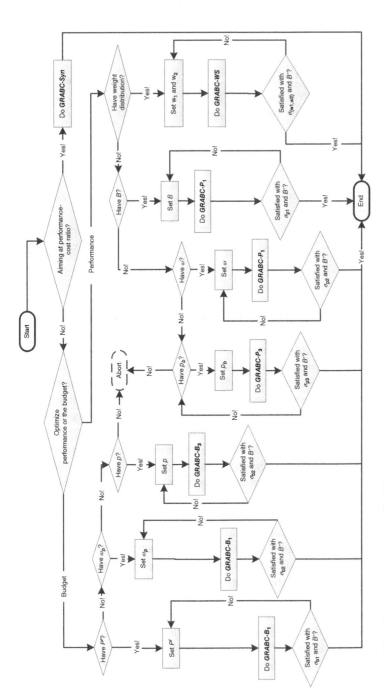

Figure 7.2 The flowchart to solve GRABC problems.

of a team simply requires that (i) each member should be assigned to more than one position without any guarantee of team robustness. Role transfer (Chapter 9) clarifies the reason why a group is not guaranteed to be workable (Zhu and Zhou 2009, 2012). The other situations may require that (ii) their team members should be assigned to as many positions as possible. In fact, solution (i) may be insufficient to make the team robust and solution (ii) is definitely unnecessary. Simply stated, the question is: "How many" is good enough? This section will provide an assured answer to such a question. This section analyzes the key scenarios of the GMEO problem. Based on this work, critical team members can be avoided by following the GMEO principle and setting the necessary number of backups. The results obtained in this section can be applied in many applications, such as administration, management, robot teams, and highly available computing platforms.

7.2.1 A Real-World Scenario

In company X, Ann, the Chief Executive Officer (CEO), has just signed a contract, the value of which is one million dollars. She asks Bob, the Human Resources (HR) officer, to organize a team from the employees of the company. Bob drafts a position list as shown in the second line of Table 7.6 for the team and a candidate staff shortlist as shown as the left-most column in Table 7.6. Next, Bob sends out a survey email to let the candidates express their preferences for the positions as shown

Table 7.6 Job preferences.

	Positions			
Candidates	**Project manager**	**Senior programmer**	**Programmer**	**Tester**
Adam	0.18	0.82	0.29	0.01
Brian	0.35	0.80	0.58	0.35
Chris	0.84	0.85	0.86	0.36
Doug	0.96	0.51	0.45	0.64
Edward	0.22	0.33	0.68	0.33
Fred	0.96	0.50	0.10	0.73
George	0.25	0.18	0.23	0.39
Harry	0.56	0.35	0.80	0.62
Ice	0.49	0.09	0.33	0.58
Joe	0.38	0.54	0.72	0.20
Kris	0.91	0.31	0.34	0.15
Larry	0.85	0.34	0.43	0.18
Matt	0.44	0.06	0.66	0.37

Table 7.7 Initial assignment plan of Bob.

Candidates	Positions			
	Project manager	Senior programmer	Programmer	Tester
Adam	0	1	0	0
Brian	0	1	0	0
Chris	0	0	1	0
Doug	0	0	0	1
Edward	0	0	1	0
Fred	0	0	0	1
George	0	0	0	0
Harry	0	0	1	0
Ice	0	0	0	0
Joe	0	0	1	0
Kris	1	0	0	0
Larry	0	0	0	0
Matt	0	0	0	0

in Table 7.6. Based on Table 7.6, Bob establishes an assignment plan as shown in Table 7.7 by applying the GRA algorithm (Zhu et al. 2012b) to maximizing the sum of all the assigned preference values when $L = [1\ 2\ 4\ 2]$. Ann tells Bob that it is necessary to have backups in the assignment plan to guarantee that no critical persons exist, i.e. the team is still in a good state even if a person cannot work for some reasons under the least redundant assignment (expressed by the number of 1s in Table 7.7) while considering the candidates' preferences.

This is not a trivial decision-making problem and requires time to be solved. Evidently, we assume that the backup is mainly reflected by the number of extra assignments of agents to roles.

Ann's request is necessary because Bob's previous assignment in Table 7.7 makes everybody in the team a critical member, i.e. if anyone is unavailable for work, then there is a situation where at least one position of the project has insufficient agents. This is, in fact, an instance of the GMEO problem. As we can see, the GMEO problem is complex and requires a thorough investigation.

7.2.2 Problem Formalizations

To ease the description of the GMEO problem, we use Q to express the preferences rather than the qualifications and follow the same style as previous chapters in formalizations. Table 7.6 forms a preference matrix. Table 7.7 forms an assignment matrix that makes the group with Q in Table 7.6 work when vector $L = [1\ 2\ 4\ 2]$. The sum of the assigned values is 6.96. In the following sections, we assume that

$$m \geq \sum_{j=0}^{n-1} L[j]$$ unless special cases are clearly specified.

GRA (Chapter 5) and GRA^+ (Chapter 6) support the solution of the "Expert in One (EO)" part, i.e. GRA assigns the agents who are "Experts in One." This section concentrates on the "Good at Many things (GM)" part of the GMEO problem. To understand the problem, we may need a new special matrix, called capability matrix denoted as $\overline{Q}$.

Definition 7.17 A *capability matrix* $\overline{Q}$ is an $m \times n$ matrix, where $\overline{Q}[i,j]\in\{0,1\}$, $\overline{Q}[i,j] = 1$ indicates that agent $i\in\mathcal{N}$ $(0 \leq i < m)$ is capable of playing (or simply, being assigned to) role $j\in\mathcal{N}$ $(0 \leq j < n)$ and $\overline{Q}[i,j] = 0$ indicates otherwise. We also call $\overline{Q}$ an *assignment plan matrix*, the 1s in $\overline{Q}$ are called *assignment units*, and the tuple $<i,j>$ making $\overline{Q}[i,j] = 1$ is called *a planned assignment*, or simply, *an assignment*.

For example, the numbers of Table 7.7 can form a capability matrix $\overline{Q}$. Note that if $\sum_{i=0}^{m-1}\sum_{j=0}^{n-1}\overline{Q}[i,j] > \sum_{j=0}^{n-1}L[j]$, then $\overline{Q}$ is a *redundant assignment* matrix. Different from the assignment matrix T, $\overline{Q}$ may be redundant but T may not.

Definition 7.18 A *group* expressed by $\overline{Q}$ and L is in a *good state iff* there exists a workable assignment matrix T, where "*iff*" means "if and only if."

Definition 7.19 The *required agent number* of a group is defined as $n_a = \sum_{i=0}^{n-1} L[j]$.

Definition 7.20 The *number of assigned agents* in a group is defined as $n_d = \sum_{i=0}^{m-1}\left\lceil\sum_{j=0}^{n-1}\overline{Q}[i,j]/n\right\rceil$, where $\lceil x \rceil$ means the whole number greater than or equal to x. That is, n_d is the number of rows that have at least one 1 in $\overline{Q}$.

Definition 7.21 The *number of total assigned units* (or planned assignments) in a group is defined as $n_t = \sum_{i=0}^{m-1}\sum_{j=0}^{n-1}\overline{Q}[i,j]$.

Definition 7.22 A *capability matrix after removing agent* i_1 $\overline{Q_{-i_1}}$ is an $(m-1)\times n$ matrix where row i_1 is removed from $\overline{Q}$, i.e.

$$\overline{Q_{-i_1}}[i,j] = \begin{cases} \overline{Q}[i,j] & (i < i_1) \\ \overline{Q}[i+1,j] & (i_1 \leq i < m-1) \end{cases} \quad (0 \leq i_1 < m).$$

Definition 7.23 An *assignment matrix after removing agent* $i_1 \overline{T_{-i_1}}$ $(0 \le i_1 < m)$ is an $(m-1) \times n$ matrix derived from $\overline{Q_{-i_1}}$.

$\overline{Q_{-i_1}}$ is not used in specifying the problems but it is used in the proof of related theorems.

Definition 7.24 Agent i_1 is *critical* in a group expressed by $\overline{Q}$ and L *iff* no $\overline{T_{-i_1}}$ $(0 \le i_1 < m)$ is workable.

Definition 7.25 Given a group expressed by Q and L, the GMEO problem is to find the best assignment plan matrix $\overline{Q}$ such that no critical agent exists in the group, i.e.

$$max \sum_{i=0}^{m-1} \sum_{j=0}^{n-1} Q[i,j] \times \overline{Q}[i,j]$$

and

$$min \sum_{i=0}^{m-1} \sum_{j=0}^{n-1} \overline{Q}[i,j]$$

subject to

$$\overline{Q}[i,j] \in \{0,1\} \quad (0 \le i < m, 0 \le j < n) \tag{7.12}$$

$$\overline{T_{-i_1}}[i,j] \in \{0,1\} \quad (0 \le i_1 < m, 0 \le i < m-1, 0 \le j < n) \tag{7.13}$$

$$\overline{T_{-i_1}}[i,j] \le \begin{cases} \overline{Q}[i,j] & (i < i_1) \\ \overline{Q}[i+1,j] & (i_1 \le i < m-1) \end{cases} \quad (0 \le i_1 < m) \tag{7.14}$$

$$\sum_{i=0}^{m-2} \overline{T_{-i_1}}[i,j] = L[j] \quad (0 \le i_1 < m, 0 \le j < n) \tag{7.15}$$

$$\sum_{j=0}^{n-1} \overline{T_{-i_1}}[i,j] \le 1 \quad (0 \le i_1 < m, 0 \le i < m-1) \tag{7.16}$$

where constraints (7.12) and (7.13) inform that agent i can or cannot be assigned to role j; constraint (7.14) states that $\overline{T_{-i_1}}$ is derived from $\overline{Q}$ or, more exactly, $\overline{Q_{-i_1}}$; constraint (7.15) expresses that the number of assigned agents for a role in $\overline{T_{-i_1}}$ must follow L; and constraint (7.16) specifies that each agent can only be assigned to one role in $\overline{T_{-i_1}}$.

Evidently, the GMEO problem is complex. Note that if $m = \sum_{i=0}^{n-1} L[j]$, every agent is a critical one. An extreme situation (**Extreme** 1) is that if $m = n_a + 1$ and $\overline{Q}[i,$

$j] = 1 \, (0 \leq i < m, 0 \leq j < n)$, then it is a team without critical agents. However, this is costly to implement because $n_t = (n_a + 1) \times n$. There also might be another extreme case (**Extreme** 2) where if $m = n_a + 1$, ONE agent (i.e. agent i', $0 \leq i'$ $< m$) is assigned as $\overline{Q}[i', j] = 1 \, (0 \leq j < n)$ as a backup agent, and other n_a agents meet the requirement of L, then there would be no critical agents in the group. However, this solution is not practical because agent i' is overloaded. That is, agent i' is not critical in *Definition 7.24* but potentially critical in some sense.

7.2.3 A Solution with CPLEX

Note that the GMEO problem defined in *Definition 7.25* is a dual-objective optimization problem that does not have a feasible solution in general. We have to modify it into different ILP problems (Burkard et al. 2009; Hansen et al. 1993; Papadimitriou 1981; Wolsey and Nemhauser 1999) to obtain feasible solutions. The most popular methods are the weighted sum and goal programming methods (Rardin 1997). The weighted sum method is not viable because the two objectives are not compatible, i.e. the second objective is a whole number, which expresses how many elements in the capability matrix are assigned but the first objective is a property to express the preference values of all the assigned people. Goal programming is a better option. We let $n_t (= \sum_{i=0}^{m-1} \sum_{j=0}^{n-1} \overline{Q}[i,j])$ be set by an administrator. Then the GMEO problem becomes a solvable ILP problem.

Definition 7.26 Given a group expressed by Q, L, and n_t, *the GMEO problem with a limited number of assignment units* (GMEO-1) aims to find the best assignment plan matrix $\overline{Q}$ such that no critical agent exists in the group, i.e. obtain

$$\sigma_{g1} = max \sum_{i=0}^{m-1} \sum_{j=0}^{n-1} Q[i,j] \times \overline{Q}[i,j]$$

subject to (7.12)–(7.16), and

$$\sum_{i=0}^{m-1} \sum_{j=0}^{n-1} \overline{Q}[i,j] \leq n_t \tag{7.17}$$

where constraint (7.17) limits the largest number of assignment units. We name the problem in *Definition 7.26* as GMEO-1 to be consistent with the definition of a critical agent in *Definition 7.24*.

Now, the GMEO-1 problem becomes an ILP problem even though it has a large search space. Therefore, we can use an ILP solver to solve it. In this section, we implement the solution by using the CPLEX (IBM 2017) package.

With the above CPLEX solution, the problem described in Section 7.2.1 ($L = [1\ 2\ 4\ 2]$ and Q is obtained from Table 7.6) can be solved with $n_t = 13$ because **Extreme** 2 shows that there is a feasible solution when $n_t = 13$ (Table 7.8, i.e. $\overline{Q}$).

It is notable that the problem in *Definition 7.26* has no feasible solution if $n_t < 13$. Therefore, Table 7.8 is the best assignment in terms of agents' preferences for the GMEO-1 problem and $\sigma_{g1} = 10.05$ (the sum of the underlined values in Table 7.6). Compared with Table 7.7, there is only one more agent used in Table 7.8. This solution is much better than **Extreme** 2. In this assignment, even though some agents could do more jobs than others, every agent is equally important in terms of critical agents defined in *Definition 7.24*.

One may argue that every agent is a critical agent after one agent leaves. This is a new GMEO problem, i.e. how to form a robust team that can afford the loss of two members at the same time. This problem may be called GMEO-2. Note that GMEO-2 has a much larger search space than GMEO-1 for any ILP problem solver because it involves $\binom{m}{2} = m \times (m-1)/2$ assignment matrices (with $m - 2$ rows and n columns) other than m $\overline{T_{-i_1}}$s in similar constraints (2)–(5). The following discussions concentrate only on the GMEO-1 problem and the GMEO-2 problem will be left for future investigation.

Table 7.8 The revised assignment plan ($n_t = 13$).

Candidates	Positions			
	Project manager	Senior programmer	Programmer	Tester
Adam	0	1	0	0
Brian	0	1	0	0
Chris	0	1	1	0
Doug	1	0	0	1
Edward	0	0	1	0
Fred	0	0	0	1
George	0	0	0	0
Harry	0	0	1	1
Ice	0	0	0	0
Joe	0	0	1	0
Kris	1	0	0	0
Larry	0	0	0	0
Matt	0	0	1	0

To discern the problem above more carefully, we find that n_t can be enumerated if it is not very large. In practice, it is reasonable to assume that n_t is not large. Therefore, it is possible for us to find an economical solution for the GMEO-1 problem without a human being's interventions. If we name the initial CPLEX solution as GMEO-1(Q, $\overline{Q}$, $\overline{T_{-i_1}}$, L, n_t), an upgraded algorithm GMEO-U is obtained as follows using n_a as the initial number for n_t:

Algorithm 7.2 GMEO-U

```
Input: m, n, Q, L, nₐ
Output: Q̄, nₜ
begin
  nₜ=nₐ;
  do
      nₜ = nₜ₊₁;
      succ=GMEO-1(Q, Q̄, T₋ᵢ₁, L, nₜ);
  while (!succ);
  return nₜ;//The assignment plan is in Q̄ and the
           //returned number is the least nₜ.
end
```

With the GMEO-U algorithm, it seems that we have solved the GMEO-1 problem. The solution to the problem in Section 7.2.1 is the same as that shown in Table 7.8, where $n_a = 9$, $n_d = 10$, and $n_t = 13$.

7.2.4 Performance Experiments and Improvements

With the platform shown in Table 7.5, the time used by GMEO-U to solve the problem presented in Section 7.2.1 is 93 seconds. In another case, with $m = 20$, $n = 5$, $L = [1\ 2\ 3\ 4\ 5]$, and Q with randomly generated values in the range of $[0, 1]$, no result was generated even after 10 hours. These two cases show that the GMEO-U algorithm based on CPLEX is still not practical for teams that do not have many members, i.e. $m \leq 20$.

We created alternative experiments with algorithm GMEO-U by setting different initial values for CPLEX to search for a solution, i.e. n_a, $n_a + 1$, $n_a + 2$, $n_a + 3$, and $n_a + 4$. We notice that CPLEX consumes considerable searching time when there is no feasible solution, i.e. when $n_t = n_a + 1$, $n_a + 2$, and $n_a + 3$. Based on this fact, we recognize an opportunity to improve the GMEO-U algorithm, i.e. to find the smallest number of n_t that guarantees a feasible solution. We find the following Lemma and Theorems:

Lemma 7.1 The necessary condition for GMEO-1 to have a feasible solution is $\sum_{i=0}^{m-1} \overline{Q}[i,j] \geq L[j] + 1 (0 \leq j < n)$.

[**Proof**]: Suppose that there is one role j_1, $\sum_{i=0}^{m-1} \overline{Q}[i, j_1] = L[j] < L[j] + 1$. Now, any agent i_1 that has $\overline{Q}[i_1, j_1] = 1$ is a critical agent because if agent i_1 is removed, no workable $\overline{T_{-i_1}}$ exists. That is, for every $\overline{T_{-i_1}}$, $\sum_{i=0}^{m-2} \overline{T_{-i_1}}[i, j_1] < L[j_1]$. This violates Constraint (5).

Lemma 7.1 is proved.∎

Theorem 7.9 GMEO-1 has a feasible solution *iff* $n_t \geq n_a + n$, i.e. $\sum_{i=0}^{m-1}\sum_{j=0}^{n-1} \overline{Q}[i,j] \geq n_a + n$.

[**Proof**]: We first prove that $n_t \geq n_a + n$ if GMEO-1 has a feasible solution.

From Lemma 7.1, $n_t \geq \sum_{j=0}^{n-1}(L[j] + 1) = \left(\sum_{j=0}^{n-1} L[j]\right) + n = n_a + n$.

The necessary condition part is proved.

Then, we prove that GMEO-1 has a feasible solution *if* $n_t \geq n_a + n$.

Suppose $n_t = n_a + n$. We do have a feasible solution with **Extreme** 2.

Theorem 7.9 is proved. ∎

Based on Theorem 7.9, we may formulate the revised algorithm GMEO-CPLEX as follows. By searching only one case of n_t, we save considerable time by eliminating useless work for CPLEX.

Algorithm 7.3 GMEO-CPLEX

```
Input: m, n, Q, L, nₐ
Output: Q̄, n_d
begin
  n = sizeof(L);
  nₜ=nₐ+n;
  succ=GMEO-1(Q, Q̄, T̄₋ᵢ₁, L, nₜ);
```

$$n_d = \sum_{i=0}^{m-1}\left\lceil \sum_{j=0}^{n-1} \overline{Q}[i, j]/n \right\rceil;$$

(Continued)

Algorithm 7.3 (Continued)

```
if (succ)
      return n_d;//The assignment plan is in Q̄ and the
                 //returned number is the least n_d.
else
      return -1;// To mean failing to find a solution.
endif
end
```

For problem GMEO-1, the number of assigned agents is a side product, i.e. n_d. The ideal situation is the scenario where the less the n_d, the better the assignment plan is, under the same σ_{g1} and n_t. From the above algorithm, we acquire Theorem 7.10.

Theorem 7.10 GMEO-1 has a feasible solution *iff* $n_d \geq n_a + 1$.
[**Proof**]: We at first prove that GMEO-1 has a feasible solution if $n_d \geq n_a + 1$.
Suppose $n_d \geq n_a + 1$, or $n_d = n_a + 1$.
We use **Extreme** 2 to set $\overline{Q}[n_a, j] = 1$ $(0 \leq j < n)$. Now, $n_t \geq n_a + n$.
From Theorem 7.9, GMEO-1 has a feasible solution.
The first part is proved.
Now we prove that $n_d \geq n_a + 1$ if GMEO-1 has a feasible solution.
Suppose that GMEO-1 has a feasible solution $\overline{Q}$.
We obtain $n_t \geq n_a + n$.
Because n_a is required by L, there is at least one more agent in $\overline{Q}$ to be the backup for one agent to leave. That is, $n_d \geq n_a + 1$.
Theorem 7.10 is proved. ■

Based on the above discussions, $n_d \geq n_a + 1$ and $n_t \geq n_a + n$ illustrate an important result, particularly for administrators. For a team with n positions requiring n_a people, we need $n_a + 1$ people and each position must have $L[j] + 1$ assigned people to guarantee that there are no critical people in the team. In other words, ONE backup person is required and you do not have to assign each person with two positions. Note that the theorems only state the conditions for feasible solutions, not the optimal solutions. That is why, in the simulations, some cases obtain $n_d = n_a + 2$. From now on, we assume that $m \geq n_a + 1$ unless stated otherwise.

With the new GMEO-CPLEX algorithm, we test 10 cases of different sizes. In each case, m (=10, 20, ...100), n (=3, 5, 6, 8,..., 20) are determined with corresponding n_as, n_ts, and n_ds, and L $(1 \leq L[j] \leq 5, 0 \leq j < n)$ and $Q \in [0, 1]$ are created randomly. Because this experiment aims to acquire the scales of the consumed time, we use one random group in each test. The results show that GMEO-CPLEX is much better than the initial solution, i.e. GMEO-U has an improved time efficiency in solving the GMEO-1 problem. It is a practical solution that can be applied to the real world. We conducted another experiment where $n = 6$ and $n_a = 8$ were kept unchanged ($1 \leq L[j] \leq 2$) and other settings are similar to those in the above experiment. The result shows that the time used is shorter than that in the above experiment because n and n_a are kept constant (Zhu 2020a).

7.2.5 A Simple Formalization of GMEO with an Efficient Solution

Even though GMEO-CPLEX seems useful, Section 7.2.4 demonstrates that the CPLEX-based solution to the GMEO-1 problem still consumes a lot of time for large groups. Now we investigate if we can simplify the problem and find a more efficient solution. Suppose that we do not consider the preferences of agents or we assume that the efforts of assigning all agents are the same, i.e. Q is ignored or all the elements of Q are 1 s, then the GMEO problem can be simplified. The simplified GMEO problem is denoted as GMEO-S. Similarly, we use GMEO-S$_1$ to mean the critical agents defined in *Definition 7.24*. GMEO-S$_2$ will be left for future research.

Definition 7.27 Given a group expressed by m and L, the *simple GMEO problem* (GMEO-S$_1$) aims to find the best assignment plan matrix $\overline{Q}$ such that no critical agent exists in the group, i.e.

$$min \sum_{i=0}^{m-1} \sum_{j=0}^{n-1} \overline{Q}[i,j]$$

subject to (7.12)–(7.16).

Even though GMEO-S$_1$ is simplified in formalizations, the CPLEX based solution, i.e. GMEO-CPLEX with $Q[i, j] = 1$ ($0 \leq i < m, 0 \leq j < n$), named GMEO-S$_1$-CPLEX, consumes even more time compared with solving GMEO-1. With the same setting of experiments, to solve a problem of $m = 50$ and $n = 10$, the CPLEX-based solution cannot obtain a result after 10 hours of running. The reason is that the search algorithm of CPLEX has to compare all the same values to find the minimum one. This results in useless computations that consume considerable time (Zhu 2020b).

As a matter of fact, GMEO-S$_1$ is simple because the matrix of preference Q is ignored. Based on Theorem 7.9, there is a feasible solution when $n_t = n_a + n$. We only need to set up a $\overline{Q}$ that meets the constraints of L, T, and n_t. Such a $\overline{Q}$ is undoubtedly the best because there is no difference among the agents for all roles. We propose a simple algorithm GMEO-S$_1$-S to solve GMEO-S$_1$ as follows:

Algorithm 7.4 GMEO-S$_1$-S

```
Input: m, n, L, n_a
Output: Q̄, n_d
begin
  st = 0;
  Q̄={0};//Initialize all the elements of Q̄ with 0s.
  for (0≤j<n)
        for(st≤k≤st+L[j])
            Q̄[k,j]=1;
        endfor
        st=st+L[j];
  endfor
```

$$n_d = \sum_{i=0}^{m-1} \left\lceil \sum_{j=0}^{n-1} \overline{Q}[i,j]/n \right\rceil ;$$

```
  return n_d;//The assignment plan is in Q̄ and the
            //returned number is n_d.
end
```

Evidently, algorithm GMEO-S$_1$-S is much more efficient than the CPLEX-based one (GMEO-CPLEX) because the complexity of GMEO-S$_1$-S is $O(n_a + n)$. Now we need to prove that algorithm GMEO-S$_1$-S is correct!

Theorem 7.11   Algorithm GMEO-S$_1$-S is correct.

[**Proof**]: From algorithm GMEO-S$_1$-S, we can obtain a matrix $\overline{Q}$, where Agents L [0], $L[0] + L[1]$,..., and $\sum_{k=0}^{n-2} L[k]$ are assigned to two roles.

To check if there is any critical agent in the obtained matrix, we need to remove agent $i_1 = \sum_{k=0}^{j-1} L[k](0 \le i_1 < m)$, we get $\overline{Q_{-i_1}}$. We check if a workable $\overline{T_{-i_1}}$ exists. The worst case is that the removed agent i_1 has two assigned roles. Now,

$$\text{Let } \overline{T_{-i_1}}[i, j_1] = \begin{cases} \overline{Q_{-i_1}}[i, j_1] & \text{if}\left(i \neq \sum_{k=0}^{j_1} L[k]\right) \\ 0 & \text{if}\left(i = \sum_{k=0}^{j_1} L[k]\right) \end{cases} \quad (0 \le i < i_1, 0 \le j_1 < j);$$

$$\overline{T_{-i_1}}[i, j_1] = \overline{Q_{-i_1}}[i, j_1](0 \le i < i_1, j \le j_1 < n);$$

$$\overline{T_{-i_1}}[i, j_1] = \overline{Q_{-i_1}}[i, j_1](i_1 \le i < m-1, 0 \le j_1 \le j); \text{ and}$$

$$\overline{T_{-i_1}}[i, j_1] = \begin{cases} \overline{Q_{-i_1}}[i, j_1] & \text{if}\left(i \neq \sum_{k=0}^{j_1-1} L[k] - 1\right) \\ 0 & \text{if}\left(i = \sum_{k=0}^{j_1-1} L[k] - 1\right) \end{cases} \quad (i_1 \leq i < m-1, j+1 \leq j_1 < n).$$

$\overline{T_{-i_1}}$ is workable because

$$\sum_{i=0}^{m-2} \overline{T_{-i_1}}[i, j] = L[j] \quad (0 \leq i_1 < m, 0 \leq j < n), \text{ and}$$

$$\sum_{j=0}^{n-1} \overline{T_{-i_1}}[i, j] \leq 1 \quad (0 \leq i_1 < m, 0 \leq i < m-1).$$

Because agent i_1 is not specific, we are sure that there is no critical agent in the group expressed by $\overline{Q}$ and L.

For other cases, suppose that agent i_1 with one assigned role j is removed. We can follow a similar method to obtain a workable $\overline{T_{-i_1}}$: $\overline{T_{-i_1}}[i, j] = \overline{Q_{-i_1}}[i, j] (0 \leq i < m-1)$;

$$\overline{T_{-i_1}}[i, j_1] = \begin{cases} \overline{Q_{-i_1}}[i, j_1] & \text{if}\left(i \neq \sum_{k=0}^{j_1} L[k]\right) \\ 0 & \text{if}\left(i = \sum_{k=0}^{j_1} L[k]\right) \end{cases} \quad (0 \leq i < m-1, 0 \leq j_1 < j); \text{ and}$$

$$\overline{T_{-i_1}}[i, j_1] = \begin{cases} \overline{Q_{-i_1}}[i, j_1] & \text{if}\left(i \neq \sum_{k=0}^{j_1-1} L[k] - 1\right) \\ 0 & \text{if}\left(i = \sum_{k=0}^{j_1-1} L[k] - 1\right) \end{cases} \quad (0 \leq i < m-1, j+1 \leq j_1 < n).$$

Theorem 7.11 is proved. ∎

Note that any feasible solution is an optimal one when dealing with GMEO-S₁ because Q is ignored. Algorithm GMEO-S₁-S and Theorem 7.11 together form an easy-to-follow organizational method, which makes a team that requires n_a agents for n different roles and has no critical agents, as follows:

1) Assign $L[j]$ agents to only role j ($0 \leq j < n-1$) and each agent is assigned at most one role;
2) Choose one extra agent $n_a + 1$ with initial role $j_1(=0)$;
3) Choose one agent among those assigned with role j_1 with role $j_1 + 1$ ($j_1 < n-1$);
4) Repeat 3) until $j_1 = n-2$.

Because the complexity of algorithm GMEO-S₁-S is $O(n_a + n)$, or simply $O(m)$ since we suppose $m \geq n_a + n$, algorithm GMEO-S₁-S is a very fast solution. With the same experiment platform, we implemented algorithm GMEO-S₁-S. The first case we test is $m = 30$, $n = 6$, and $L = [2\ 4\ 4\ 1\ 5\ 3]$ ($n_a = 19$, $n_t = 25$, and $n_d = 20$). The CPLEX-based solution to GMEO-S_1 consumes 11.46 seconds but the implemented algorithm GMEO-S_1-S uses <1 millisecond (ms) and the result of $\overline{Q}$ is the same as that from the CPLEX-based solution. We even tested 100 random cases with $m = 600$, $n = 50$; n_a between 273 and 386, and the GMEO-S_1-S algorithm solved each one in less than 1 ms.

7.2.6 A More Efficient Solution for GMEO-1

From Section 7.2.4, it seems that we have obtained a practical solution to GMEO-1. However, for large groups, GMEO-CPLEX is still time-consuming, we need to check if there is a better solution. GRA has been solved by adapting the Kuhn-Munkres algorithm, also called the Hungarian algorithm (Kuhn 1955; Munkres 1957). The complexity of the GRA algorithm is $O(m^3)$. If we could use GRA to solve GMEO-1, we can design an even more efficient solution.

Theorem 7.12 A necessary condition for **GMEO-1** to have a feasible solution is that in $\overline{Q}$, no two agents are assigned to the same two roles.

[**Proof**]: Suppose that two agents, i_1 and i_2 are assigned to play the same roles j_1 and j_2 in $\overline{Q}$, i.e.

$\overline{Q}[i_1, j_1] = \overline{Q}[i_1, j_2] = \overline{Q}[i_2, j_1] = \overline{Q}[i_2, j_2] = 1$. Now we consider $\overline{Q_{-i_1}}$.

$$\because \sum_{i=0}^{m-1} \overline{Q}[i,j] = L[j] + 1 (0 \leq j < n);$$

$$\therefore \sum_{i=0}^{m-2} \overline{Q_{-i_1}}[i,j_1] = L[j_1] \text{ and } \sum_{i=0}^{m-2} \overline{Q_{-i_1}}[i,j_2] = L[j_2];$$

To form a workable $\overline{T_{-i_1}}$ from $\overline{Q_{-i_1}}$, we have to let $\overline{T_{-i_1}}[i, j_1] = \overline{Q_{-i_1}}[i, j_1]$ $(0 \leq i < m-1)$ to follow constraint (7.4), i.e. $\sum_{i=0}^{m-2} \overline{T_{-i_1}}[i,j] = L[j]$ $(0 \leq i_1 < m, 0 \leq j < n)$.

At the same time, we have to let $\overline{T_{-i_1}}[i, j_2] = \overline{Q_{-i_1}}[i, j_2]$ $(0 \leq i < m-1)$.

Now $\overline{T_{-i_1}}[i_2, j_1] = \overline{T_{-i_1}}[i_2, j_2] = 1$.

$$\because \overline{T_{-i_1}}[i_2,j_1] + \overline{T_{-i_1}}[i_2,j_2] = 2 > 1,$$

$\therefore \overline{T_{-i_1}}$ violates constraint (5), i.e.

$$\sum_{j=0}^{n-1} \overline{T_{-i_1}}[i,j] \leq 1 (0 \leq i_1 < m, 0 \leq i < m-1).$$

$\therefore$ No workable $\overline{T_{-i_1}}$ exists.

Theorem 7.12 is proved. ∎

From the properties of the GMEO problem and Theorem 7.12, we propose a new algorithm as follows. The basic idea is: (i) to use algorithm GRA to obtain the best assignment matrix T for L; and (ii) to add n more assignments to T to form $\overline{Q}$ in consideration of the constraints.

Algorithm 7.5 GMEO-GRA

```
Input: m, n, Q, L, nₐ
Output: Q̄, n_d
begin
        declare Q₁ an m×n matrix;
        n_t=n_a+n;
        if (n_t>m) return 0;// No solution!
        endif
        declare T an m×n matrix;
        GRA(L, Q, T, m, n, 0);
        Initialize Q₁ with Q by setting negative values to
the assigned units to avoid being considered,
```

$$\text{i.e., } Q_1[i, j] = \begin{cases} Q[i, j] & if(T[i, j] \neq 1) \\ -m \times n & if(T[i, j] = 1) \end{cases};$$

```
        declare S as a set of role numbers;
        for (0≤j<n)S = S∪{j};//Initialize the set.
        endfor
        declare p as an assignment <i, j>;
        // To record the indices of an assignment;
        while (S≠Φ)
           Search for the maximum unassigned value amax from
           Q₁ and record the indices, i.e.,
           amax=max{ Q₁ [i, j]|0≤i<m, 0≤j<n-1}; and
           p=<i', j'>, where Q₁ [i', j']=amax.
           if (SatisfiedWithConstraints(T, L, p, m, n))
                       T[p.i, p.j]=1;
                       Q₁[p.i, p.j]= -m×n;
                       S=S-{p.j};
           else //Discard assignment p;
                       Q₁[p.i, p.j]= -m×n;
           endif
        endwhile
        Q̄=T;
        Count the 1s in Q̄ to get n_d;
        return n_d;//The results are in Q̄.
end
```

Algorithm 7.6 Satisfied With Constraints

```
- - - - - - - - - - - - - - - - - - - - - - - - - - - - - - - -
Input: m, n, T, L, p //p is an assignment, i.e., <i,j>.
Output: True or False
begin
     sum= The total number of assigned roles for agent p.i;
     if (sum≥2)
             return False;
     endif
     if (sum==1)
             Record the assigned role as j;
             Check if there is an agent assigned to roles
p.j and j;
             if (there is such an agent)
                     return False;
             endif
     endif
     Count the number of agents assigned to role p.j to x;
     if (x = L[p.j]+1)
             return False;
     endif
     return True;
end
```

Algorithm 7.5 is used to obtain $\overline{Q}$, and Algorithm 7.6 is to check if an assignment follows the constraints.

The most complex part in GMEO-GRA, other than GRA, is the **while** loop whose complexity is at worst $O(m^3)$ because the complexity of the part to search the maximum unassigned Q value is $O(m^2)$ at worst, and **SatisfiedWithConditions** has the complexity of $O(m)$. Therefore, the total complexity of GMEO-GRA is $O(m^3)$.

GMEO-GRA is correct because it follows all the constraints in GMEO-1. The only concern is its optimality expressed by the group performance σ and the number of assigned agents, n_d. Please note that both algorithms are trying to optimize $\sum_{i=0}^{m-1} \sum_{j=0}^{n-1} Q[i,j] \times \overline{Q}[i,j]$ and n_d is just a side-product.

From GMEO-GRA, let $\sigma_{g2} = max \sum_{i=0}^{m-1} \sum_{j=0}^{n-1} Q[i,j] \times \overline{Q}[i,j]$.

Compared with the GMEO-CPLEX solution, we notice that GMEO-GRA may not obtain the globally optimal result, because it uses two locally optimal results, i.e. T and n more assignments, to compose the final result.

We need to check not only the time but also the optimality $(\sigma_{g2}/\sigma_{g1})$ of the new GMEO-GRA algorithm to verify its practicability. The results are very exciting. The first experiment is set as $m = 20$ and $n = 4$. We randomly create 100 groups with different Ls $(1 \leq L[j] \leq 5)$ and $Qs(Q[i, j] \in [0, 1], 0 \leq i < m, 0 \leq j < n)$, and n_a is between 6 and 19. The time range of GMEO-GRA is between 0 and 3 ms (0 means less than 1 ms) and that of GMEO-CPLEX is between 119 and 564 ms. The optimality is ranging from 98%–100%. There are exceptions in the experiments because these four cases have no solutions for both algorithms due to $m < n_t$. The other parameters produced by both algorithms, i.e. n_ds, do not have many differences (Zhu 2020b).

Another experiment is set with $m = 50$ and $n = 10$. We randomly create 100 groups with different Ls $(1 \leq L[j] \leq 5, 0 \leq j < n)$ and Qs $(Q[i, j] \in [0, 1], 0 \leq i < m, 0 \leq j < n)$, and set n_a between 21 and 40. The time range of GMEO-GRA is between 0 and 24 ms (0 means less than 1 ms) and that of GMEO-CPLEX is between 7.6 and 3365 s. The optimality $(\sigma_{g2}/\sigma_{g1})$ is 99%–100%. Only a few n_ds have a small difference. In this experiment, there are also four exceptions for those cases that have no feasible solutions (Zhu 2020b).

The reason why we do not provide other comparisons when m is large, e.g. 100 or more, is that the time consumed by the GMEO-CPLEX solution is too much to acquire 100 results in an acceptable time. The time used by GMEO-GRA is between 100 and 165 ms for another experiment with 100 random groups in the settings of $(m = 200, n = 30, 1 \leq L[j] \leq 5$, and $Q[i, j] \in [0, 1], 0 \leq i < m, 0 \leq j < n)$. The results show that GMEO-GRA is highly practical. It can be well applied in dealing with large real-world teams (Zhu 2020b).

7.3 Related Work

Budget constraints are important for optimization problems in the industry. There have been many researchers studying budget constraints in various fields and applications (Andelman and Mansour 2004; Bhattacharya and Dupas 2012; Cao et al. 2014; Karabakal et al. 2000; Sakellariou et al. 2007; Yu and Buyya 2006). However, the applications of the budget constraints vary considerably depending on specific requirements. No solution can be applied directly to solving the proposed GRABC problems.

Andelman and Mansour (2004) studied combinatorial auctions with budget constraints. They discussed the exact and approximate solutions. Note that the budget constraints are on the individual payers' side. In GRA, the budget is for the team.

It is exciting for us to recognize the work of Bhattacharya and Dupas (2012) because their work extended the potential application of our proposed problem in this section. Their work aimed to allocate a binary treatment among a target population based on observed covariates. Their goals include: (i) to maximize the mean social welfare arising from an eventual outcome distribution, when a budget constraint limits what fraction of the population can be treated and (ii) to infer the dual value, i.e. the minimum resources needed to attain a specific level of mean welfare via efficient treatment assignment. Their methodology was verified by a case study in Western Kenya.

Cao et al. (2014) focused their work on developing workflow-scheduling algorithms considering both budget and throughput constraints. The objectives of their problem include (i) maximizing throughput under a budget constraint and (ii) minimizing the execution cost under the minimum throughput constraint. Their work also presented a potential application field for our proposed problems and solutions.

Karabakal et al. (2000) investigated the asset replacement problem with budget constraints. Due to the lack of efficient ILP solvers at the time of their research, they proposed a dual heuristic method, i.e. first solve the problem without the budget constraint and then reduce budget violations by solving a Lagrangian dual problem.

Sakellariou et al. (2007) developed two workflow-scheduling approaches for Grids, i.e. LOSS and GAIN, to adjust schedules that are generated by time-optimized and cost-optimized heuristics, to meet users' budget constraints. Their algorithm is based on a Directed Acyclic Graph (DAG) scheduling algorithm that aims to minimize the makespan.

Yu and Buyya (2006) put effort into workflow-scheduling problems with considerations of budget constraints. Their application background is the workflow scheduling on the Grids. The solution minimizes the execution time while meeting a specified budget for delivering results by applying a genetic algorithm. Their work is also a potential application field for our proposed solutions.

The most related work to GMEO is found in the literature concerning cross-training in management and psychology, where cross-training means that team members are trained on each others' roles (Gorman et al. 2010; Easton 2011; Li et al. 2012; Marks et al. 2002). However, the problems, the approaches, and the objectives of such studies have significant differences from GMEO. Nobody has provided an exact answer to similar problems. GMEO expresses an entirely different problem compared with the previously investigated problems. The solution to the GMEO problem extends the application of RBC and E-CARGO.

The aforementioned research indicates a strong need to investigate the problems discussed in this chapter.

7.4 Summary

This chapter discusses two complex problems in engineering, which can be formalized into multi-objective optimization problems with the assistance of E-CARGO. The GRABC problems are discussed through formalizations, theoretical analysis, and practical solutions. We also verified the proposed solutions through simulations and experiments. The experiments indicate that the proposed solutions perform well enough for relatively large problems ($m = 400$). The various solutions are synthesized into a flowchart to guide practitioners in applying the solutions to real-world problems. A complete investigation into the GMEO (or GMEO-1) problem is described and efficient solutions are proposed with the complexity of $O(m^3)$ and $O(m)$ for the two fundamental forms of GMEO, i.e. GMEO-1 and GMEO-S$_1$.

The work of this chapter again demonstrates that RBC and E-CARGO are an excellent vehicle to convey challenging collaboration and management to a successful accomplishment. Further investigations may be required along the following directions:

1) In theory, we may analyze different forms of the GRABC problem to specify the necessary and sufficient conditions for a problem to have a feasible solution.
2) Applying the different forms of GRABC into social simulations may help solve more problems in social systems.
3) Improving the GRABC formalizations to adapt to the welfare treatment assignment problems described in (Bhattacharya and Dupas 2012; Cao et al. 2014; Yu and Buyya 2006) may find better ways to solve these problems.
4) Theoretically, we only solve the GMEO problem in the most fundamental form, i.e. a team is kept in a good state if ONE agent leaves the team, or simply, GMEO-1 (GMEO-S$_1$). A problem of GMEO may have more complex forms, such as if a team can afford TWO (GMEO-2 (GMEO-S$_2$), or even more, GMEO-k (GMEO-S$_k$, $k > 2$)) agents leaving the team at the same time. Therefore, it is important to investigate such problems.

There are many similar problems to what we have discussed in this chapter, i.e. optimization with multiple objectives when dealing with assignments. In intelligent adaptive systems (Hou et al. 2014), cloud computing (Ma et al. 2017), group decision making (Qin et al. 2017), Manufacturing (Yang et al. 2016), and intelligent transportation systems (Zuo et al. 2015), there are many GRA⁺⁺ problems. Readers can follow the methods described in this chapter to formalize and solve other complex problems in engineering.

References

Andelman, N. and Mansour, Y. (2004). Auctions with budget constraints. In: *Algorithm Theory - SWAT 2004*, Lecture Notes in Computer Science, vol. 3111 (eds. T. Hagerup and J. Katajainen), 26–38. Berlin Heidelberg: Springer.

Bhattacharya, D. and Dupas, P. (2012). Inferring welfare maximizing treatment assignment under budget constraints. *J. of Econometrics* 167 (1): 168–196.

Bradley, S.P., Hax, A., and Magnanti, T. (1977). *Applied Mathematical Programming*. Boston, MA: Addison-Wesley.

Burkard, R.E., Dell'Amico, M., and Martello, S. (2009). *Assignment Problems, Revised Reprint*. Philadelphia, PA: Siam.

Cao, F., Zhu, M.M., and Ding, D. (2014). Distributed workflow scheduling under throughput and budget constraints in grid environments. In: *Job Scheduling Strategies for Parallel Processing*, Lecture Notes in Computer Science, vol. 8429 (eds. N. Desai and W. Cirne), 62–80. Berlin Heidelberg: Springer.

Easton, F.F. (2011). Cross-training performance in flexible labor scheduling environments. *IIE Trans.* 43: 589–603.

Garey, M.R. and Johnson, D.S. (1979). *Computers and Intractability: A Guide to the Theory of NP-Completeness*. New York: W. H. Freeman and Company.

J. C. Gorman, N. J. Cooke, and P. G. Amazeen (2010), Training adaptive teams *Human Factors*, 52, 2, pp. 295–307.

Hansen, P., Jaumard, B., and Mathon, V. (1993). State-of-the-art survey -constrained nonlinear 0-1 programming. *ORSA J. on Computing* 5 (2): 97–119.

Hou, M., Banbury, S., and Burns, C. (2014). *Intelligent Adaptive Systems: An Interaction-Centered Design Perspective*. Boca Raton, FL: CRC Press, Taylor and Francis Group.

IBM (2017). ILOG CPLEX optimization studio. http://www-01.ibm.com/software/integration/optimization/cplex-optimization-studio/ (accessed 2017).

Karabakal, N., Bean, J.C., and Lohmann, J.R. (2000). Solving large replacement problems with budget constraints. *The Engineering Economist*, A J. Devoted to the Problems of Capital Investment 45 (4): 290–308.

Kuhn, H.W. (1955). The Hungarian method for the assignment problem. *Naval Research Logistic Quarterly* 2: 83–97. (Reprinted in vol. 52, no. 1, 2005, pp. 7-21.)

Kuhn, H.W. (2014). Nonlinear programming: a historical view. In: *Traces and Emergence of Nonlinear Programming* (eds. G. Giorgi and T.H. Kjeldsen), 393–414. Basel: Springer.

Li, Q., Gong, J., Fung, R.Y.K., and Tang, J. (2012). Multi-objective optimal cross-training configuration models for an assembly cell using non-dominated sorting genetic algorithm-II. *International Journal of Computer Integrated Manufacturing* 25 (11): 981–995.

Ma, H., Zhu, H., Hu, Z. et al. (2017). Multi-valued collaborative QoS prediction for cloud service via time series analysis. *Future Generation Computer Systems* 68: 275–288.

Marks, M.A., Sabella, M.J., Burke, C.S., and Zaccaro, S.J. (2002). The impact of cross-training on team effectiveness. *Journal of Applied Psychology* 87 (1): 3–13.

Munkres, J. (1957). Algorithms for the assignment and transportation problems. *J. of the Society for Industrial and Applied Mathematics* 5 (1): 32–38.

Papadimitriou, C.H. (1981). On the complexity of integer programming. *Journal of the ACM* 28 (4): 765–768.

Qin, Q., Liang, F., Li, L. et al. (2017). A TODIM-based multi-criteria group decision making with triangular intuitionistic fuzzy numbers. *Applied Soft Computing J.* 55: 93–107.

Rardin, R.L. (1997). *Optimization in Operations Research*. Upper Saddle River, NJ: Prentice Hall.

Sakellariou, R., Zhao, H., Tsiakkouri, E., and Dikaiakos, M.D. (2007). Scheduling workflows with budget constraints. In: *Integrated Research in GRID Computing* (eds. S. Gorlatch and M. Danelutto), 189–202. Springer, US.

Schreiber, C. and Carley, K.M. (2004). Key personnel: identification and assessment of turnover risk. *Proceedings of the North American Association of Computational Social and Organizational Science Conference* (27–29 June 2004). Pittsburgh, PA, July 2004. https://www.researchgate.net/profile/Kathleen_Carley/publication/228856499_Key_personnel_Identification_and_assessment_of_turnover_risk/links/00b7d517d66cda6964000000.pdf.

Wolsey, L.A. and Nemhauser, G.L. (1999). *Integer and Combinatorial Optimization*. New York: Wiley-Interscience.

Yang, C., Shen, W., Lin, T., and Wang, X. (2016). A hybrid framework for integrating multiple manufacturing clouds. *The Int'l J. of Advanced Manufacturing Technology* 86 (1): 895–911.

Yu, J. and Buyya, R. (2006). A budget constrained scheduling of workflow applications on utility grids using genetic algorithms. *Proceedings of the Workshop on Workflows in Support of Large-Scale Science/HPDC'06*, Paris, France (June 2006), 10 pages.

Zhu, H. (2016). Avoiding conflicts by group role assignment. *IEEE Trans. on Systems, Man, and Cybernetics: Systems* 46 (4): 535–547.

Zhu, H. (2020a). Maximizing group performance while minimizing budget. *IEEE Trans. on Systems, Man, and Cybernetics: Systems* 50 (2): 633–645.

Zhu, H. (2020b). Avoiding critical members in a team by redundant assignment. *IEEE Trans. on Systems, Man, and Cybernetics: Systems* 50 (7): 2729–2740.

Zhu, H. and Zhou, M.C. (2006). Role-based collaboration and its kernel mechanisms. *IEEE Trans. on Systems, Man and Cybernetics, Part C* 36 (4): 578–589.

Zhu, H. and Zhou, M. (2009). M–M role-transfer problems and their solutions. *IEEE Trans. on Systems, Man and Cybernetics, Part A: Systems and Humans* 39 (2): 448–459.

Zhu, H. and Zhou, M.C. (2012). Efficient role transfer based on Kuhn–Munkres algorithm. *IEEE Trans. on Systems, Man and Cybernetics, Part A: Systems and Humans* 42 (2): 491–496.

Zhu, H., Hou, M., Wang, C., and Zhou, M.C. (2012a). An efficient outpatient scheduling approach. *IEEE Trans. on Automation Science and Engineering* 9 (4): 701–709.

Zhu, H., Zhou, M.C., and Alkins, R. (2012b). Group role assignment via a Kuhn-Munkres algorithm-based solution. *IEEE Trans. on Systems, Man and Cybernetics, Part A* 42 (3): 739–750.

Zhu, H., Liu, D., Zhang, S. et al. (2016). Solving the many to many assignment problem by improving the Kuhn-Munkres algorithm with backtracking. *Theoretical Computer Science* 618: 30–41.

Zhu, H., Liu, D., Zhang, S. et al. (2017). Solving the group multi-role assignment problem by improving the ILOG approach. *IEEE Trans. on Systems, Man, and Cybernetics: Systems* 47 (12): 3418–3424.

Zhu, H., Sheng, Y., Zhou, X.-Z., and Zhu, Y. (2018). Group role assignment with cooperation and conflict factors. *IEEE Trans. on Systems, Man, and Cybernetics: Systems* 48 (6): 851–863.

Zuo, X., Chen, C., Tan, W., and Zhou, M.C. (2015). Vehicle scheduling of an urban bus line via an improved multiobjective genetic algorithm. *IEEE Trans. on Intelligent Transportation Systems* 16 (2): 1030–1041.

Exercises

1 What do we mean by GRA^{++}?

2 What is the definition of GRABC?

3 Specify the problems $GRABC\text{-}P_1$, $GRABC\text{-}P_2$, $GRABC\text{-}P_3$, $GRABC\text{-}B_1$, $GRABC\text{-}B_2$, and $GRABC\text{-}B_3$.

4 Specify the problems of GRABC-WS and GRABC-Syn.

5 Could you describe the flowchart for processing a GRABC problem?

6 What data do you need to collect when you hope to solve a GRABC problem?

7 What is the definition of GMEO?

8 In GMEO, what do we mean by $\overline{Q}$, $\overline{Q_{-i_1}}$, and $\overline{T_{-i_1}}$?

9 What do we mean by GMEO-1, GMEO-k, and GMEO-U?

10 What do we mean by GMEO-S, $GMEO\text{-}S_1$, and $GMEO\text{-}S_2$?

Part III

Applications

8

Solving Engineering Problems with GRA

8.1 Group Role Assignment with Agents' Busyness Degrees

Task allocations in collaboration are complex. They involve many factors that are difficult to specify and handle. Failing to consider such factors in the role assignment stage may produce obstacles that lead to collaboration failures. Such factors should be taken seriously while assigning roles. Busyness is a prominent example of such a problem. This section presents a challenging assignment problem called Group Role Assignment (GRA) with Agents' Busyness Degrees (GRAABD) (Zhu and Zhu 2017), which is an instance of the GRA with Multiple Objectives (GRA^{++}) problem. The solution to this problem aims at creating a high-performance group through role assignment with the consideration of the agents' busyness degrees.

People's abilities and energy are limited, however powerful they are. For example, every individual has the same number of hours in a day, i.e. 24. Even though different people may have different working efficiencies and may extend their usage of the 24 hours, the total time in a day is fixed. Every administrator would prefer to assign tasks to an individual with good attributes like work ethic, reputation, experience, passion, motivation, and qualification. However, such people are almost always busy. Overloading may cause highly qualified people to lower their quality of work and efficiency of task executions, i.e. overloading may negatively affect their health and task performance (Ann et al. 2010; Estes 2015; Kerfoot 2006; Ruffing 1995). Busyness, then, becomes an important factor in the task assignment process. In this section, we believe that busyness is a state of a person or an agent that consists of physical, psychological, and mental factors. By considering busyness, we will be able to avoid overloading and maintain team productivity.

E-CARGO and Role-Based Collaboration: Modeling and Solving Problems in the Complex World,
First Edition. Haibin Zhu.
© 2022 by The Institute of Electrical and Electronics Engineers, Inc.
Published 2022 by John Wiley & Sons, Inc.

8.1.1 A Real-World Scenario

In a company, Ann, the Chief Executive Officer (CEO), has just signed a million-dollar contract. She asks Bob, the Human Resources (HR) officer, to organize a team from the employees of the company. Bob drafts a position list as shown in Table 8.1 for the team and a candidate staff shortlist as shown in Table 8.2, where the numbers in parentheses are indices of positions and people. Then, Bob initiates an evaluation process and asks the branch officers to evaluate the employees for each possible position (Table 8.2).

After that, Ann and Bob have a meeting with the branch officers to assign positions to the candidate employees. The officers report that most candidates have other duties and responsibilities. Ann asks the branch officers to judge the candidates' busyness degrees using the Likert-type scale: strongly agree, agree, neutral, disagree, and strongly disagree. The results are shown in Table 8.3.

Table 8.1 The required positions.

Position	Project manager	Senior programmer	Programmer	Tester
Required Number	1	2	4	2

Table 8.2 The candidates and evaluations on positions.

	Positions			
Candidates	Project manager (0)	Senior programmer (1)	Programmer (2)	Tester (3)
Adam (0)	0.18	0.82	0.29	0.01
Bret (1)	0.35	**0.80**	0.58	0.35
Chris (2)	0.84	**0.85**	[0.86]	0.36
Doug (3)	0.96	0.51	0.45	0.64
Edward (4)	0.22	0.33	**0.68**	0.33
Fred (5)	0.96	0.50	0.10	**0.73**
George (6)	0.25	0.18	0.23	**0.39**
Harry (7)	0.56	0.35	**0.80**	0.62
Ice (8)	0.49	0.09	0.33	0.58
Joe (9)	0.38	0.54	**0.72**	0.20
Kris (10)	[0.91]	0.31	0.34	0.15
Larry (11)	**0.85**	0.34	0.43	0.18
Matt (12)	0.44	0.06	**0.66**	0.37

Table 8.3 The candidates' busyness degrees in the Likert-type scale.

Candidates	Busy?
Adam (0)	Strongly agree
Bret (1)	**Agree**
Chris (2)	**Agree**
Doug (3)	Agree
Edward (4)	**Disagree**
Fred (5)	**Strongly disagree**
George (6)	**Disagree**
Harry (7)	**Neutral**
Ice (8)	Neutral
Joe (9)	**Neutral**
Kris (10)	Agree
Larry (11)	**Neutral**
Matt (12)	**Disagree**

Subsequently, Ann tells Bob to try his best to maximize the group performance by assigning appropriate candidates to the relevant positions based on their qualifications and busyness degrees. Bob considers this for a while and then tells Ann that such a problem may require a significant amount of processing time to find a satisfactory solution. Fortunately, Ann, as a skilled administrator, allows for a reasonable response time.

In the above scenario, Ann and Bob follow the initial steps of Role-Based Collaboration (RBC) (Chapter 3; Zhu 2015; Zhu and Zhou 2006) and Bob encounters a GRA problem (Chapter 5; Zhu et al. 2012) that requires consideration of an additional constraint, i.e. the busyness degree of the agents.

8.1.2 Problem Formalization

In the previous chapters, we notice that the state of a group after role assignment is highly dependent on different situations. Therefore, prior to assignment, we need to specify the constraints, e.g. agent conflicts and cooperation factors specified in Chapter 6.

In this section, we introduce a new factor, i.e. the busyness degrees of agents. Considering this factor, a reasonable and acceptable assumption is that busy agents' real performances on roles are not as high as those shown in Q. To reflect the busyness of agents in GRA, we need to define a new vector.

Definition 8.1 A *busyness vector* is an *m*-vector $V^b[i] \in [0, 1]$ that expresses the busyness degree of agent i ($0 \le i \le m - 1$).

For example, if we use numbers in $[0, 1]$ to express the busyness degrees, Table 8.3 becomes Table 8.4, i.e. the values are assigned as: strongly agree = 0.95, agree = 0.75, netural = 0.5, disagree = 0.25, and strongly disagree = 0.05.

Definition 8.2 Given Q, L, and V^b, the *GRA with Agents' Busyness Degree* (GRAABD) problem is to find a T to:

$$max \sum_{i=0}^{m-1} \sum_{j=0}^{n-1} Q[i,j] \times T[i,j]$$

$$and \ min \sum_{i=0}^{m-1} \sum_{j=0}^{n-1} V^b[i] \times T[i,j]$$

subject to

$$T[i,j] \in \{0, 1\} \quad (0 \le i < m, 0 \le j < n), \tag{8.1}$$

$$\sum_{i=0}^{m-1} T[i,j] = L[j] \ (0 \le j < n), \tag{8.2}$$

$$\sum_{j=0}^{n-1} T[i,j] \le 1 \ (0 \le i < m) \tag{8.3}$$

Table 8.4 The candidates' busyness degrees in $[0, 1]$.

Candidates	Busy degrees?
Adam (0)	0.95
Bret (1)	0.75
Chris (2)	0.75
Doug (3)	0.75
Edward (4)	0.25
Fred (5)	0.05
George (6)	0.25
Harry (7)	0.5
Ice (8)	0.5
Joe (9)	0.5
Kris (10)	0.75
Larry (11)	0.5
Matt (12)	0.25

Evidently, GRAABD is a multi-objective optimization problem (MOOP) (Rardin 1997). Mathematically, GRAABD is equivalent to GRA with Budget Constraints (GRABC) (Chapter 7). Therefore, we can use the same methods to solve the GRAABD problem. The key point is to evaluate agents on roles to obtain a pertinent Q matrix. We can compose Q with the weighted sum method, i.e. using two objectives, the weight w parameter for performance, and 1-w for busyness.

8.1.3 Solutions

Definition 8.3 Given Q, L, V^b, and w, the *Group Role Assignment with Agents' Busyness Problem in the Weighted Sum Form* (*GRAABD-WS*) aims to find a workable T to obtain

$$\sigma_{bw} = max \left\{ w \times \sum_{i=0}^{m-1} \sum_{j=0}^{n-1} Q[i,j] \times T[i,j] - (1-w) \times \sum_{i=0}^{m-1} \sum_{j=0}^{n-1} V^b[i] \times T[i,j] \right\}$$

subject to (8.1)–(8.3).

To solve GRAABD-WS by GRA, we compose a new Q matrix based on the objective in Definition 8.3 of GRAABD-WS. By transforming the objective, we justify the following.

Because $w \times \sum_{i=0}^{m-1} \sum_{j=0}^{n-1} Q[i,j] \times T[i,j] - (1-w) \times \sum_{i=0}^{m-1} \sum_{j=0}^{n-1} V^b[i] \times T[i,j]$

$$= \sum_{i=0}^{m-1} \sum_{j=0}^{n-1} w \times Q[i,j] \times T[i,j] - \sum_{i=0}^{m-1} \sum_{j=0}^{n-1} (1-w) \times V^b[i] \times T[i,j]$$

$$= \sum_{i=0}^{m-1} \sum_{j=0}^{n-1} \left(w \times Q[i,j] - (1-w) \times V^b[i] \right) \times T[i,j], \text{ the equivalent objective becomes}$$

$$max \left\{ \sum_{i=0}^{m-1} \sum_{j=0}^{n-1} \left(w \times Q[i,j] - (1-w) \times V^b[i] \right) \times T[i,j] \right\}.$$

Therefore, the GRAABD-WS objective becomes to find a workable T to obtain

$$\sigma_{bw} = max \left\{ \sum_{i=0}^{m-1} \sum_{j=0}^{n-1} Q^{b'}[i,j] \times T[i,j] \right\}$$

subject to (8.1)–(8.3), and

$$Q^{b'}[i,j] = w \times Q[i,j] - (1-w) \times V^b[i] \quad (0 \le i < m, 0 \le j < n) \tag{8.4}$$

In (8.4), to consider the impact of busyness on the qualification values of agents on roles, we assume that an increase of busyness degrees leads to a decrease in qualification values. We use T^{bw} to express the T that makes σ_{bw}.

Using the above formalization, we propose a straightforward algorithm GRAABD-WS based on the GRA algorithm, i.e. Algorithm 5.3 (Chapter 5). The algorithm is as follows:

Algorithm 8.1 GRAABD-WS.

```
Input:
        m - the number of agents;
        n - the number of roles;
        Q - the mxn matrix of qualifications;
        L - the n-vector of role requirements;
        Vᵇ - the m-vector of agents' busyness degrees; and
        w - the weight for performance.
Output:
        Success: T′
        Fail: no workable T is obtained.
begin
   for(0≤i< m)
     for(0≤j< n)
            Q′[i, j]= w×Q[i, j] - (1- w)× Vᵇ[i]);
     endfor
   endfor
   succ=SolveRGRAP(Q′,L,T′,m,n);
   return succ;
end
```

Because the complexity of the GRA algorithm is $O(m^3)$, Algorithm 8.1 is of the same complexity, i.e. $O(m^3)$ and can be used to solve the problem in Section 8.1.1 for Bob.

The assignment result from GRAABD-WS is shown in Tables 8.2 and 8.3 in **Bold**. It is evident that this assignment accounts for the busyness degree of agents as no agent with "strongly agree (very busy)" as their busyness degree is assigned.

If the original Q matrix is used to calculate the group performance of the new assignment, we obtain a group performance of 6.48. Compared to the original group performance of 6.96 obtained from GRA (Tables 8.2–8.4, underlined numbers), the performance change is only $(6.96 - 6.48)/6.96 \approx 7\%$. That is to say, in the situation of Section 8.1.1, only a 7% decrease in group performance is needed to take care of the team members' busyness degrees. Considering the impacts of busyness on performance, this may even imply an overall performance increase. The sum of all the assigned busyness values by GRAABD-WS is 3.8 (Table 8.3, **Bold**), while the value calculated using GRA is 5.25. Thus, the reduction in busyness is $(5.25 - 3.8)/5.25 = 27.6\%$, which is much more significant than the 7% decrease in performance.

Using $w = 0.0$ to 1.0 with a step of 0.1, Figure 8.1 shows the effects of different weight distributions on the performance and busyness of the GRAABD-WS solution. Compared with GRA, GRAABD-WS obtains balanced performance and busyness (Zhu and Zhu 2017).

8.1.4 Simulations and Benefits

To verify the benefits of GRAABD-WS compared with GRA, we assume that busy and highly energetic agents can deal with their jobs more efficiently and busyness affects the performance in the actual execution of tasks. The following matrix Q^r defines the effects.

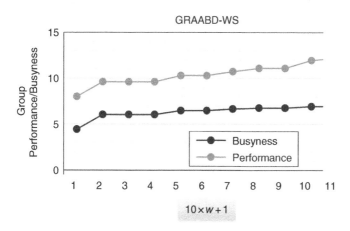

Figure 8.1 The effects of weights in GRAABD-WS.

Definition 8.4 The *actual qualification matrix* Q^r is an $m \times n$ matrix, where $Q^r \in [0, 1]$ expresses the qualification value of agent $i \in \mathcal{N}$ $(0 \leq i < m)$ for role $j \in \mathcal{N}$ $(0 \leq j < n)$ in actual task execution when considering the busyness vector V^b, such that:

$$Q^r[i,j] = \begin{cases} 3 \times (1 - V^b[i]) \times Q[i,j] & (Q[i,j] \geq 0.8) \\ 2 \times (1 - V^b[i]) \times Q[i,j] & (0.5 \leq Q[i,j] < 0.8) \\ (1 - V^b[i]) \times Q[i,j] & (Q[i,j] < 0.5) \end{cases} \qquad (8.5)$$

In (8.5), we suppose that highly qualified agents are more effective than those with lower qualifications, e.g. 3, 2, and 1. For example, if agent i's busyness is 0.8 and $Q[i, j]$ is 0.9, then the actual performance $Q^r[i, j]$ is $3 \times (1 - 0.8) \times 0.9 = 0.54$; and if agent i's busyness is 0.3, $Q[i, j]$ is 0.7, the actual performance $Q^r[i, j]$ is $2 \times (1 - 0.3) \times 0.7 = 0.98$. This implies that a person with a lower qualification value and sufficient time may perform significantly better than a highly qualified person who is too busy.

To make Q^r comparable with Q, we normalize Q^r to Q^{nr} using the following:

$$Q^{nr}[i,j] = \frac{Q^r[i,j] - \min\{Q^r[i,j]\}}{\max\{Q^r[i,j] - \min\{Q^r[i,j]\}\}} \qquad (0 \leq i < m, 0 \leq j < n). \qquad (8.6)$$

$\sigma_0 = \sum_{i=0}^{m-1} \sum_{j=0}^{n-1} Q[i,j] \times T^*[i,j]$ is the group performance with GRA based on Q (*Chapter 5*).

$\sigma_{bw} = \sum_{i=0}^{m-1} \sum_{j=0}^{n-1} Q[i,j] \times T^{bw}[i,j]$ is the group performance using GRAABD-WS $(w = 0.5)$ based on Q (Definition 8.3).

$\sigma_{b1} = \sum_{i=0}^{m-1} \sum_{j=0}^{n-1} Q^{nr}[i,j] \times T^*[i,j]$ is called the actual group performance using GRA.

$\sigma_{b2} = \sum_{i=0}^{m-1} \sum_{j=0}^{n-1} Q^{nr}[i,j] \times T^{bw}[i,j]$ is called the actual group performance considering the busyness of agents.

The comparison between the group performance obtained by *GRAABD-WS* and that by GRA is $\lambda_{b1} = \sigma_{bw}/\sigma_0$. λ_{b1} represents the rate between the group performance of the GRAABD-WS assignment and that of the GRA assignment.

The benefit of GRAABD-WS is $\lambda_{b2} = (\sigma_{b2}-\sigma_{b1})/\sigma_{b1}$. λ_{b2} shows the percent increase in the actual group performance of GRAABD-WS compared with GRA.

We use simulations to obtain λ_{b1} and λ_{b2}. The simulation platform is shown in Table 8.5.

Table 8.5 Simulation platform configuration for GRAABD.

Hardware	
CPU	**Intel core i7-4650U@1.7 GHz**
MM	8 GB
Software	
OS	Windows 7 Enterprise 64-bit
Eclipse	Version: Oxygen.3a Release (4.8.1a)
JDK	Java 8 64 bit

8.1.4.1 Simulation 1

Assume $m = 50$, $n = 10$, $w = 0.5$. Random numbers are set using parameters with the following ranges:

1) $1 \leq L[j] \leq 5$ to guarantee there are enough agents.
2) $0 \leq Q[i, j] \leq 1$ $(0 \leq i < m, 0 \leq j < n)$.
3) $0 \leq V^b[i] \leq 1$ $(0 \leq i < m)$.

We simulate 100 random groups and collect σ_0, σ_{bw}, σ_{b1}, σ_{b2}, λ_{b1}, and λ_{b2}.

The results show that the performance (λ_{b1}) of GRAABD-WS compared with GRA is relatively stable and fluctuates only between 91% and 98%. The increase in the actual group performance of GRAABD-WS (λ_{b2}) compared to GRA varies considerably from 9% to 56% (Zhu and Zhu 2017).

8.1.4.2 Simulation 2

To verify the result obtained in Simulation 1, we conduct another simulation to observe the benefits in different groups sizes.

In this simulation, we use m ranging from 10 to 200 with steps of 10, $n = m/5$, and $w = 0.5$. The ranges of other parameters are identical to Simulation 1. For each m, we create 100 random groups and collect the maximum, minimum, and average λ_{b2}. The result shows that GRAABD-WS has clear benefits in terms of the actual group performances compared to GRA, especially when $m \geq 20$.

Note that when $m = 10$ and $n = 2$, the average benefit is positive but some cases present losses. In the simulation, there is a special case of GRAABD-WS that has losses, where Figure 8.2a is Q, Figure 8.2b is Q^{nr}, Figure 8.2c is T^{bw}, Figure 8.2d is T^*, $L = [5\ 2]$, $V^b = [0.03\ 0.35\ 0.03\ 0.94\ 0.11\ 0.82\ 0.11\ 0.13\ 0.51\ 0.13]$, to illustrate a negative case with $\sigma_1 = 3.61$, and $\sigma_2 = 3.57$ for comparison. In such a case, GRAABD-WS results in a poorer assignment because GRA (Figure 8.2d) chooses agents with better qualifications (0.02 and 0.07 in Q^{nr}). In contrast, GRAABD-WS (Figure 8.2c) chooses agents (0.03 and 0.02 in Q^{nr}) considering busyness.

(a)   (b)   (c)   (d)

$$
\begin{bmatrix}
0.85 & 0.37 \\
0.22 & 0.29 \\
0.86 & 0.42 \\
0.52 & 0.44 \\
0.12 & 0.28 \\
0.57 & 0.13 \\
0.55 & 0.63 \\
0.08 & 0.42 \\
0.10 & 0.74 \\
0.12 & 0.84
\end{bmatrix}
\begin{bmatrix}
0.98 & 0.14 \\
0.05 & 0.07 \\
1.00 & 0.15 \\
0.02 & 0.00 \\
0.03 & 0.09 \\
0.07 & 0.00 \\
0.39 & 0.45 \\
0.02 & 0.14 \\
0.01 & 0.28 \\
0.03 & 0.87
\end{bmatrix}
\begin{bmatrix}
1 & 0 \\
0 & 0 \\
1 & 0 \\
0 & 0 \\
1 & 0 \\
0 & 0 \\
1 & 0 \\
1 & 0 \\
0 & 1 \\
0 & 1
\end{bmatrix}
\begin{bmatrix}
1 & 0 \\
0 & 0 \\
1 & 0 \\
1 & 0 \\
0 & 0 \\
1 & 0 \\
1 & 0 \\
0 & 0 \\
0 & 1 \\
0 & 1
\end{bmatrix}
$$

Figure 8.2 The matrices of a negative case. (a) Q. (b) Q^{nr}. (c) T^{bw}. (d) T^{*}.

Fortunately, such cases are rare and exist predominately in small groups ($m = 10$). Over 80% of the simulated GRAABD-WS cases perform better than GRA.

8.1.4.3 Simulation 3

Using Definition 8.4, we propose a new solution to the GRAABD problem in order to address the ineffective GRAABD-WS cases in simulation 2.

Definition 8.5   Given Q, L, V^{b}, and w, the *Group Role Assignment with Agents' Busyness Problem in the Synthesized Form (GRAABD-Syn)* aims to find a workable T to obtain

$$
\sigma_{syn} = max \left\{ \sum_{i=0}^{m-1} \sum_{j=0}^{n-1} Q^{nr}[i,j] \times T[i,j] \right\}
$$

subject to (8.1)–(8.3), (8.5), and (8.6).

Compared with GRAABD-WS, GRAABD-Syn optimizes role assignment using Q^{nr}, which represents the actual qualification Q matrix that considers the busyness of agents, i.e. V^{b}. The simulation shows that GRAABD-Syn surpasses GRA in every case and demonstrates that GRAABD-Syn is the best role assignment methodology if the impact of busyness on agent qualification is known (Zhu and Zhu 2017).

8.2 Group Multi-Role Assignment with Coupled Roles

There are a variety of constraints when assigning agents to roles within a group. One such constraint is role coupling. Role coupling is a complex problem that cannot be solved without appropriate modeling tools. Assigning components to programmers is a common task undertaken by software project managers prior to the coding phase. From the context of software engineering (Pressman and Maxim

2014), the components assigned to the software team staff should be highly coherent and lowly coupled. That is, a software developer should be assigned highly coherent modules that are lowly coupled with the modules assigned to other developers. Such a coherence/coupling idea is easy to understand but difficult to accomplish. This is due to the many factors that affect the manager's ability to accomplish this assignment. The lack of modeling methodology is one of those factors.

In the software development life cycle (Pressman and Maxim 2014), software engineers emphasize the process of requirement analysis, design, coding, and maintenance. Advanced software development methodologies concentrate on software design and design tools that support software development but neglect the crucial issue of staff management. Even though coupling is an important aspect of software product development, there are limited methodologies that directly support component assignment. This section illustrates that component assignment is an important and complex problem that has been neglected in existing software development research and investigation.

Component assignment should be accomplished between the steps of design and coding in the software development life cycle. Such a task is normally accomplished subjectively by a project manager using traditional management skills, such as programming skills, experience, and historical data. This situation is made more difficult due to the lack of available tools and models in formalizing such a complex problem.

This section systematically analyzes the Group Multi-Role Assignment with Coupled Roles (GMRACR) problem (Zhu 2019a) and formalizes it as an optimization problem. After formalization, we solve the problem using the IBM ILOG CPLEX Optimization Platform (CPLEX) (IBM 2019). With the proposed solutions, the GMRACR problem can be solved with numerical exactness. We believe that the results obtained in this section can be applied to many other applications, such as administration, management, robot teams (Mosteo et al. 2017), highly available computing platforms, and high-performance services.

8.2.1 A Real-World Scenario

In an Information Technology (IT) company, Ann, the project manager of a software product, has led the completion of the design of a software product and prepares to start coding. She is concerned with the coupling of the component assignments because she knows that if the components are not carefully assigned, she will face challenges arising from possible arguments among the programmers, especially those who have modules tightly coupled with other programmers' modules. She lists all the components and their couplings in Table 8.6, where the numbers express the degree of coupling, i.e. larger numbers indicate more coupling between the two components.

The qualifications of each programmer for each component are shown in Table 8.7. They are primarily based on historical data about the programmers.

Table 8.6 The components and their couplings.

Components	User interaction	Visualization	Data collection	Data analysis	Data sharing	Database management	Communication	System logging	Extensive computation
User interaction	0	2	3	2	1	2	1	2	1
Visualization	2	0	2	4	2	1	2	1	3
Data collection	3	2	0	3	3	3	2	1	1
Data analysis	2	4	3	0	2	3	1	1	3
Data sharing	1	2	3	2	0	3	4	4	1
Database management	2	1	3	3	3	0	1	1	1
Communication	1	2	2	1	4	1	0	1	1
System logging	2	1	1	1	4	1	1	0	0
Extensive computation	1	3	1	3	1	1	1	0	0

Table 8.7 The qualifications.

Components / Programmers	User interaction	Visualization	Data collection	Data analysis	Data sharing	Database management	Communication	System logging	Extensive computation
Adam	0.85	**0.82**	0.60	0.65	0.45	0.40	0.30√	0.75√	0.40
Brian	0.75	0.80√	0.58	**0.76**	0.58	0.65	0.45	0.40	**0.85**√
Chris	0.40	0.45	**0.86**	0.36√	**0.76**√	0.80√	0.76	0.85	0.45
Doug	**0.96**√	0.56	0.45√	0.64√	0.33	0.68	**0.75**	0.78	0.35
Edward	0.22	0.33	0.68√	0.33	0.65	**0.70**	0.56	**0.76**	0.45

Ann is concerned with maximizing group performance while considering coupling effects. This is because conventional wisdom in software engineering suggests that the coupling between components hampers software development (Pressman and Maxim 2014). Note that Ann's problem here only occurs when the number of components is larger than the number of people.

The challenge for Ann is how to assign programmers to obtain the highest possible total qualification value while minimizing the interactions among programmers.

Evidently, this assignment is a GMRACR problem that cannot be accomplished manually. One can imagine how a project manager can accomplish it without the assistance of computers and optimization tools. One may argue that Ann can assign the programmers mainly by her impressions of how programmers can work. However, this is not objective decision making and implies bias and ambiguity. What we pursue in this section is an objective solution with the optimization of specific objectives.

8.2.2 The Problem Specification

Definition 8.6 A *role coupling matrix* expressed by C^R is defined as an $n \times n$ matrix, where $C^R[j_1, j_2] \in \mathcal{N}$ means the number of couplings between two roles. Without loss of generality, we also define $C^R[j_1, j_1] = 0 (0 \leq j_1 < n)$.

Note C^R can be extracted from the message sets of a role in the original Environments – Classes, Agents, Roles, Groups, and Objects (E-CARGO) model, i.e. $\mathcal{M}_{in}$ and $\mathcal{M}_{out}$ (Chapter 4). Figure 8.3 shows an example of matrix C^R for $n = 9$, i.e. Table 8.6. By the above definitions, we can define a new abstract assignment problem, i.e. the GMRACR problem.

Definition 8.7 Given Q, C^R, L and L^a, the *Group Multi-Role Assignment with Coupled Roles (GMRACR)* problem aims to find a workable T to obtain

$$max \sum_{i=0}^{m-1} \sum_{j=0}^{n-1} Q[i,j] \times T[i,j], \text{ and}$$

$$min \sum_{i_1=0}^{m-2} \sum_{i_2=i_1+1}^{m-1} \sum_{j_1=0}^{n-1} \sum_{j_2=0}^{n-1} C^R[j_1, j_2] \times T[i_1, j_1] \times T[i_2, j_2]$$

subject to (7.1)–(7.3).

Solving the GMRACR problem is not a trivial task because it is a dual-objective nonlinear optimization problem, which may not have a feasible solution due to conflicts between the two objectives. Furthermore, the second objective is a nonlinear expression. There is no generalized way to solve such an optimization problem (Rardin 1997).

$$\begin{bmatrix} 0 & 2 & 3 & 2 & 1 & 2 & 1 & 2 & 1 \\ 2 & 0 & 2 & 4 & 2 & 1 & 2 & 1 & 3 \\ 3 & 2 & 0 & 3 & 3 & 3 & 2 & 1 & 1 \\ 2 & 4 & 3 & 0 & 2 & 3 & 1 & 1 & 3 \\ 1 & 2 & 3 & 2 & 0 & 3 & 4 & 4 & 1 \\ 2 & 1 & 3 & 3 & 3 & 0 & 1 & 1 & 1 \\ 1 & 2 & 2 & 1 & 4 & 1 & 0 & 1 & 1 \\ 2 & 1 & 1 & 1 & 4 & 1 & 1 & 0 & 0 \\ 1 & 3 & 1 & 3 & 1 & 1 & 1 & 0 & 0 \end{bmatrix}$$

Figure 8.3 The matrix C^R for Table 8.6.

Using a similar method to what we used in Section 6.3.2 (Chapter 6), we can write $T[i_1, j_1] \times T[i_2, j_2]$ into a linear expression by logical expression transformations because $T[i, j] \in \{0, 1\}$, $T[i_1, j_1] \times T[i_2, j_2] = 1$ if $T[i_1, j_1] = 1$, and $T[i_2, j_2] = 1$ and $T[i_1, j_1] \times T[i_2, j_2] = 0$ otherwise.

We may introduce additional variables $T'[i_1, j_1, i_2, j_2] = T[i_1, j_1] \times T[i_2, j_2]$ $(0 \leq i_1, i_2 < m, 0 \leq j_1, j_2 < n)$. The constraints are shown as follows:

$$T'[i_1, j_1, i_2, j_2] \in \{0, 1\} \quad (0 \leq i_1, i_2 < m, 0 \leq j_1, j_2 < n) \tag{8.7}$$

$$2T'[i_1, j_1, i_2, j_2] \leq T[i_1, j_1] + T[i_2, j_2] \quad (0 \leq i_1, i_2 < m, 0 \leq j_1, j_2 < n) \tag{8.8}$$

$$T[i_1, j_1] + T[i_2, j_2] \leq T'[i_1, j_1, i_2, j_2] + 1 \quad (0 \leq i_1, i_2 < m, 0 \leq j_1, j_2 < n) \tag{8.9}$$

Now the GMRACR problem becomes a problem to find a workable T to obtain

$$max \sum_{i=0}^{m-1} \sum_{j=0}^{n-1} Q[i,j] \times T[i,j], \text{ and}$$

$$min \sum_{i_1=0}^{m-2} \sum_{i_2=i_1+1}^{m-1} \sum_{j_1=0}^{n-1} \sum_{j_2=0}^{n-1} C^R[j_1, j_2] \times T'[i_1, j_1, i_2, j_2]$$

subject to (7.1)–(7.3), (8.7)–(8.9).

Note that the revised form of the GMRACR problem is still a linear dual-objective optimization problem. No algorithms and tools are available to solve it directly. To solve such a problem, there are two common ways: one is the weighted sum and the other is goal programming (Rardin 1997).

In applying the weighted sum method, we need to make Q and C^R comparable, i.e. we need to normalize C^R into $C^{R'}$, i.e. $C^{R'}[j_1, j_2] = C^R[j_1, j_2]/\max\{C^R[j_1, j_2](0 \leq j_1, j_2 < n)\}$.

If we use w_1 and w_2 to express the weight of the two objectives, respectively, then we have a new definition.

Definition 8.8 Given Q, $C^{R'}$, L, L^a, w_1 and w_2, *the weighted sum form of the GMRACR (GMRACR-WS) problem is to find a workable T to obtain*

$$max \left\{ w_1 \times \sum_{i=0}^{m-1} \sum_{j=0}^{n-1} Q[i,j] \times T[i,j] - w_2 \times \sum_{i_1=0}^{m-2} \sum_{i_2=i_1+1}^{m-1} \sum_{j_1=0}^{n-1} \sum_{j_2=0}^{n-1} \right.$$

$$\left. C^{R'}[j_1, j_2] \times T'[i_1, j_1, i_2, j_2] \right\}$$

subject to (7.1)–(7.3), (8.7)–(8.9).

Now GMRACR-WS becomes a solvable problem. However, it is still NP-hard (Papadimitriou 1981) because it can be categorized as a Quadratic Assignment Problem (QAP) (Burkard 1984; Burkard et al. 2009). The following CPLEX solutions and experiments confirm this statement. We use σ_{ws}^c and γ_{ws}^c to expresses the $\sum_{i=0}^{m-1} \sum_{j=0}^{n-1} Q[i,j] \times T_{ws}^*[i,j]$, and $\sum_{i_1=0}^{m-2} \sum_{i_2=i_1+1}^{m-1} \sum_{j_1=0}^{n-1} \sum_{j_2=0}^{n-1} C^R[\,j_1,j_2] \times T_{ws}^*[i_1,j_1]$
$\times T_{ws}^*[i_2,j_2]$, respectively, where T_{ws}^* is the T obtained by Definition 8.8.

8.2.3 The Solutions with CPLEX and Initial Results

With the CPLEX package, we need to prepare objective function coefficients, constraint coefficients, right-hand side constraint values, and upper and lower bounds. Then, we need to transform the objective and the constraints into Linear Programming (LP) formats with one-dimensional (1-D) vectors, i.e. Q, $C^{R'}$, L, and L^a 1-D vectors, in order to define an LP problem. In this case, Q provides the coefficients, T and T' are the variables, and the upper and lower bounds of $T[i,j]$ and $T'[i_1,j_1,i_2,j_2]$ $(0 \le i_1, i_2 < m, 0 \le j_1, j_2 < n)$ are 1 and 0.

In the implementation of the CPLEX solution for GMRACR-WS, we need a transformation from T', an $m \times n \times m \times n$ 4-D matrix, to a 1-D vector Y with $m \times n \times m \times n$ elements.

The formula is as follows:

$$Y[i_1 \times m \times n \times n + i_2 \times n \times n + j_1 \times n + j_2] = T'[i_1,j_1,i_2,j_2]$$
$$(0 \le i_1, i_2 < m, 0 \le j_1, j_2 < n).$$

With the above implementation, we use $w_1 = 0.8$ and $w_2 = 0.2$ to solve the problem mentioned in Section II and obtain the result T^*(underlined numbers) as shown in Table 8.7. Compared with the result obtained by GMRA (**Bolded** numbers), we find that the two assignments are different. The GMRACR-WS evidently produces a lower group performance of $\sigma_{ws}^c = 4.37$ compared with the GMRA result of $\sigma_1 \left(= \sum_{i=0}^{m-1} \sum_{j=0}^{n-1} Q[i,j] \times T^*[i,j], T^* \text{ is the } T \text{ obtained by GMRA} \right) = 8.12$.
However, the new assignment (with $\gamma_{ws}^{c*} = 0$ couplings) avoids coupling between people (agents) compared with the γ^c $\left(= \sum_{i_1=0}^{m-2} \sum_{i_2=i_1+1}^{m-1} \sum_{j_1=0}^{n-1} \sum_{j_2=0}^{n-1} C^R[\,j_1,j_2] \right.$
$\left. \times T^*[i_1,j_1] \times T^*[i_2,j_2] \right) = 40$ couplings in the GMRA result.

The group performance cost for removing all the couplings is $(8.12 - 4.37)/8.12 = 40.3\%$.

Now, we attempt another way to solve GMRACR, i.e. goal programming. We set up one objective as a constraint and pursue the other optimization objective. Because we are more concerned with the couplings, we can establish group performance as a constraint, and the GMRA result σ_1 can be taken as a threshold to set the constraint. We use λ to express the percentage of the group performance obtained by GMRA.

Definition 8.9 Given Q, C^R, L, L^a and λ, the goal programming form of the GMRACR (GMRACR-GP) problem is to find a workable T to obtain

$$min \sum_{i_1=0}^{m-2} \sum_{i_2=i_1+1}^{m-1} \sum_{j_1=0}^{n-1} \sum_{j_2=0}^{n-1} C^R[j_1, j_2] \times T'[i_1, j_1, i_2, j_2]$$

subject to (7.1)–(7.3), (8.7)–(8.9), and

$$\sum_{i=0}^{m-1} \sum_{j=0}^{n-1} Q[i,j] \times T[i,j] \geq \lambda \times \sigma_1 \tag{8.10}$$

where σ_1 means the group performance obtained by GMRA (Definition 6.2, Chapter 6). We use $\sigma_{GP}^c = \sum_{i=0}^{m-1}\sum_{j=0}^{n-1}Q[i,j] \times T_{GP}^*[i,j]$ and $\gamma_{GP}^c = \sum_{i_1=0}^{m-2}\sum_{i_2=i_1+1}^{m-1} \sum_{j_1=0}^{n-1}\sum_{j_2=0}^{n-1}C^R[j_1,j_2] \times T_{GP}^*[i_1,j_1] \times T_{GP}^*[i_2,j_2]$ to express the number of couplings in the result. With CPLEX, we implement the solution of GMRACR-GP. We obtain the result for the problem in Section 8.2.1 by setting $\lambda = 90\%$: $\gamma_{GP}^c = 13$, T_{GP}^* is shown in Table 8.7 (the numbers with $\sqrt{}$s), and $\sigma_{GP}^c = 6.54$. Compared with the result obtained by GMRA, we use the cost of 10% of group performance to gain the benefit of $(\gamma^c - \gamma_{GP}^c)/\gamma^c = (40 - 13)/40 = 67.5\%$ of couplings.

8.2.4 Verification Experiments

We have used a special example (Section 8.2.3) to show that we can obtain a better assignment in the sense of couplings by solving the GMRACR problem. In this section, we conduct experiments to verify the benefits of GMRACR-WS and GMRACR-GP compared with GMRA. The experiment is set as $m = 7$, $n = 11$, and $Q[i, j]$ is randomly created in $[0, 1]$, $L[j] = 1$, $L^a[i] = 2$, $1 \leq C^R[j_1, j_2] \leq 4$, $w_1 = 0.8$, and $w_2 = 0.2$ ($0 \leq i < m$, $0 \leq j, j_1, j_2 < n$). The test ran for 100 random cases, each of which is solved by GMRA and GMRACR-WS, respectively.

In the experiment, we collect the number of couplings and the group performances of both GMRACR-WS and GMRA, and the time used by GMRACR-WS to solve each random case. The platform used in the experiment is shown in Table 8.5.

The numbers of couplings collected by GMRA are between 6 and 33 (most are between 10 and 33), and those collected by GMRACR-WS are between 0 and 30 (most are between 0 and 15). The group performances for GMRA are in [8.43, 10.18] and those of GMRACR-WS are in [4.94, 9.67].

The results confirm that GMRACR-WS obtains benefits in the coupling numbers with some group performance loss. Note that in our example and experiments, we set C^R as a symmetric matrix, i.e. $C^R[j_1, j_2] = C^R[j_2, j_1]$, $(0 \leq j_1, j_2 < n)$. However, our solutions are not limited by this condition, i.e. C^R does not have to be symmetric. The experimental results show that most gains are in [0.5, 1] and most losses are in [0, 0.5].

With similar settings and $\lambda = 90\%$, we conduct 100 random cases for GMRACR-GP and GMRA. Note that the performance loss is determined in GMRACR-GP, i.e. $1 - \lambda = 10\%$.

In the experiment, we collect the numbers of couplings and the group performances of both GMRACR-GP and GRA, and the time used by GMRACR-GP to solve each random case on the platform in Table 8.5. The results show that at the loss of 10% constraint, the gains are in [0.16, 0.93] and most are in [0.4, 0.8].

From the performance experiment, we may understand the complexity of the GMRACR problem, both the CPLEX solutions use seconds (for GMRACR-WS, in [1, 12], and for GMRACR-GP, in [1,7]) to solve a 5×11 ($m = 5, n = 11$) problem. For problems at the scale of $10 \times 20 (m = 10, n = 20)$, the used time by GMRACR-GP ranges from minutes to hours from another experiment of 100 random cases on a similar laptop. Note that CPLEX is believed to be one of the few powerful optimization platforms in the world (IBM 2019). Therefore, it is beneficial to investigate more efficient algorithms for the GMRACR problem. The visual results of these experiments can be found in (Zhu 2019a).

8.3 Most Economical Redundant Assignment

Team performance and efficiency can suffer due to the unavailability of its members (Alidaee et al. 2011; Gardner et al. 2017). Redundancy is normally used to address this situation and conduct successful adaptive collaborations (Sheng et al. 2016). However, redundancy requires additional resources, i.e. the budget is not a trivial expenditure. Therefore, it is a big challenge for an administrator to balance the management budget and the robustness of a team. Conventional assignment methods are based on the experience, wisdom, logical analysis, or intuitive judgments of administrators. Such a method cannot consistently guarantee the most economical role assignment in a team, especially if the team size is relatively large.

The Most Economical Redundant Assignment (MERA) problem is complex because there are many different situations to consider when organizing a team. The "most economical" requirement reflects two aspects: the number of redundant assignments and the training budget of the assignments. Conventional solutions are either insufficient at making a robust team or produce a lot of waste. Simply stated, the question is: "How many" is enough for redundancy? This section provides an assured answer to such a problem.

After the Good at Many and Expert in One (GMEO) problem (Chapter 7; Zhu 2020) is solved, the MERA problem becomes less complex and solvable because the formalization method of GMEO can be well applied to MERA. By introducing a new data structure to express the different training costs of roles, we can formalize the MERA problem concisely. The solution to GMEO reveals that an exact number of redundant assignments can keep a team in a workable state if one member leaves. This section asserts that the most economical way for such a redundant assignment exists. The results obtained in this section can be applied in many applications as mentioned in Section 8.2.

8.3.1 A Real-World Scenario

Continue the scenario of Section 8.1.1. Ann signs a new contract and recognizes a new situation. Busyness is not her major concern now. However, she is concerned about the training and the team's robustness. With the same evaluation of Table 8.1, she tells Bob that it is necessary to have backup assignments to guarantee that no critical person exists, i.e. the team remains viable even though one member is unavailable for some reason, and thus fulfilling the GMEO requirement. Ann also requires Bob to conduct training for members with qualification values less than 0.85. Bob composes the training costs as shown in Table 8.8 (in \$K).

Besides not meeting the requirement of redundancy, the GRA result may make many highly qualified people be trained unnecessarily, i.e. the assigned people with the evaluation value ≥ 0.85 as shown in Table 8.2 (Note: Read it by using the header lines of Table 8.9.) using the square bracketed numbers. This situation is, in fact, an instance of the MERA problem. Obviously, the MERA problem is complex and needs thorough investigation.

Table 8.8 The training costs for different courses (in \$K).

Training Courses	Distributed systems	Block chain	User interface	Artificial intelligence
Cost	3	5	4	2

8.3.2 Problem Formalizations

Using the E-CARGO model (Chapter 4), the GMEO problem was formalized in Chapter 7, and $m, n, L, T, n_a, n_t, \sigma,$ and $\overline{T_{-i_1}} (0 \leq i_1 < m)$ have been defined. Here, we restate some concepts, revise others, and define new ones to facilitate the formalization of the MERA problem.

Definition 8.10 An assignment matrix T is *redundant* if it keeps:

$$\sum_{i=0}^{m-1} T[i,j] \geq L[j] \quad (0 \leq j < n), \tag{8.11}$$

$$n_t = \sum_{i=0}^{m-1} \sum_{j=0}^{n-1} T[i,j] > n_a. \tag{8.12}$$

With GRA, GRA with constraints (GRA$^+$) (Chapter 6), or GRA with Multiple Objectives (GRA^{++}) (Chapter 7), we can obtain an optimized assignment matrix. If we consider the qualification threshold τ, combining matrices Q and T, we know that some assigned agents may not be qualified to play some roles. Training is required to help these unqualified agents become qualified. Furthermore, training is needed only for those agents that are assigned but not qualified.

Definition 8.11 A role training *cost vector* V^c is an n-vector where $V^c[j] \in (0, +\infty)$ expresses the cost for an agent to be trained for role $j \in \mathcal{N} (0 \leq j < n)$.

In Definition 8.11 we assume that the training cost of agents for the same role is identical, e.g. the tuition of a training course is the same for every enrolled student.

Definition 8.12 The *capability matrix* $\overline{Q}$ is a derived matrix from Q, where

$$\overline{Q}[i,j] = \begin{cases} 1 & (Q[i,j] \geq \tau) \\ 0 & (Q[i,j] < \tau) \end{cases} \quad (0 \leq i < m, 0 \leq j < n). \tag{8.13}$$

Note that $\overline{Q}[i,j] = 1$ indicates that agent $i \in \mathcal{N} (0 \leq i < m)$ is qualified to play the role $j \in \mathcal{N} (0 \leq j < n)$ and $\overline{Q}[i,j] = 0$ indicates otherwise.

Definition 8.13 The *required training matrix* $\overline{T}$ is a derived matrix defined by T and $\overline{Q}$, where:

$$\overline{T}[i,j] = \begin{cases} 1 & (T[i,j] - \overline{Q}[i,j] = 1) \\ 0 & (T[i,j] = 0) \end{cases} \quad (0 \leq i < m, 0 \leq j < n). \tag{8.14}$$

Definition 8.14 The *training cost* of a group is defined as $\sum_{i=0}^{m-1} \sum_{j=0}^{n-1} V^c[i] \times \overline{T}[i,j]$.

From the Definition of $\overline{T}$, we may understand that only the assigned and unqualified agents need training and we assume that agents are qualified after training. We can describe $\overline{T}$ with T and $\overline{Q}$ in a linear expression as:

$$\overline{T}[i,j] = T[i,j] \times \left(1 - \overline{Q}[i,j]\right) \quad (0 \le i < m, 0 \le j < n). \tag{8.15}$$

Definition 8.15 Given a group expressed by Q, L, V^c, and τ, the MERA problem is to find the most economical redundant assignment matrix T and the training plan matrix $\overline{T}$ such that no critical agent exists in the group, i.e.

$$max \quad \sum_{i=0}^{m-1} \sum_{j=0}^{n-1} Q[i,j] \times T[i,j]$$

and

$$min \quad \sum_{i=0}^{m-1} \sum_{j=0}^{n-1} V^c[i] \times \overline{T}[i,j]$$

subject to (8.1), (8.13), (8.15), and

$$\overline{T_{-i_1}}[i,j] \in \{0,1\} \quad (0 \le i_1 < m, 0 \le i < m-1, 0 \le j < n), \tag{8.16}$$

$$\overline{T_{-i_1}}[i,j] \le \begin{cases} \overline{Q}[i,j] & (i < i_1) \\ \overline{Q}[i+1,j] & (i_1 \le i < m-1) \end{cases} \quad (0 \le i_1 < m), \tag{8.17}$$

$$\sum_{i=0}^{m-2} \overline{T_{-i_1}}[i,j] = L[j] \quad (0 \le i_1 < m, 0 \le j < n), \tag{8.18}$$

$$\sum_{j=0}^{n-1} \overline{T_{-i_1}}[i,j] \le 1 \quad (0 \le i_1 < m, 0 \le i < m-1), \tag{8.19}$$

where constraints (8.13) and (8.15) express how $\overline{Q}$ and $\overline{T}$ are formed; (8.16) informs that agent i can be assigned or not assigned to role j; (8.17) informs that $\overline{T_{-i_1}}$ is an $(m-1) \times n$ assignment matrix derived from $\overline{Q}$ where agent $i_1(0 \le i_1 < m)$ has been removed; (8.18) expresses that the number of assigned agents for a role in $\overline{T_{-i_1}}$ must follow L; and (8.19) specifies that each agent can only be assigned to one role in $\overline{T_{-i_1}}$.

Evidently, the MERA problem is a GRA^{++} problem, which has no generalized solution. As a result, we have to use special methods to solve it.

8.3.3 A Solution with CPLEX

The MERA problem (Definition 8.15) is a dual-objective optimization problem that does not have a general feasible solution. We should adapt it into an Integer Linear Programming (ILP) problem (Rardin 1997; Wolsey and Nemhauser 1999) to gain feasible solutions. The most popular methods to do this are weighted sum (WS) and goal programming (GP) (Rardin 1997). GP, i.e. setting one objective as a constraint and optimizing the other objective, is a better option for the MERA problem due to the features of MERA. For MERA, we can either set the group performance as the constraint or set the training budget as the constraint.

Definition 8.16 Given a group expressed by Q, L, V^c, and τ, the *MERA problem with Goal Programming* (MERA$_{GP}$) with the required group performance $\bar{\sigma}$ (MERA$_{GPP}$) is to find the best training plan matrix $\bar{T}$ such that no critical agent exists in the group, i.e. obtain

$$min \quad \sum_{i=0}^{m-1} \sum_{j=0}^{n-1} V^c[i] \times \bar{T}[i,j]$$

subject to (8.1), (8.13), (8.15)–(8.19), and

$$\sum_{i=0}^{m-1} \sum_{j=0}^{n-1} Q[i,j] \times T[i,j] \geq \bar{\sigma}, \tag{8.20}$$

where constraint (8.20) sets the least group performance.

If we have an appropriate number for $\bar{\sigma}$, then the MERA$_{GPP}$ problem becomes a solvable ILP problem.

Definition 8.17 Given a group expressed by Q, L, V^c, and τ, the *MERA problem with Goal Programming* (MERA$_{GP}$) on the limited Budget $\bar{C}$ (MERA$_{GPB}$) is to find the best assignment matrix T such that no critical agent exists in the group, i.e. obtain

$$max \quad \sum_{i=0}^{m-1} \sum_{j=0}^{n-1} Q[i,j] \times T[i,j]$$

subject to (8.1), (8.13), (8.15)–(8.19), and

$$\sum_{i=0}^{m-1} \sum_{j=0}^{n-1} V^c[i] \times \bar{T}[i,j] \leq \bar{C}, \tag{8.21}$$

where constraint (8.21) limits the total training cost.

Now, we need to estimate and set up relevant values for $\bar{\sigma}$ and $\bar{C}$ to solve MERA$_{GPP/GPB}$ (to mean MERA$_{GPP}$ or MERA$_{GPB}$). In the MERA$_{GPP/GPB}$ problem, because the assignment is redundant and the number of training assignments is determined by Q, $\bar{Q}$, L, and τ, we can estimate $\bar{\sigma}$ and $\bar{C}$ using the GRA framework (Chapter 5; Zhu et al. 2012). Let T^* be the T obtained by GRA, then we obtain $\sigma_0 = \sum_{i=0}^{m-1} \sum_{j=0}^{n-1} Q[i,j] \times T^*[i,j]$.

Definition 8.18 Let $L'[j] = \sum_{i=0}^{m-1} \sum_{j=0}^{n-1} \bar{Q}[i,j] \times T^*[i,j]$ $(0 \leq j < n)$ mean the number of the assigned qualified agents by GRA. We define $L''[j] = L[j] - L'[j]$ $(0 \leq j < n)$ to mean the number of unqualified agents assigned by GRA.

Lemma 8.1 The necessary condition for MERA$_{GPP/GPB}$ to have a feasible solution is $\sum_{i=0}^{m-1} T[i,j] \geq L[j] + 1$ $(0 \leq j < n)$.

[Proof]: Ref. Lemma 7.1 (Chapter 7) and replace $\bar{Q}$ with T.∎

Theorem 8.1 MERA$_{GPP/GPB}$ has a feasible solution iff $n_t \geq n_a + n$, i.e. $\sum_{i=0}^{m-1}$ $\sum_{j=0}^{n-1} T[i,j] \geq n_a + n$.

[Proof]: Ref. Theorem 7.9 (Chapter 7) and replace $\overline{Q}$ with T.∎

From Theorem 8.1, if all the redundant assignments need training, we obtain that the cost $\sum_{j=0}^{n-1}(L''[j] + 1) \times V^c[j]$. Therefore, we can set $\overline{C} = \sum_{j=0}^{n-1}$ $(L''[j] + 1) \times V^c[j]$.

To provide a solution to MERA$_{GPP/GPB}$ by using the CPLEX package, the four elements (i.e. objective function coefficients, constraint coefficients, right-hand side constraint values; and upper and lower bounds) to define a Linear Programming (LP) problem in CPLEX must be specified. In this case, Q provides the coefficients, T and $\overline{T_{-i_1}}$ $(0 \leq i_1 < m)$ are the variables, and τ, L, and L'' are constraints. The upper and lower bounds of T and $\overline{T_{-i_1}} (0 \leq i_1 < m)$ are 1 and 0.

We solve the problem presented in Section 8.3.2 with $L = [1\,2\,4\,2]$, $V^c = [3\,5\,2\,4]$, and $\tau = 0.85$.

The results of MERA$_{GPP}$ and MERA$_{GPB}$ are shown in Tables 8.9 and 8.10, respectively, where the 1s inform the position assignments and the bolded 1s are the training plans (MERA$_{GPP}$: $\sigma^* = 13.84$ and $C^* = \$38K$; and MERA$_{GPB}$: $\sigma^* = 9.36$ and $C^* = \$30K$, here σ^* and C^* are computed by the T and $\overline{T}$ obtained by MERA$_{GPP}$ and MERA$_{GPB}$, respectively).

Table 8.9 The assignment (1s) and training plan (the **Bold** 1s) for the problem in Section 8.3.1 (MERA$_{GPP}$).

Training Courses Candidates	Distributed systems	Block chain	User interface	Artificial intelligence
Adam	0	0	**1**	0
Brian	0	0	**1**	0
Chris	0	1	1	0
Doug	1	0	0	0
Edward	0	0	0	0
Fred	1	0	0	0
George	0	0	0	1
Harry	0	1	0	0
Ice	0	0	0	1
Joe	0	1	0	0
Kris	1	0	0	1
Larry	1	0	**1**	0
Matt	0	0	**1**	0

Table 8.10 The assignment (1s) and training plan (the **Bold** 1s) for the problem in Section 8.3.1 (MERA$_{GPB}$).

Training courses Candidates	Distributed systems	Block chain	User interface	Artificial intelligence
Adam	0	**1**	0	0
Brian	0	**1**	**1**	0
Chris	**1**	1	1	0
Doug	1	0	0	**1**
Edward	0	0	**1**	0
Fred	1	0	0	**1**
George	0	0	0	0
Harry	**1**	0	**1**	**1**
Ice	0	0	0	0
Joe	0	0	**1**	0
Kris	1	0	0	0
Larry	1	0	0	0
Matt	0	0	**1**	0

8.3.4 A New Form of the MERA Problem and a More Efficient Solution

Examining the solutions in Tables 8.9 and 8.10, we find that there are more than enough redundant assignments, e.g. <Larry, Distributed Systems> is an unnecessary assignment. Using GRA to check the 13 cases where one agent is removed, we find that none of the optimal GRA solutions uses the above assignment. Note that GRA seeks to maximize group performance by satisfying the constraints expressed by L (Zhu et al. 2012). That is to say, the pursued maximum group performance with MERA$_{GPP/GPB}$ is useless. This extra assignment only increases the cost of team management. Therefore, we must admit that the MERA$_{GPP/GPB}$ algorithms are impractical. We need to revise the form of the MERA problem to develop more effective algorithms.

Definition 8.19 Given a group expressed by Q, L, and τ, the *MERA$_{NEW}$ problem* is to find the most economical redundant assignment matrix T and the training plan matrix $\overline{T}$ such that no critical agent exists in the group, i.e.

$$max \sum_{i=0}^{m-1} \sum_{j=0}^{n-1} Q[i,j] \times T[i,j],$$

$$min \sum_{i=0}^{m-1} \sum_{j=0}^{n-1} V^c[i] \times \overline{T}[i,j], \text{ and}$$

$$min \sum_{i=0}^{m-1} \sum_{j=0}^{n-1} T[i,j]$$

subject to (8.1), (8.11), (8.13)–(8.19).

To solve the MERA$_{NEW}$ problem, we have to modify it into an ILP problem.

Definition 8.20 Given a group expressed by Q, L, and τ, the *new MERA problem with Goal Programming on a limited budget $\overline{C}$* (MERA$_{NEWGPB}$) is to find the best assignment matrix T and the training plan matrix $\overline{T}$ such that no critical agent exists in the group, i.e. obtain

$$\sigma_{GPB} = max \sum_{i=0}^{m-1} \sum_{j=0}^{n-1} Q[i,j] \times T[i,j]$$

subject to (8.11), (8.13)–(8.19), and

$$\sum_{i=0}^{m-1} \sum_{j=0}^{n-1} T[i,j] = n_t. \tag{8.22}$$

Based on *Lemma 8.1*, we created the following algorithm. The basic idea is to set up an initial n_t and finally search for a feasible solution by adding n_t by 1, where MERA$_{GPB}$-CPLEX expresses the CPLEX solution for MERA$_{GPB}$.

Algorithm 8.2 MERA$_{NEWGPB}$

```
Input: m, n, Q, L, τ, Vᶜ
Output: T, T̄
begin
nₜ = nₐ +n;
succ=False;
while (succ≠True)
        succ = MERAGPB-CPLEX (m, n, Q, L, τ, Vᶜ)  with
        constraint ∑ᵢ₌₀ᵐ⁻¹∑ⱼ₌₀ⁿ⁻¹T[i, j] = nₜ;
nₜ = nₜ+1;
endwhile
```

(Continued)

Algorithm 8.2 (Continued)

```
if (succ)
      T̄[i, j] = T[i, j] × (1 - Q̄[i, j]);
      C* = ∑ⁿ⁻¹ⱼ₌₀ (L″[j] + 1) × Vᶜ[j];

      //The assignment matrix is T, the training one T̄,
      //and the cost is in C*.
endif
end
```

With Algorithm 8.2, the problem in Section 8.3.1 is solved and the result is shown in Table 8.11 ($\sigma_{GPB} = 10.05$ and $C_{GPB} = \$30K$). We omit the description of MERA$_{NEWGPP}$ for three reasons: (i) to save space; (ii) it is similar to MERA$_{NEWGPB}$; and (iii) The σ of MERA$_{NEWGPP}$ is not as good as that of MERA$_{NEWGPB}$.

With the platform shown in Table 8.12, to solve the problem presented in Section 8.3.2 with MERA$_{NEWGPB}$, the time used is 270 milliseconds (ms) (Table 8.11). In another case, we set $m = 20$, $n = 5$, $L = [1\ 2\ 3\ 4\ 5]$, $V^c = [3\ 1\ 4\ 1\ 5]$

Table 8.11 The assignment (1s) and training plan (the **Bold** 1s) for the problem in Section 8.3.1 with MERA$_{NEWGPB}$.

Training courses Candidates	Distributed systems	Block chain	User interface	Artificial intelligence
Adam	0	**1**	0	0
Brian	0	**1**	0	0
Chris	0	1	1	0
Doug	1	0	0	1
Edward	0	0	1	0
Fred	0	0	0	1
George	0	0	0	0
Harry	0	0	1	1
Ice	0	0	0	0
Joe	0	0	1	0
Kris	1	0	0	0
Larry	0	0	0	0
Matt	0	0	1	0

Table 8.12 Test platform configuration for MERA.

CPU	Intel Core i7-2677U CPU @1.8 GHz 1.80 GHz
MM	6 GB
OS	Windows 7 Enterprise
Eclipse	Version: Version: Neon.1a Release (4.6.1)
JDK	Java 1.8.0_91

and Q with random values in [0, 1]. The result is obtained in 393 ms. To verify the practicability of the MERA$_{\text{NEWGPB}}$ algorithm for different teams, we also conducted performance experiments.

In the experiment, we set $\tau \in [0.6, 1]$, $m = 40$, $n = 8$; we simulated 100 random groups with randomly generated (uniform distributions) Q, L, and V^c where $0 \leq Q$ $[i, j] \leq 1$, $1 \leq L[j] \leq 5$, and $1 \leq V^c[j] \leq 5$ $(0 \leq i < m, 0 \leq j < n)$. The time (in seconds) used to solve each random case is in [1.7, 12.07]. The MERA$_{\text{NEWGPB}}$ algorithm is practical for small problems, i.e. $m \leq 40$. From MERA$_{\text{NEWGPB}}$, its redundancy should be emphasized. Note that GRA can obtain an assignment matrix that can be made redundant.

Theorem 8.2 A necessary condition for the MERA$_{\text{NEW}}$ problem is $m \geq n_a + 1$, where $n_a = \sum_{j=0}^{n-1} L[j]$.

[Proof]: Suppose we have only $n_a = \sum_{j=0}^{n-1} L[j]$ agents. From constraint (8.19), i.e., $\sum_{j=0}^{n-1} \overline{T_{-i_1}}[i,j] \leq 1$. We have at most $\left(\sum_{j=0}^{n-1} L[j]\right) - 1$ assignments. That means we have at least one role j ($0 \leq j < n$) such that $\sum_{i=0}^{m-2} \overline{T_{-i_1}}[i,j] < L[j]$. This does not satisfy constraint (8.18).

Therefore, we must have $m \geq n_a + 1$.

Theorem 8.2 is proved. ∎

Theorem 8.2 shows that if we only have $n_a = \sum_{j=0}^{n-1} L[j]$ agents and one agent is unavailable, then we do not have enough agents to meet the requirement of L because each agent can only be assigned one role in GRA (Zhu et al. 2012).

Theorem 8.3 A necessary condition for MERA$_{\text{NEW}}$ to have a feasible solution is that in T, no two agents are assigned to the same two roles.

[Proof]: Ref. Theorem 7.12 (Chapter 7) and replace $\overline{Q}$ with T. ∎

From the properties of the MERA problem and Theorem 8.3, we can apply the GMEO-GRA algorithm (Chapter 7; Zhu 2019b) to solve MERA (MERA$_{\text{GRA}}$). The basic idea is to use the GRA algorithm to obtain the best assignment matrix T^* for L and then add n more assignments to T^* to form T in consideration of constraints like Theorem 8.3.

Figure 8.4 The solution for the problem in Section 8.3.1. (a) The assignment matrix. (b) The training matrix.

(a)

$$\begin{bmatrix} 0 & 1 & 0 & 0 \\ 0 & 1 & 0 & 0 \\ 0 & 1 & 1 & 0 \\ 0 & 0 & 0 & 1 \\ 0 & 0 & 1 & 0 \\ 0 & 0 & 0 & 1 \\ 0 & 0 & 0 & 0 \\ 0 & 0 & 1 & 1 \\ 0 & 0 & 0 & 0 \\ 0 & 0 & 1 & 0 \\ 1 & 0 & 0 & 0 \\ 1 & 0 & 0 & 0 \\ 0 & 0 & 1 & 0 \end{bmatrix}$$

(b)

$$\begin{bmatrix} 0 & 1 & 0 & 0 \\ 0 & 1 & 0 & 0 \\ 0 & 0 & 0 & 0 \\ 0 & 0 & 0 & 1 \\ 0 & 0 & 1 & 0 \\ 0 & 0 & 0 & 1 \\ 0 & 0 & 0 & 0 \\ 0 & 0 & 1 & 1 \\ 0 & 0 & 0 & 0 \\ 0 & 0 & 1 & 0 \\ 0 & 0 & 0 & 0 \\ 0 & 0 & 0 & 0 \\ 0 & 0 & 1 & 0 \end{bmatrix}$$

Using $MERA_{GRA}$, the problem in Section 8.3.1 is solved under the conditions of $L = [1\ 2\ 4\ 2]$, $V^c = [3\ 5\ 2\ 4]$, and $\tau = 0.85$. The result is shown in Figure 8.4, where Figure 8.4a shows the assignment matrix and Figure 8.4b shows the training matrix. Note, σ^* (=0.94) is 98.9% of the σ^* (=10.05) obtained by $MERA_{NEWGPB}$, and $C^* = \$30\,K$ for both. The processing time is 4 ms using the platform in Table 8.12. Evidently, $MERA_{GRA}$ follows all the constraints in Definition *8.15* and guarantees the requirement of redundancy. The only concern is its optimality expressed by the group performance.

The proposed solution has been verified on the platform in Table 8.12. For more details, please refer to (Zhu 2019b).

8.4 Related Work

All the problems, GRAABD, GMRACR, and MERA in this chapter, are important, challenging, and complex problems. However, to the author's knowledge, there is a lack of fundamental research on these issues due to the lack of abstraction models other than E-CARGO (Liu et al. 2018; Sheng et al. 2016; Zhu 2019a,b; Zhu and Zhou 2006; Zhu and Zhou 2008; Zhu et al. 2012).

Most of the related works are found in the literature of assignment problems, (Burkard et al. 2009; Kuhn 2010; Kumar et al. 2013; Mosteo et al. 2017; Munkres 1957; Rico et al. 2012; Stone et al. 2013) but these studies have significant differences compared to GRAABD, GMRACR, and MERA.

Assignments have been considered mathematical problems and many different forms of assignments have been discovered (Burkard et al. 2009; Kuhn 2010; Kumar et al. 2013; Mosteo et al. 2017; Munkres 1957; Rico et al. 2012; Stone et al. 2013). However, the GMRACR problem in this chapter has not been discussed, because mathematicians used different modeling and discovery methodologies. This situation demonstrates that the RBC and E-CARGO model (Liu et al. 2018; Sheng et al. 2016; Zhu 2019a,b; Zhu and Zhou 2006; Zhu and Zhou 2008; Zhu et al. 2012) are promising frameworks in the investigation of assignment problems and contribute to the literature in assignment research.

There is also other research in role assignments that deal with different applications, such as wireless sensor networking (Bhardwaj and Chandrakasan 2002), robot teams (Mosteo et al. 2017; Stone and Veloso 1999; Vail and Veloso 2003), workflow management (Kumar et al. 2013; Shen et al. 2003), multi-agent systems (Dastani et al. 2003), and task assignment (Durfee et al. 2014). They have not yet considered the busyness constraints in solving their problems.

Some research works are related to busyness (Kerfoot 2006; Ruffing 1995). Note that, busyness is different from the workload concept in that busyness is a synthesized factor and a soft constraint, while workload (Estes 2015; Wickens 2008) has a determined number of limits and is a hard constraint.

Ann et al. (2010) argue that multiple directorships affect the quality of managerial oversight and negatively influence agency conflicts in acquisition decisions. Their results support the notion that multiple directorships overstretch directors' time, resulting in less effective monitoring. Therefore, their work supports the assumption that busyness affects the efficiency and quality of task execution. It also supports the significance of our proposed formalization and solution. Estes (2015) studies the workload curve and concludes that the subjective rate changes with the workload in an S shape, i.e. nonlinearly. The subjective rate is similar to the meaning of busyness. Ruffing (1995) expresses busyness as the fourth demon in the sense of spiritual life. Her argument informs that busyness is not a concrete indicator of overloading but an abstract, mental, and psychological one that is very different from the term "workload." Young et al. (2015) review the mental workload (MWL) research in ergonomics. They summarize the current state of affairs regarding the understanding, measurement, and application of MWL in the design of complex systems. The conclusion is that real-world problems, particularly in transportation, are the major applications of MWL research results, and new developments in neuropsychological measurement techniques offer promise in quantifying both the physical and mental workload.

Even though the MERA problem is significant, there are few systematic investigations of it because it is complex and challenging. Another reason for the lack of research on MERA is the absence of modeling tools to support researchers in their studies. Most of the related works are found in the literature of assignment (Burkard et al. 2009; Kuhn 1955, 2010; Kumar et al. 2013; Mosteo et al. 2017; Munkres 1957; Rico et al. 2012; Stone et al. 2013). However, the problems, approaches, and objectives in these studies are different from MERA.

No literature has provided an exact answer to the MERA problem, including the well-known book (Burkard et al. 2009) on assignment problems. This book (Burkard et al. 2009) provides a comprehensive treatment of the theoretical, algorithmic, and practical development of assignment problems from its conceptual beginnings in the 1920s to the date of publication. The topics covered include bipartite matching algorithms, linear assignment problems, quadratic assignment problems, multi-index assignment problems, and many variations of these problems. This fact again demonstrates the value of RBC and E-CARGO in helping discover new problems from the viewpoint of collaboration and complex systems.

Mathematicians investigating graph theory (Dourado 2016) define a k-role assignment of a graph G as a surjective function $r: V(G) \rightarrow \{1, ..., k\}$, where $V(G)$ is the set of vertices of graph G, $\{r(u'):u' \in N(u)\} = \{r(v'):v' \in N(v)\}$; ($N(u)$ is the set of the neighbor nodes of node u) for every pair u. There are instances of such a concept in social networks. We believe that there are theoretical connections between k-role assignment and MERA, but these are out of the scope of this section and will be investigated in future studies.

The aforementioned research indicates a strong need to fundamentally investigate the variations of GRA, particularly the problems discussed in this section.

8.5 Summary

From the examples in this chapter, we may find that collaboration and team management are complex. This chapter presents several examples of real-world problems in management. These examples again demonstrate the modeling power of RBC, E-CARGO, and GRA.

The GRAABD problem is a dual-objective optimization one that has no feasible solutions. We use the weighted sum approach to specify a GRAABD-WS problem that is solvable. Compared with GRA, GRAABD-WS produces assignments that do not decrease much of the ideal group performance. The benefit obtained by GRAABD approaches an average of 37% as the group size increases. If the impact of busyness on personal qualifications is quantified, then GRAABD-Syn can be used to obtain a considerably better assignment compared to both GRAABD-WS and GRA.

We obtain more gains in coupling and less loss in group performance by solving the GMRACR problem. The CPLEX solutions to the *GMRACR problem* are acceptable for small groups ($m \times n \leq 200$).

Based on the formalization of the MERA problem and continuous investigations, we discover an important conclusion: $MERA_{GRA}$ is an efficient algorithm to be used in common scenarios to find the training scheme $\overline{T}$ for a team that has $Q, L, V^c, and \tau$.

GRA is also a method to investigate new ways to solve real-world problems. The steps to solve such a problem are as follows:

1) Clarify requirements as roles and supplies as agents.
2) Evaluate each agent on each role to form the qualification matrix.

3) Analyze and formalize constraints with the symbols of RBC, E-CARGO, and new symbols if necessary.
4) Specify the problem as a GRA^+ or GRA^{++} problem.
5) If the problem is solvable by an optimization platform within an acceptable time, the problem is solved.
6) If the problem is not solvable by an optimization platform, try to revise the formalization by changing nonlinear equations to linear equations; or changing the multiple objectives into a single objective; or adding more constraints to conduct goal programming.
7) If both 5) and 6) are not satisfied and the program has low efficiency, trying to find necessary, sufficient, or necessary and sufficient conditions for the problem to be feasible may allow you to obtain a more efficient solution.

From this chapter, further investigations may be conducted along the following directions:

1) Studying innovative ways to express the busyness degrees of agents.
2) Composing better methods to form Q with consideration of busyness degrees.
3) Investigating more efficient algorithms for the GMRACR problem.
4) Modeling and solving GRA problems to avoid problematic agents, e.g. adding a new component to express the problematic members and compose a new constraint for GRA.
5) Applying the proposed algorithms in highly available systems that need backups, such as robot teams, service computing platforms, and cloud computing environments.
6) Applying the proposed methods in management and industry, such as adaptive systems (Sheng et al. 2016), cloud computing (Gardner et al. 2017; Queiroz et al. 2017; Yang et al. 2016), agent teams (Stone et al. 2013), Cyber-Physical Systems (Queiroz et al. 2017), and Manufacturing (Yang et al. 2016). Then collecting and evaluating the actual team performance or productivity of a team and verifying the proposed method by real-world practice.
7) Applying RBC and E-CARGO to the frontier topics, such as cloud computing (Yang et al. 2016), cyber-physical systems, robot teams (Mosteo et al. 2017), and intelligent manufacturing.

References

Alidaee, B., Wang, H., and Landram, F. (2011). On the flexible demand assignment problems: case of unmanned aerial vehicles. *IEEE Trans. on Automation Science and Engineering* **8** (4): 865–868.

Ann, S., Jiraporn, P., and Kim, Y.S. (2010). Multiple directorships and acquirer returns. *Journal of Banking & Finance* **34** (9): 2011–2026.

Bhardwaj, M. and Chandrakasan, A.P. (2002), Bounding the lifetime of sensor networks via optimal role assignments, *Proceeding of Twenty-First Annual Joint Conference of the IEEE Computer and Communications Societies (INFOCOM 2002),* New York, USA (June 2002), vol. 3, pp. 1587- 1596.

Burkard, R.E. (1984). Quadratic assignment problems. *European Journal of Operational Research* **15** (3): 283–289.

Burkard, R.E., Dell'Amico, M., and Martello, S. (2009). *Assignment Problems, Revised Reprint.* Philadelphia, PA: Siam.

Dastani, M., Dignum, V. and Dignum, F. (2003) Role-assignment in open agent societies, *Proceedings of the Second Int'l Joint Conference on Autonomous Agents and Multiagent Systems,* Melbourne, Australia (July 2003), pp. 489–496.

Dourado, M.C. (2016). Computing role assignments of split graphs. *Theoretical Computer Science* **635**: 74–84.

Durfee, E.H., Boerkoel, J.C. Jr., and Sleight, J. (2014). Using hybrid scheduling for the semi-autonomous formation of expert teams. *Future Generation Computer Systems* **31**: 200–212.

Estes, S. (2015). The workload curve: subjective mental workload. *Human Factors* **57** (7): 1174–1187.

Gardner, K., Harchol-Balter, M., Scheller-Wolf, A., and Van Houdt, B. (2017). A better model for job redundancy: decoupling server slowdown and job size. *IEEE/ACM Trans. on Networking* **25** (6): 3353–3367.

IBM (2019), ILOG CPLEX optimization studio. http://www-01.ibm.com/software/integration/optimization/cplex-optimization-studio/ (accessed 10 August 2020).

Kerfoot, K. (2006). Beyond busyness: creating slack in the organization. *Nursing Economic* **24** (3): 168–170.

Kuhn, H.W. (1955). The Hungarian method for the assignment problem. *Naval Research Logistic Quarterly* **2**: 83–97. (Reprinted in vol. 52, no. 1, 2005, pp. 7 – 21.

Kuhn, H.W. (2010). The Hungarian method for the assignment problem. In: *50 Years of Integer Programming 1958-2008* (eds. M. Jünger, T.M. Liebling, D. Naddef, et al.), 29–48. Berlin Heidelberg: Springer-Verlag.

Kumar, A., Dijkman, R., and Song, M. (2013). Optimal resource assignment in workflows for maximizing cooperation. *Lecture Notes in Computer Science* **8094**: 235–250.

Liu, D., Yuan, Y., Zhu, H. et al. (2018). Balance preferences with performance in group role assignment. *IEEE Trans. on Cybernetics* **48** (6): 1800–1813.

Mosteo, A.R., Montijano, E., and Tardioli, D. (2017). Optimal role and position assignment in multi-robot freely reachable formations. *Automatica* **81**: 305–313.

Munkres, J. (1957). Algorithms for the assignment and transportation problems. *J. of the Society for Industrial and Applied Mathematics* **5** (1): 32–38.

Papadimitriou, C.H. (1981). On the complexity of integer programming. *Journal of the ACM* **28** (4): 765–768.

Pressman, R. and Maxim, B. (2014). *Software Engineering: A Practitioner's Approach*, 8e. Columbus, OH: McGraw-Hill Education.

Queiroz, J., Leitão, P., and Oliveira, E. (2017). Industrial cyber physical systems supported by distributed advanced data analytics. In: *Service Orientation in Holonic and Multi-Agent Manufacturing* (eds. T. Borangiu, D. Trentesaux, A. Thomas, et al.), 47–59. Cham: Springer.

Rardin, R.L. (1997). *Optimization in Operations Research*. Upper Saddle River, NJ: Prentice Hall.

Rico, R., Antino, M., and Lau, D. (2012). Bridging team faultiness by combining task role assignment and goal structure strategies. *Journal of Applied Psychology* **97** (2): 407–420.

Ruffing, J. (1995). Resisting the demon of busyness. *Spiritual life* **41** (2): 79–89.

Shen, M., Tzeng, G.-H., Liu, D.-R. (2003) Multi-criteria task assignment in workflow management systems, *Proceedings of the 36th Annual Hawaii Int'l Conference on System Sciences*, Hawaii, USA (6–9 January 2003), pp. 1–9.

Sheng, Y., Zhu, H., Zhou, X., and Hu, W. (2016). Effective approaches to adaptive collaboration via dynamic role assignment. *IEEE Trans. on Systems, Man, and Cybernetics: Systems* **46** (1): 76–92.

Stone, P. and Veloso, M. (1999). Task decomposition, dynamic role assignment, and low-bandwidth communication for real-time strategic teamwork. *Artificial Intelligence* **110**: 241–273.

Stone, P., MacAlpine, P., and Barrera, F. (2013). Positioning to win: a dynamic role assignment and formation positioning system. In: *RoboCup 2012*, LNCS, vol. **7500** (eds. X. Chen, P. Stone, L.E. Sucar and T. van de Zant), 190–121. Heidelberg: Springer.

Vail, D. and Veloso, M. (2003). Multi-robot dynamic role assignment and coordination through shared potential fields. In: *Multi-Robot Systems* (eds. A. Schultz, L. Parkera and F. Schneider), 87–98. Kluwer.

Wickens, C.D. (2008). Multiple resources and mental workload. *Human Factors*, June 2008 **50** (3): 449–455.

Wolsey, L.A. and Nemhauser, G.L. (1999). *Integer and Combinatorial Optimization*. New York: Wiley-Interscience.

Yang, C., Shen, W., Lin, T., and Wang, X. (2016). A hybrid framework for integrating multiple manufacturing clouds. *The Int'l J. of Advanced Manufacturing Technology*, Sept. 2016 **86** (1): 895–911.

Young, M.S., Brookhuis, K.A., Wickens, C.D., and Hancock, P.A. (2015). State of science: mental workload in ergonomics. *Ergonomics* **58** (1) 2015: 1–17.

Zhu, H. (2015). Role-based collaboration and the E-CARGO: revisiting the developments of the last decade. *IEEE Systems, Man, and Cybernetics Magazine* **1** (3): 27–35.

Zhu, H. (2019a), Group multi-role assignment with coupled roles, *The 16ᵗʰ IEEE Int'l Conference on Networking, Sensing, and Control*, Banff, Canada (9–11 May 2019), pp.281–286.

Zhu, H. (2019b), The most economical redundant assignment, *The IEEE Int'l Conference on Systems, Man and Cybernetics* (SMC'19), Bari, Italy, (6–9 October 2019), pp. 146–151.

Zhu, H. (2020). Avoiding critical members in a team by redundant assignment. *IEEE Trans. on Systems, Man, and Cybernetics: Systems* **50** (7) July 2020: 2729–2740.

Zhu, H. and Zhou, M.C. (2006). Role-based collaboration and its kernel mechanisms. *IEEE Transactions on Systems, Man and Cybernetics, Part C* **36** (4): 578–589.

Zhu, H. and Zhou, M. (2008). Role transfer problems and algorithms. *IEEE Trans. on Systems, Man and Cybernetics, Part A: Systems and Humans (IF: 2.03)* **38** (6): 1442–1450.

H. Zhu, and Y. D. Zhu (2017), Group role assignment with agents' busyness degrees, *The IEEE Int'l Conference on Systems, Man and Cybernetics* (SMC'17), Banff, Canada, (5–8 October 2017), pp. 3201–3206.

Zhu, H., Zhou, M.C., and Alkins, R. (2012). Group Role Assignment via a Kuhn-Munkres Algorithm-based Solution. *IEEE Transactions on Systems, Man and Cybernetics, Part A: Systems and Humans* **42** (3): 739–750.

Exercises

1 Describe a scenario that matches each of the GRAABD, GMRACR, and MERA problems.

2 Write the formalizations of all the discussed problems in this chapter, i.e. GRAABD, GMRACR, and MERA problems.

3 Describe the steps involved in solving a real-world problem with RBC, E-CARGO, and GRA.

4 List examples of factors that may affect GRA.

5 Based on E-CARGO and the discussion of relations in Chapter 5, there are other factors that affect GRA. Can you think of some examples?

6 In this chapter, we made many assumptions. Do you agree with these assumptions? Why or why not? Can you provide methods to obtain the assumed data, e.g. busyness, and couplings? Give examples.

7   This chapter again demonstrates the complexity of collaboration and team management. Compose a GRA problem with several mentioned factors combined, e.g. busyness, or couplings together and try to solve them.

8   We listed a few conditions for the problems discussed in this chapter to have a feasible solution. Could you provide additional conditions, e.g. sufficient, necessary, or necessary and sufficient conditions?

9   The GRA problems discussed in this chapter may also help deal with the problems in the other phases of Role-Based Collaboration. Discuss them.

10   The research methods in this chapter can also be applied to cutting-edge fields of engineering and industry. Can you investigate some?

9

Role Transfer

9.1 Role Transfer Problems

Role transfer is a common phenomenon in the real world. For example, on a battlefield, if a high-ranking officer is disabled, a similar or lower rank officer is needed to play their role (Zhu and Zhou 2006b). The role played by that similar or lower rank officer may need to be played by another, etc. A high availability computing system should meet a similar requirement for computers, subsystems, or components to transfer their functions (or roles) to recover from a faulty state. We generally need to duplicate them to tolerate their faults.

There are two scenarios to consider in role transfer research. One is that there are sufficient agents compared to roles and the other is when there are not. Different methods are required to resolve each scenario. For the latter, we introduce a special type of role transfer that uses time intervals to solve the problem. We call this temporal role transfer.

In this chapter, to ease the clarification of role transfer problems, we assume that each agent can only have one current role, i.e. $\forall a \, (|a.\mathcal{R}_c| = 1)$. Therefore, $\mathcal{R}_c$ is simplified to r_c.

Definition 9.1 Role r is *current* to agent a if a is currently playing r, i.e. $(r \in a.\mathcal{R}_c)$ $\bigwedge (|\mathcal{R}_c| = 1)$, or simply, $r = a.r_c$ and a is called a current agent of r, i.e. $a \in r.\mathcal{A}_c$. Role r is *potential* to a if a is qualified to play r but is not currently playing it, i.e. $r \in a.\mathcal{R}_p$; and a is called a potential agent of r, i.e. $a \in r.\mathcal{A}_p$. The current role and all potential roles of an agent are called its role repository, i.e. $\{r_c\} \cup \mathcal{R}_p$. Role r is *critical* if it has only enough agents currently playing it, i.e. $\exists g(< r, g, \mathcal{D}_o > \in g.e.\mathcal{B}) \wedge (| r.\mathcal{A}_c | = g.\ell)$.

E-CARGO and Role-Based Collaboration: Modeling and Solving Problems in the Complex World, First Edition. Haibin Zhu.

Definition 9.2   Group g is *workable* if there are enough agents to play each role r, i.e. $\forall r \in \mathcal{R} (\exists g (< r, g, D_o > \in g.e.\mathcal{B}) \rightarrow (| r.A_e | \geq g.\ell))$. If a role loses its current agent(s), g redistributes agents to play this role. Such redistribution is called *role transfer*. A role transfer is *successful* if g is workable after the transfer.

Definition 9.3   An agent is *critical* if there is no successful role transfer to make group g workable once the agent leaves g. When an agent leaves g, g loses this agent and this agent becomes a *lost agent* of g. Group g is *critical* if every agent in it is critical. A *level-n emergency* occurs if g has lost n (≥ 1) agents. g is *level-n strong* if it is workable via successful role transfer after a level-n emergency occurs.

Note that $a.\mathcal{R}_p$ is a set of all the potential roles of an agent a. The agent can only directly serve the requests relevant to its current role. For an agent, a potential role can become current and vice versa. An agent can have only one current role. That is to say, an agent can hold many potential roles at the same time but only one current role at a time. By holding only one current role, we can avoid role-role conflicts (Bostrom 1980). If a role transfer for agent a occurs, $a.r_c$ is swapped with an r in $a.\mathcal{R}_p$. The repository of an agent includes all the roles that the agent is qualified to play.

In a crisis, people must concentrate on their roles all the time. Every role must have enough players at any time. A critical group defined in Definition 9.3 mandates that no agent can be lost to maintain its workability.

Role transfer problems can be understood by the following examples. For simplicity, we assume that a role must have at least one current agent, i.e. $\ell \geq 1$.

Case 1: Figure 9.1a shows a workable group. Suppose that, as the current player of r_7, a_8 leaves the group. Is the group still workable by role transfer? r_7 can be played by a_7. Hence, let a_7 play r_7. Consequently, r_6 has no current agent. We can let a_6 play r_6, then there is no agent to play r_5. Therefore, the group shown in Figure 9.1a is not workable, after a_8 leaves.

Case 2: Figure 9.1b shows a group similar to Figure 9.1a. The only difference is that the agents' qualifications are more than those in Figure 9.1a. After a_8 leaves the group, a_7 can play r_7, a_6 can play r_6, and a_5 can play r_5. Although a_4 can play r_4, we cannot find another qualified agent to play r_3. Therefore, the group is not workable after a_8 leaves.

Case 3: Figure 9.2a shows a rearrangement of Figure 9.1a. If a_8 leaves the group, we can make it a workable group by letting a_3 play r_7, and a_2 playing r_3. That is to say, a_8 is not *a critical agent*.

Case 4: Figure 9.2b shows a group similar to Figure 9.1b. The only difference is that the agents' qualifications are rearranged. After a_8 leaves the group, we can let a_7 play r_7, a_3 play r_6, and a_2 play r_2. Then, the problem is solved. This means that a_8 is not a critical agent for the group in Figure 9.2b.

We can define the role transfer problem by introducing an additional matrix.

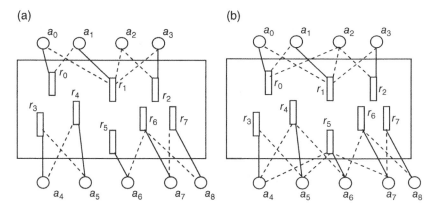

Figure 9.1 Two workable assignments with different potential roles. (a) A workable group. (b) A workable group with more potential roles assigned.

Definition 9.4 A *potential role matrix* is defined as $Q_p: \mathcal{A} \times \mathcal{R} \in \{0,1\}$, where $Q_p[i, j] = 1$ expresses $a_i \in r_j.\mathcal{A}_\phi$, and $Q_p[i, j] = 0$ means $a_i \notin r_j.\mathcal{A}_\phi$.

Definition 9.5 (Zhu and Zhou 2008a)
Role transfer is a process in which 1s of T are exchanged with Q_p, i.e. if there is an agent i, and roles j_1 and j_2, $T[i, j_1] = 1$, $Q_p[i, j_1] = 0$, $T[i, j_2] = 0$, and $Q_p[i, j_2] = 1$, a role transfer occurs when $T[i, j_1] := 0$, $Q_p[i, j_1] := 1$, $T[i, j_2] := 1$, and $Q_p[i, j_2] := 0$

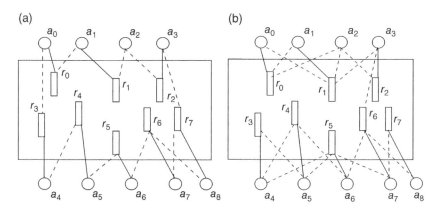

Figure 9.2 The re-arrangement of Figure 9.1. (a) A workable group. (b) A workable group with more potential roles assigned.

$(0 \leq i < m, 0 \leq j_1, j_2 < n)$; on the other hand, if there is a role j, and agents i_1 and i_2, $T[i_1, j] = 1$, $Q_p[i_1, j] = 0$, $T[i_2, j] = 0$, and $Q_p[i_2, j] = 1$, a role transfer also occurs when $T[i_1, j] := 0$, $Q_p[i_1, j] := 1$, $T[i_2, j] := 1$, $Q_p[i_2, j] := 0$ $(0 \leq i_1, i_2 < m, 0 \leq j < n)$.

Definition 9.6 Given m, n, L, Q_p, and T that is not workable, the *role transfer problem (RTP)* is to find a workable T through role transfers.

9.2 The M-M Role Transfer Problems

This section demonstrates that a practical role-transfer algorithm can help group managers understand which agents in the group are critical. In reality, there are even more complicated problems. Suppose that there is no critical agent in one group. We still need to know if some agent combinations are critical to the group. By a critical agent set, we mean that a group becomes unworkable when all the agents in the set leave the group, i.e. no successful role transfer exists. In such a case, we need to develop algorithms to check the agent combinations' critical properties by dealing with role transfer problems when some roles lose their current agents. Also, in a group, one role may need more than one current agent to work on. This factor makes role transfer problems even more complicated.

Definition 9.7 Suppose that group g composed of m (≥ 1) agents and n (≥ 1) roles is unworkable, where each of M ($1 \leq M \leq n$) unworkable roles requires M' ($1 \leq M' \leq m$) extra current agents. This is an *M-M role transfer problem* that aims to find a successful role transfer, if g has one.

Furthermore, *1-1, 1-M, and M-1 role transfer problems* can be defined by assigning a specific M or M' with 1.

Section 9.1 only deals with a special case of a 1-1 problem, i.e. in a group, only one role loses one current agent and this role only requires one more current agent to work. In general, a role in a group has a corresponding range $[\ell, \alpha]$ to express the role's workable state (Zhu and Zhou 2006a, 2008a, b, 2009, 2012; Zhu et al. 2008, 2012). That is to say, a role is workable only if it has at least l current agents and a group is workable only if every role is workable. Because we are mainly concerned with the workable state of a group, we only consider the lower bound of a role range, i.e. ℓ. Note that α is used to express the maximum number of agents that can play the relevant role in a group.

In the following figures, a circle is used to express an agent; a box, a role; a grey box, a role without enough current agents; a solid line, a current agent (role), and a dashed line, a potential agent (role). We also use $r_0, r_1, \ldots$ to express role names and $a_0, a_1, \ldots$ agent names.

The 1-1 problem states that there is only one unworkable role that requires only one extra agent to make it work. In Figure 9.3, the number in parentheses in each box expresses the required number (i.e. ℓ) of current agents for the role represented by the box, i.e. $L = [1\ 4\ 3\ 3]$. Note: a_6 is idle but cannot play r_3. Because r_3 has only two current agents, it requires a third one. Thus, the group is not in a workable state and requires role transfers. Figure 9.4 is a resulting state after multiple successful role transfers. In Figure 9.4, all roles have sufficient current agents.

In reality, Figures 9.3 and 9.4 can be used to express a soccer team with 11 soccer players. In this scenario, L would represent the formation of the team. To make this group workable for the formation specified, there should be 1 *goalie* (r_0), 4 *defenders* (r_1), 3 *middle fields* (r_2), and 3 *forwards* (r_3). In Figure 9.3, player a_6 has not yet been assigned the position (role) but it is qualified to play a *defender* or *middle field*. Figure 9.3 is not workable for the formation of 4-3-3 but Figure 9.4 is.

Figure 9.5 expresses the current role matrices of Figures 9.3 and 9.4, respectively. Figure 9.6 shows the potential role matrices for Figures 9.3 and 9.4, respectively.

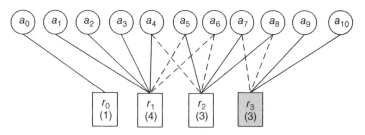

Figure 9.3 An unworkable group due to that r_3 needs 3 agents but has only 2.

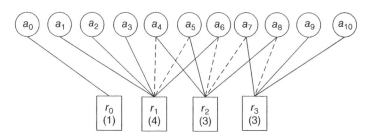

Figure 9.4 A workable state of the group in Figure 9.3.

(a)

```
[1 0 0 0]
[0 1 0 0]
[0 1 0 0]
[0 1 0 0]
[0 1 0 0]
[0 0 1 0]
[0 0 0 0]
[0 0 1 0]
[0 0 1 0]
[0 0 0 1]
[0 0 0 1]
```

(b)

```
[1 0 0 0]
[0 1 0 0]
[0 1 0 0]
[0 1 0 0]
[0 0 1 0]
[0 0 1 0]
[0 1 0 0]
[0 0 0 1]
[0 0 1 0]
[0 0 0 1]
[0 0 0 1]
```

Figure 9.5 The current role matrices of Figures 9.3 and 9.4. (a) The current role matrix for Figure 9.3. (b) The current role matrix for Figure 9.4.

(a)

```
[0 0 0 0]
[0 0 0 0]
[0 0 0 0]
[0 0 0 0]
[0 0 1 0]
[0 1 0 0]
[0 1 1 0]
[0 0 0 1]
[0 0 0 1]
[0 0 0 0]
[0 0 0 0]
```

(b)

```
[0 0 0 0]
[0 0 0 0]
[0 0 0 0]
[0 0 0 0]
[0 1 0 0]
[0 1 0 0]
[0 0 1 0]
[0 0 1 0]
[0 0 0 1]
[0 0 0 0]
[0 0 0 0]
```

Figure 9.6 The potential role matrices for Figures 9.3 and 9.4. (a) The potential role matrix for Figure 9.3. (b) The potential role matrix for Figure 9.4.

9.2.1 M-1 Problem

By using the same example of a soccer team, we may have an M-1 problem shown in Figure 9.7. In Figure 9.7, the required formation is 3-3-4. Agents a_4 and a_6 have not yet decided their positions (roles). Therefore, Figure 9.7 is not workable for the formation, where two roles (r_2 and r_3) require one more current agent to play. Figure 9.8 is a workable state for the formation after transferring roles from Figure 9.7.

9.2.2 1-M Problem

We may have a 1-M problem shown in Figure 9.9. Figure 9.9 is not workable because one role (r_2) requires two more current agents.

9.2.3 M-M Problem

In Figure 9.10, $L = [1, 5, 2, 2]$, r_1 has two current agents and requires three more, and r_3 has one current agent and requires one more. Therefore, the group is unworkable and requires role transfers. Figure 9.11 shows a workable state of the group in Figure 9.10 after successful role transfers. In Figure 9.11, all roles have sufficient current agents.

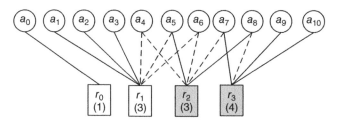

Figure 9.7 An unworkable group with 2 roles, each of which requires 1 extra current agent.

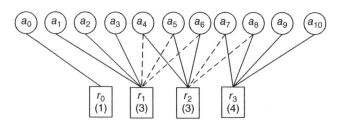

Figure 9.8 A workable group.

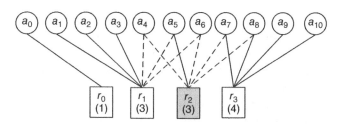

Figure 9.9 An unworkable group with 1 role requiring 2 more current agents.

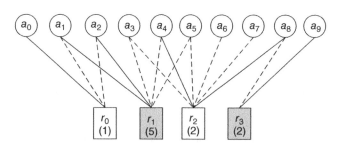

Figure 9.10 An M-M problem.

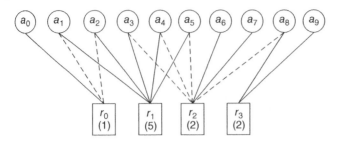

Figure 9.11 A solution for Figure 9.10.

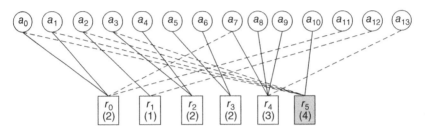

Figure 9.12 A 1-M problem.

Figures 9.10 and 9.11 can also be used to represent a computer science department with 10 staff. To make this group workable, there should be 1 *chair* (r_0), 5 *professors* (r_1), 2 *academic advisors* (r_2), and 2 *secretaries* (r_3). The group in Figure 9.10 is unworkable because there are insufficient *professors* and *secretaries*. In Figure 9.10, a_5 has not yet taken its position (role), and a_6 and a_7 have not yet been offered the *academic advisor* position (role). Figure 9.11 shows a workable group formed by transferring roles among the staff from Figure 9.10.

The 1-M problem may become a new M-M problem after one role transfer is conducted, i.e. making the unworkable role workable. To understand this situation, we can consider the example in Figure 9.12. This figure, at first, seems to be a 1-M problem, i.e. r_5 requires 3 more current agents. However, when making r_5 workable, we need 3 potential agents of r_5 to take role r_5. Then, we can ignore r_5 and its associated agents to form a new situation shown in Figure 9.13. The problem in Figure 9.13 is now an M-M problem, i.e. two roles, r_0 and r_1, require 2 and 1 more current agents, respectively. Note that in Figure 9.12, after making r_5 workable, r_5, a_0, a_1, a_2, and a_{10} will not be considered in the following role transfers.

The problems in Figures 9.12 and 9.13 have their instances in the real world. Suppose that Figure 9.12 represents a fire brigade trying to put out a fire in a building. r_5 here represents a role to rescue the people on the fifth floor, and this

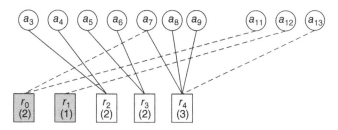

Figure 9.13 An M-M problem.

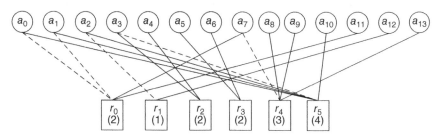

Figure 9.14 The solution of the problem in Figure 9.12.

requires 4 firefighters. The commander discovers that role r_5 is short of firefighters. Thus, he commands some of those who can climb the building to rescue the people on the fifth floor. However, after he has issued this command, he finds that both r_0 (*pumper operator*) and r_1 (*fire truck driver*) do not work as well. Now it is even worse, i.e. two roles cannot work properly. He needs to issue more commands. Fortunately, he finds a solution as shown in Figure 9.14.

The M-M RTPs can be solved with an exhaustive search method (Zhu and Zhou 2009). However, the method in (Zhu and Zhou 2009) has high complexity and is not practical for large problems. M-M RTPs can also be transferred to an assignment problem and solved with the GRA (Zhu and Alkins 2009b) algorithm discussed in Chapter 5.

9.3 From M-M RTPs to Role Assignment Problems

Based on the definitions of Sections 9.1 and 9.2, the M-M RTP can be defined as follows.

Definition 9.8 The *M-M RTP* aims to find a new T to make group g workable given T and Q_p of group g.

At first glance, GRAP and RTP are very different. However, if we examine the simple GRAP (Chapter 5) and RTP (Zhu and Zhou 2008a, 2009), it is easy to see their similarity.

Definition 9.9 We call role repository matrix Q a *simple qualification matrix*, i.e. $Q := T + Q_p$ in M-M role transfer problems.

It is called the qualification matrix in role assignment (Chapter 5), and it is also called the repository role matrix in role transfer (Section 9.1).

Definition 9.10 Given a simple qualification matrix Q, the *Simple Group Role Assignment Problem (SGRAP)* aims to find a role assignment matrix T that makes g workable.

For example, suppose that Q is shown in Figure 9.20a and $L = [1\ 4\ 3\ 3]$. One solution T for the GRAP is shown in Figure 9.15b.

Now an RTP can be converted to an SGRAP. Let $Q := T + Q_p$. Then, an M-M RTP is actually an SGRAP (Chapter 5). For example, Figure 9.16a is a T, Figure 9.16b a Q_p, and Figure 9.16c a Q.

(a)

$$\begin{bmatrix} 1 & 0 & 0 & 0 \\ 0 & 1 & 0 & 0 \\ 0 & 1 & 0 & 0 \\ 0 & 1 & 0 & 0 \\ 0 & 1 & 1 & 0 \\ 0 & 1 & 1 & 0 \\ 0 & 1 & 1 & 0 \\ 0 & 0 & 1 & 1 \\ 0 & 0 & 1 & 1 \\ 0 & 0 & 0 & 1 \\ 0 & 0 & 0 & 1 \end{bmatrix}$$

(b)

$$\begin{bmatrix} 1 & 0 & 0 & 0 \\ 0 & 1 & 0 & 0 \\ 0 & 1 & 0 & 0 \\ 0 & 1 & 0 & 0 \\ 0 & 1 & 0 & 0 \\ 0 & 0 & 1 & 0 \\ 0 & 0 & 1 & 0 \\ 0 & 0 & 1 & 0 \\ 0 & 0 & 0 & 1 \\ 0 & 0 & 0 & 1 \\ 0 & 0 & 0 & 1 \end{bmatrix}$$

Figure 9.15 A simple qualification matrix and the assignment matrix. (a) The simple Q. (b) The T.

(a)

$$\begin{bmatrix} 1 & 0 & 0 & 0 \\ 0 & 1 & 0 & 0 \\ 0 & 1 & 0 & 0 \\ 0 & 1 & 0 & 0 \\ 0 & 1 & 0 & 0 \\ 0 & 0 & 1 & 0 \\ 0 & 0 & 0 & 0 \\ 0 & 0 & 1 & 0 \\ 0 & 0 & 1 & 0 \\ 0 & 0 & 0 & 1 \\ 0 & 0 & 0 & 1 \end{bmatrix}$$

(b)

$$\begin{bmatrix} 0 & 0 & 0 & 0 \\ 0 & 0 & 0 & 0 \\ 0 & 0 & 0 & 0 \\ 0 & 0 & 0 & 0 \\ 0 & 0 & 1 & 0 \\ 0 & 1 & 0 & 0 \\ 0 & 1 & 1 & 0 \\ 0 & 0 & 0 & 1 \\ 0 & 0 & 0 & 1 \\ 0 & 0 & 0 & 0 \\ 0 & 0 & 0 & 0 \end{bmatrix}$$

(c)

$$\begin{bmatrix} 1 & 0 & 0 & 0 \\ 0 & 1 & 0 & 0 \\ 0 & 1 & 0 & 0 \\ 0 & 1 & 0 & 0 \\ 0 & 1 & 1 & 0 \\ 0 & 1 & 1 & 0 \\ 0 & 1 & 1 & 0 \\ 0 & 0 & 1 & 1 \\ 0 & 0 & 1 & 1 \\ 0 & 0 & 0 & 1 \\ 0 & 0 & 0 & 1 \end{bmatrix}$$

Figure 9.16 Examples of T, Q_p, and Q. (a) T. (b) Q_p. (c) Q.

The following algorithm applies K-M to solve the SGRAP. The details of adaptation from a GRA problem (GRAP) to a GAP (Kuhn 1955; Munkres 1957; Toroslu and Üçoluk 2007) can be found in (Zhu et al. 2012). Several key points are described as follows:

(a)   (b)

$$\begin{bmatrix} 1 & 1 & 0 \\ 0 & 1 & 1 \\ 1 & 0 & 1 \\ 1 & 1 & 0 \\ 1 & 0 & 1 \end{bmatrix} \quad \begin{bmatrix} 1 & 1 & 1 & 0 & 1 \\ 0 & 1 & 1 & 1 & 1 \\ 1 & 0 & 0 & 1 & 1 \\ 1 & 1 & 1 & 0 & 1 \\ 1 & 0 & 0 & 1 & 1 \end{bmatrix}$$

Figure 9.17 Square matrix from Q. (a) A Q. (b) The square matrix for Figure 9.17a.

1) Form a square matrix, i.e. duplicating the columns whose corresponding role range is greater than 1 ($L[j] > 1$) and adding columns ($m - \sum_{j=0}^{n-1} L[j]$) for non-existing roles. Please note that $m \geq \sum_{j=0}^{n-1} L[j]$ is a necessary condition for a successful role transfer.

2) Adjust the problem from maximization to minimization, i.e. $Q'[i, j] = 1 - Q[i, j]$, $(0 \leq i, j \leq m - 1)$.

For example, Q shown in Figure 9.17a ($L = [1\ 2\ 1]$) can be changed to a square matrix shown in Figure 9.17b, where the last column represents a null role.

From the above assignment algorithm GRA (Chapter 5), a role transfer algorithm can be given as follows:

Algorithm 9.1 MMRoleTransfer

```
Input:
  • L : an n-dimensional role's lower bound vector;
  • T: an m×n current role matrix; and
  • Qp: an m×n potential role matrix.
Output:
  • Success: An m×n current role matrix T' in which for
    all columns j (j= 0, …, n-1), ∑ᵢ₌₀ᵐ⁻¹T'[i, j] ≥L[j]. The
    new potential role matrix is in Qp'.
  • Failure:  An m×n current role matrix T' in which there
    is at least one column j (j= 0, …, n-1), ∑ᵢ₌₀ᵐ⁻¹T'[i, j] <
    L[j]. The new potential role matrix is in Qp'.
begin
  Step 1: Q := T + Qp;
  Step 2: result := GRA(L, Q, T', m, n);
  Step 3: Extract Qp' from T' and Qp;
  Step 4: Return result;
End
```

The complexity of Algorithm 9.1 is the same as GRA, which is $O(m^3)$. Therefore, the M-M role transfer problems are solved. As mentioned in Chapter 5, the SGRA problem even has a simpler solution, which is left for exercises.

9.4 Temporal M-M Role Transfer Problems

In the above discussion, we assume that there are sufficient people (agents) to avoid a crisis. In an emergency, scarcity is often more frequent than sufficiency. We may need to repeatedly assign roles to an agent in different time segments due to the scarcity of agents.

Sometimes, there are many (>1) roles, each of which loses many (>1) current agents; at the same time, $|\mathcal{A}| < |\mathcal{R}|$. This is a temporal M-M role transfer problem. This problem has two cases:

- Case 1: Strong restriction. We need to find a scale s and a role transfer scheme to guarantee that in any period $\delta = s \times \nu$ in consideration, every role must have enough current agents in at least a segment $\nu = \delta/s$. For example, to play the role *piano mover*, four players are required. Hence, the *piano mover* role only works when all four players are currently playing it.
- Case 2: Weak restriction. We need to find a scale s and a role transfer scheme to guarantee that in any period of time $\delta = s \times \nu$ in consideration, every role j must be played in $L[j]$ intervals. For example, the role *technical school instructor* requires 4 people to offer 4 courses to students. The role works when 2 of them repeatedly offer 2 different courses. It is fine with the students as long as the schedule is not in conflict. Clearly when 4 courses must be offered at the same time, the restriction becomes strong.

9.4.1 Temporal Transfer with Weak Restriction

In consideration of the temporal transfer with weak restriction, we observe an interesting phenomenon, i.e. separating a period δ into s parts is similar to copying the agents $s - 1$ times. Virtually, we finally have $s \times |\mathcal{A}|$ agents to play the original roles. In fact, we can provide a simple algorithm to solve this problem. However, we need to keep in mind the following requirements:

- Requirement 1: the s times of the number of all agents should be larger than the sum of the number of the required agents for each role, i.e. $s \times |\mathcal{A}| \geq \sum_{j=0}^{n-1} L[j]$.

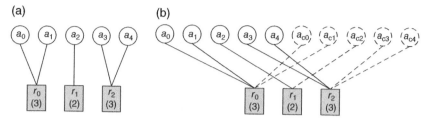

Figure 9.18 The group is not workable. (a) An unworkable group. (b) A possible solution by temporal transfer under the weak restriction.

- Requirement 2: when all the agents are copied $s - 1$ times, $Q = T + Q_p$, and $\sum_{i=0}^{m-1} Q[i,j] \geq L[j]$, i.e. each role obtains sufficient current agents.

Definition 9.11 Suppose that $m < n$, a *Temporal M-M Role Transfer Problem with Weak Restriction (WRTP)* aims to find an $m \times n$ matrix sequence: C^0, C^1,..., C^{s-1} and $C'[i,j] := \sum_{k=0}^{s-1} C^k[i,j]$ $(0 \leq i < m, 0 \leq j < n)$ to make group g workable given the T and Q_p of group g.

For example, Figure 9.18a shows an unworkable group that requires temporal role transfer. Figure 9.18b shows a role transfer scheme if we introduce a scale of $s = 2$. The dashed circles in Figure 9.18b are used to indicate agent copies that are available for temporal role transfer. Note that current roles become potential roles for the agent copies at a different time. In fact, Figure 9.18b is an M-M problem. In Figure 9.18b, if we can assign sufficient current agents to each role (Figure 9.19a), then we obtain the successful temporal role transfer scheme shown in Figures 9.19b and 9.19c. The scheme is obtained by splitting the original agents and their copies. For any period, we can separate it into two parts: t_0 and t_1. Figure 9.19b is the group's working state during t_0 and Figure 9.19c is the working state during t_1.

To understand Figure 9.19, we can consider the situation where 5 *professors* offer 3 courses (8 sections) to students. We assign *professors* a_0 and a_1 to teach course (role) r_0, *professor* a_2 to teach course (role) r_1, and *professors* a_3 and a_4 to teach course (role) r_2. A professor cannot teach two courses at the same time, but they can teach the students at different times. Therefore, Figure 9.19 represents a possible workable schedule in this scenario.

To find a successful role transfer scheme, we need to introduce a new operation for matrices "$\Vdash$" called merging by rows. Suppose X is an $m \times n$ matrix and Y is an $o \times n$ matrix, $X \Vdash Y$ creates a new $(m + o) \times n$ matrix Z, where $Z[i, j] := X[i,j], 0 \leq i < m$; and $Z[i, j] := Y[i - m, j], m \leq i < m + o$.

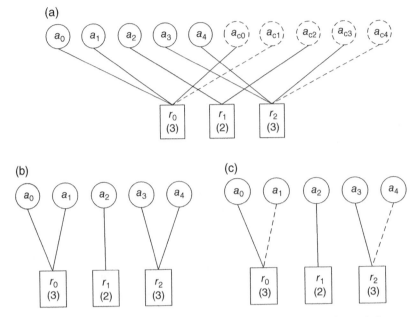

Figure 9.19 A temporal transfer scheme is obtained under the weak restriction. (a) A workable state. (b) The working state in one time slot. (c) The working state in in the other time slot.

The algorithm can be described as follows:

Algorithm 9.2 ProcessWRTP

- -

Input:
- L: an n-dimensional role's lower bound vector;
- T: an m×n current role matrix; and
- Q_p: an m×n potential roles matrix.

Output:
- **Success:** A sequence of matrices C^0, C^1, ..., C^{s-1} and a scale s.
- **Failure:** A report of an unworkable group.

begin
    //Note: ":=" means assignment from right to left and
    //"=" means that the left is equal to the right.
    Step1: Initialization and preparation.
    s:=1; C^0:=T; Q^0=Q_p; Q':= T+Q_p;
    ZERO is set as an m×n matrix with all elements being zeros.

(Continued)

Algorithm 9.2 (Continued)

---

if (there is one role that has neither current roles
nor potential roles, i.e. in matrix Q', there is one
column without 1s) *return* failure;

$$p := \sum_{j=0}^{n-1} L[j];$$

Step 2: Duplicate the agents and check if it works.
while (s<p)
//The case when (s = p) is a definite solution after
//passing the above requirements, i.e., one agent
//can play every role at least one time in the p
//intervals.

  Step 2.1: Duplicate the agents.
  s := s+1;
  $C^s := C^{s-1} \| \mathrm{ZERO}$; //The current role matrix is expanded.
  $Q^s := Q^{s-1} \| Q'$; //The potential role matrix is
                        //expanded.

  Step 2.2: Check if it works.
  result := **MMRoleTransfer**(L, C^s, C', Q^s, Q', m×s, n);

  Step 2.3: Establish the successful scheme or try more.
    if (result = success)
        Split the *(s×m)* ×n matrix C' into *s* m×n
        matrices C'^0, C'^1, ..., C'^{s-1}, i.e.,
        for (each i<m×s)
          for (each j <n)
              $C'^{i/m}$ [i%m, j] := C^s [i, j] ;
              //*i*/*m* means an integer quotient
              //when *i* is divided by *m*;
              //*i*%*m* means a remainder when *i*
              //is divided by *m*;
          endfor
        endfor
            return s +1 (scale) and all the
            matrices C'^0, C'^1, ..., C'^s;
      else
      s=s+1;
      endif
    endwhile
end

9.4.2 Temporal Transfer with Strong Restriction

The strong restriction presents a different requirement in temporal role transfer. Suppose that Figure 9.20 expresses a situation that 5 professors (agents) teach students but the professors are required to offer 3 *courses* (roles) (r_0, r_1, and r_2) for 8 sections of students. Course r_0 has 3 sections at the same time, r_1 has 2, and r_2 has 3. Evidently, a professor cannot offer the different sections of the same *course* at the same time segment.

By assigning current and potential courses to professors as shown in Figure 9.20, we can create the workable group shown in Figure 9.21 through temporal role transfers. That is to say, in one time segment, professors a_0, a_1, and a_2 are assigned to teach 3 sections of *course* r_0, respectively, and professors a_3 and a_4 are assigned to teach 2 sections of *course* r_2. In another segment, professors a_1, a_3, and a_4 are assigned to teach 3 sections of *course* r_2.

This is the most complicated case of role transfer discussed so far. In a role transfer scheme, we must guarantee that all the current agents are bound to their relevant role in a time segment.

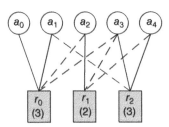

Figure 9.20 The group is not workable under strict restriction.

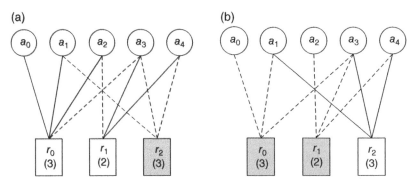

Figure 9.21 A temporal transfer scheme is obtained under strict restriction. (a) The working state in one time slot. (b) The working state in in the other time slot.

To guarantee that there is a successful role transfer scheme, a new type of requirement must be met: Suppose that T is the current assignment matrix, Q_p is the potential assignment matrix, and $Q = T + Q_p$. For a group to be workable, the number of 1s in each column should not be less than the relevant number in L, i.e. $\sum_{i=0}^{m-1} Q[i,j] \geq L[j]$ ($j = 0, 1, ..., n - 1$). That is to say, for each role j, if the number of all the potential agents and current agents is less than $L[j]$, the group is not workable because it is impossible for the role to have enough agents to play it.

The above requirement is much stronger than those in the WRTP. For example, the group in Figure 9.18 does not satisfy this requirement. Figure 9.19 shows an unworkable group for the new restriction. Figure 9.21 is a solution with a scale of 2, i.e. in any period, we can separate it into two parts: t_0 and t_1. The group's state is Figure 9.21a during t_0 and Figure 9.21b during t_1.

Definition 9.12 Suppose that $m < n$, and T and Q_p of group $\mathcal{g}$ are given, a *Temporal M-M Role Transfer Problem with Strong Restriction (SRTP)* aims to find an $m \times n$ matrix sequence: $C^0, C^1, ..., C^{s-1}$ such that every role is workable in at least one interval expressed by $C^k(0 \leq k \leq s - 1)$, i.e. $\forall j \exists k \ni (L[j] \leq \sum_{i=0}^{m-1} C^k[i,j])(0 \leq k \leq s - 1)$.

Now, we cannot use the concept of copying agents in Section 9.4.1. By observation, introducing a scale s means that we can separate the role set $\mathcal{R}$ into s nonempty subsets $\mathcal{R}_i \neq \emptyset$ ($i = 0, 1, ..., s - 1$), $\cup_{i=0}^{s-1}\mathcal{R}_i = \mathcal{R}$, $\mathcal{R}_i \cap \mathcal{R}_j = \emptyset$ ($i, j = 0, 1, ..., s - 1, i \neq j$). For each of the s parts in a period, one set of roles is played by enough current agents. For each set of roles, it is an M-M role transfer problem. This method needs to solve a recursive combination and permutation problem (Gilleland 2007). It is too complicated to apply it in an algorithm.

Therefore, we try reverse thinking: from a definite solution to a better one. We observe that a definite solution separates $\mathcal{R}$ into $n = |\mathcal{R}|$ subsets, each of which contains only one role. If these requirements are met, then the solution is set, i.e. $s = |\mathcal{R}|$. The solution is that in each interval, all the potential agents are ready to play one specific role. Evidently, this is not an optimal solution. We need to merge some role sets to obtain a better one, step by step. To simplify the discussion, we assume that no agent is free. In fact, it is easy to meet this assumption because the number of agents is less than the number of required agents.

There are many possibilities to set up an initial definite solution. To form the first n sets with one role, we set $s = n$ initially. In each $m \times n$ matrix C^k ($k = 0$, 1, ..., $s - 1$), there might be many combinations of choices. Suppose that there are $v^j = \sum_{i=0}^{m-1} C^k[i,j](j = 0, 1, ..., n - 1)$ agents, there are $\binom{v^j}{L[j]}$ possibilities for each

C^k to make role j work. From this, we can conclude that the total number of initial definite solutions is $\prod_{j=0}^{n-1} \binom{v^j}{L[j]}$.

To obtain the temporal transfer scheme, we need a kernel algorithm to merge the matrices obtained from the initial solution (a list of $m \times 1$ matrices). The key point is to start from a pair of matrices and try to merge them; repeat this process until no more merging is successful; record the scheme. Repeat these steps on other pairs of matrices and compare the schemes with what has been recorded, keeping the better one. In the end, only the best one is kept, i.e. there is no better scheme with a smaller scale.

The temporal role transfer scheme is obtained by Algorithm 9.3, which finds all the possible initial solutions.

Algorithm 9.3 SRTPKernel

```
Input:
    • L: an n-dimensional role's lower bound vector;
    • s: the number of the matrices to be merged;
    • CL: the current role matrix list with s matrices;
    • QL: the potential role matrix list with s matrices;
    • m: the agent number.
Output:
    • s: a scale;
    • CL: a transfer scheme in a matrix list; and
    • QL: a corresponding potential matrix list.
//Note: Vⁱ means the ith item of the vector V and "-"
//means an operation to remove the item on its right
//from its left vector.
begin
    Step 1: Initialize the vector and compute the
    combination number.
    A := {0, 1, ..., s-1};

    C = (s over 2) = s!/(2(s-2)!)=s(s-1)/2.

    Step 2: Establish a cx2 matrix V listing all the
    combinations taking 2 items from A, where V[i] is a
    pair of indices of matrices in the matrix lists CL
    and QL.
```

(Continued)

Algorithm 9.3 (Continued)

```
Step 3: Initialize the result matrix lists and the scale.
RCL := CL; RQL:=QL; result := s;
Step 4: Check if merging is possible
    for (each pair in V)
    Step 4.1: Reserve the initial matrix lists.
        TCL := CL; TQL:=QL; ts := s
        r₁ := the number of the columns of TCL^V[i, 0];
        r₂ := the number of the columns of TCL^V[i, 1];
    Step 4.2: Merge two matrices.
        ZERO:= an m×r₂ matrix with all elements being
        zeros;
        C := TCL^V[i, 0] ‖ ZERO;// ‖ means to merge by columns.
        Q := TQL^V[i, 0] ‖ (TCL^V[i, 1] + TQL^V[i, 1]);
        //+ means matrix add.
    Step 4.3: Check if this merging is OK.
        result :=MMRoleTransfer(L, C, C′, Q, Q′, m, r₁+r₂);
    Step 4.4: Try more merging if successful.
        if (Two matrices can be merged, i.e., result =
        success)
            CL′ := TCL - TCL^V[i, 1];
            QL′ := TQL - TQL^V[i, 1];
            C′^V[i, 0] := C′;
            Q′^V[i, 0] := Q′;
            ts :=s-1;
            kernel_result:= SRTPKernel(L, CL′, QL′,
            ts, m);
            if (A better solution is obtained, i.e.,
            kernel_result<ts)
                Copy CL′ to TCL, QL′ to TCL, and
                kernel_result to ts;
            endif
        endif    //Two Matrices can be merged.
    Step 4.5: Collect the result matrix list.
        if (A better solution is obtained in this
        loop, i.e., ts<result)
            Copy TCL to RCL, TQL to RQL, and ts to result;
        endif
    endfor //Each pair.
    return RCL to CL, RQL to QL, and result to s;
end
```

Please note that Algorithm 9.4 involves a loop with $p = \prod_{j}^{n-1}\binom{v^j}{L[j]}$, where v^j is the number of potential agents to play role j. Therefore, the complexity of Algorithm 9.4 is high, i.e. $O\left(\prod_{j}^{n-1}\binom{v^j}{L[j]}\right)$, and not practical.

Algorithm 9.4 ProcessSRTP

```
Input:
   • An n-dimensional role's low bound vector L;
   • An mxn current role matrix T; and
   • An mxn potential roles matrix Qp.
Output:
   • Success: A sequence of matrices CL and a scale s.
   • Failure: A report for an unworkable group.
   //Note: ":=" means assignment from right to left and
   "=" means that the left is equal to the right.
begin
 Step 1: Initialize and check initial conditions.
   Q := T+Qp;
   if (in matrix Q, there is one column with a number of
 1s less than the relevant number in the role low range
```
vector L, i.e., $\sum_{i=0}^{m-1} Q[i, j] < L[j]$ ($j = 0, 1, ..., n-1$))
```
      return failure;
   endif
   s[0..n-1] := {n};//All the values are n.
```
$p = \prod_{j}^{n-1}\binom{v^j}{L[j]}$, where v^j is the number of potential
```
   agents to play role j;
 Step 2: Try each possible scheme of agent assignment.
   for (0≤k<p)
      Build the kth pair of CL[k] and QL[k] as {C⁰, C¹, ...,
      and, Cⁿ⁻¹} and {Q⁰, Q¹, ..., and, Qⁿ⁻¹};
      Cʲ (j = 0, ..., n-1) := an mx1 current role matrix,
      role rᵢ is assigned with the required number (i.e.,
```

(Continued)

Algorithm 9.4 (Continued)

```
        L[i]) of agents;
        Q^j := an mx1 potential role matrix for role j;
        result := SRTPKernel (L, CL[k], QL[k], s[k], m);
     endfor
  Step 3: Select the best scheme with the least scale s.
     s=n; index =0;
     for (0≤k<p)
          if (s[k]<s)
                   s = s[k]; index = k;
          endif
     endfor
  return success with s, CL[index], and QL[index];
end
```

- -

9.4.3 A Near-Optimal Solution to SRTP with the Kuhn-Munkres Algorithm

With Algorithm 9.1, WRTP is solved quickly in the same complexity as that of Algorithm 9.1. We need to improve Algorithm 9.4 because it requires a combinational number: $\prod_{j=0}^{n-1} \binom{v^j}{L[j]}$, where v^j is the number of qualified agents to play role j, i.e. $v^j = \sum_{i=0}^{m-1} Q[i,j]$ ($0 \leq j \leq n-1$). The goal of this section is to find a near-optimal solution in the polynomial complexity.

Similar to solving the GRA problem (Chapter 5) with the Kuhn-Munkres Algorithm (Baker 2008; Bourgeois and Lassalle 1971; Kuhn 1955; Munkres 1957; Zhu and Alkins 2009a; Zhu et al. 2012), we discuss a near-optimal solution to the SRTP. Because the Kuhn-Munkres algorithm solves square matrices, we need to prepare matrices that can be transformed into square ones. However, if matrix Q for the SRTP is expanded, the number of rows (agents) of the matrix is less than the number of columns (roles). To apply the simple assignment algorithm adapted from the Kuhn-Munkres algorithm, we needed to split Q into a group of matrices in which the number of columns (agents) is less than the number of rows (roles). For example, Figure 9.22 shows T, Q_p, Q, and $L = [3, 2, 2]$.

(a)

$$\begin{bmatrix} 1 & 0 & 0 \\ 1 & 0 & 0 \\ 0 & 1 & 0 \\ 0 & 0 & 1 \\ 0 & 0 & 1 \end{bmatrix}$$

(b)

$$\begin{bmatrix} 0 & 0 & 0 \\ 0 & 0 & 1 \\ 1 & 0 & 0 \\ 1 & 1 & 0 \\ 0 & 1 & 0 \end{bmatrix}$$

(c)

$$\begin{bmatrix} 1 & 0 & 0 \\ 1 & 0 & 1 \\ 1 & 1 & 0 \\ 1 & 1 & 1 \\ 0 & 1 & 1 \end{bmatrix}$$

Figure 9.22 Three matrices for a group. (a) T. (b) Q_p. (c) Q.

$$\begin{bmatrix} 1 & 1 & 1 & 0 & 0 & 0 & 0 \\ 1 & 1 & 1 & 0 & 0 & 1 & 1 \\ 1 & 1 & 1 & 1 & 1 & 0 & 0 \\ 1 & 1 & 1 & 1 & 1 & 1 & 1 \\ 0 & 0 & 0 & 1 & 1 & 1 & 1 \end{bmatrix}$$

Figure 9.23 The expanded matrix from Figure 9.22c for the Kuhn-Munkres algorithm.

Based on Q and L, an expanded $m \times n'$ ($n' = \sum_{j=0}^{n-1} L[j]$) matrix (called flat matrix, denoted as Q^f) is obtained (Figure 9.23), i.e. column j is copied $L[j]$ ($0 \le j \le n-1$) times and an n' dimensional vector L' is used to record the indexes of L in Q^f. For Figure 9.23, $L' = [0, 0, 0, 1, 1, 2, 2]$. Note that matrix Q^f is not a square one because in the SRTP, $m < n'$. It is necessary to split Q^f into several square matrices and make the SRTP solvable.

To accomplish this split, a scheme must be found first. In other words, it is necessary to obtain a list of indices to split Q^f into a sequence of sub matrices. Based on the assumption of SRTP that $\max\{L[j]\ (0 \le j \le n-1)\} \le m$, the major task before the split is to find sub vectors whose sum is just less than m.

Given a list of natural numbers L and a positive integer m, the problem is to find the least number of sub lists $L^0, L^1, ..., L^{s-1}$ and $\cup_{i=0}^{s-1} L^i = L$.

Note that, in a split, one role should not be split into two different matrices. This is the restriction of SRTP. The basic idea for a split is, through multiple passes of searching and tagging, to form groups of elements in L and make the sum of them less than the count of available agents and record them until all the roles have been considered.

The split algorithm can be described as follows using two separate algorithms, **ObtainSplitScheme** and **Splitting**:

Using the example shown in Figure 9.22c with $s = 2$, the above algorithms generated the two matrices shown in Figure 9.24.

The matrices after the expansion step of the simple assignment algorithm are shown in Figure 9.25.

Algorithm 9.5 ObtainSplitScheme

Input:
- *L: an n-dimensional role's lower bound vector; and*
- *m: the number of agents.*

Output:
- **Success:** *a scale s and the splitting scheme S that is an s dimensional vector of roles.*

begin
    s := 0;
    Step 1: Start to check from the first role;
    Step 2: Check if all the roles are tagged, if yes go to 6;
    Step 3: Select the role that is not tagged, put this role
            into the scheme S[s] and tag it;
    Step 4: Start from the next untagged role;
        If the sum of its range and the total range of
    scheme S[s] is not larger than m, put this role
    into S[s] and tag the role. Repeat Step 4 until the
    sum exceeds m.
    Step 5: s:=s+1, go to **Step 2**;
    Step 6: exit;
end

Algorithm 9.6 Splitting

Input:
- *L: an n-dimensional role's lower bound vector;*
- *s: scale;*
- *S: the splitting scheme that is an s dimensional vector, each item of which is a list of roles; and*
- *Q: an m×n repository role matrix.*

Output:
- **Success:** A sequence of matrices Q^0, Q^1, ..., and Q^{s-1}, a sequence of vectors L^0, L^1, ..., and L^{s-1}, the dimension of Q^k is m×S[k].length, and L^k is S[k].length ($0 \leq k \leq s-1$).

(Continued)

Algorithm 9.6 (Continued)

```
begin
  for (0≤k≤s-1)
    for (0≤i≤S[k].length-1)
      for (0≤q≤m-1)
                Qᵏ[q, i]:= Q[q, S[k][i]];
      endfor
      Lᵏ[i]:=L[S[k][i]];
    endfor
  endfor
  return Qᵏ, and Lᵏ (0≤k≤s-1);
end
```
- -

(a)

$$\begin{bmatrix} 1 & 0 \\ 1 & 0 \\ 1 & 1 \\ 1 & 1 \\ 0 & 1 \end{bmatrix}$$

(b)

$$\begin{bmatrix} 0 \\ 1 \\ 0 \\ 1 \\ 1 \end{bmatrix}$$

Figure 9.24 The split matrices from Figure 9.22c. (a) The first part. (b) The second part.

After the splitting step, we generate a solution for the SRTP shown in Figure 9.26, where Figure 9.26a is for the first half and Figure 9.26b is for the second half.

The final C^0 and C^1 are shown in Figure 9.27, where Figure 9.27a is for the first half and Figure 9.27b is for the second half.

It is required to check a necessary condition for matrix Q^f to be split:

$$L[j] \le \sum_{i=0}^{m-1} Q^f[i,j](0 \le j \le n-1).$$

Even with the above condition, some split schemes may not result in a successful solution. For example, two roles at one time may compete for qualified agents. That is why we use exhaustive search (Zhu and Zhou 2008a, b).

To avoid exponential complexity, a near-optimal, but practical, solution is presented. The algorithm is near-optimal because it does not pursue the optimal scale number. It tries to find a successful scale by limiting the available agent count when one scale is not successful. In fact, with each failure, the agent count is decreased by 1 (Step 6.2 of Algorithm 9.7). This trick keeps the split

Figure 9.25 The expanded matrices from Figure 9.24 for the Kuhn-Munkres algorithm. (a) An expanded square matrix for Figure 9.24a. (b) An expanded square matrix for Figure 9.24b.

(a)

$$\begin{bmatrix} 1 & 1 & 1 & 0 & 0 \\ 1 & 1 & 1 & 0 & 0 \\ 1 & 1 & 1 & 1 & 1 \\ 1 & 1 & 1 & 1 & 1 \\ 0 & 0 & 0 & 1 & 1 \end{bmatrix}$$

(b)

$$\begin{bmatrix} 0 & 0 & 0 & 0 & 0 \\ 1 & 1 & 1 & 0 & 0 \\ 0 & 0 & 0 & 0 & 0 \\ 1 & 1 & 1 & 0 & 0 \\ 1 & 1 & 1 & 0 & 0 \end{bmatrix}$$

Figure 9.26 A possible solution for Figure 9.24. (a) The Kuhn-Munkres *solution* for Figure 9.25a. (b) The Kuhn-Munkres solution for Figure 9.25b.

(a)

$$\begin{bmatrix} 1 & 0 & 0 & 0 & 0 \\ 0 & 1 & 0 & 0 & 0 \\ 0 & 0 & 1 & 0 & 0 \\ 0 & 0 & 0 & 1 & 0 \\ 0 & 0 & 0 & 0 & 1 \end{bmatrix}$$

(b)

$$\begin{bmatrix} 0 & 0 & 0 & 0 & 0 \\ 1 & 0 & 0 & 0 & 0 \\ 0 & 0 & 0 & 0 & 0 \\ 0 & 1 & 0 & 0 & 0 \\ 0 & 0 & 1 & 0 & 0 \end{bmatrix}$$

Figure 9.27 The required solution for Figure 9.21. (a) The T for Figure 9.26a. (b) The T for Figure 9.26b.

(a)

$$\begin{bmatrix} 1 & 0 & 0 \\ 1 & 0 & 0 \\ 1 & 0 & 0 \\ 0 & 1 & 0 \\ 0 & 1 & 0 \end{bmatrix}$$

(b)

$$\begin{bmatrix} 0 & 0 & 0 \\ 0 & 0 & 1 \\ 0 & 0 & 0 \\ 0 & 0 & 1 \\ 0 & 0 & 1 \end{bmatrix}$$

algorithms (Algorithms 9.5 and 9.6) effective. For Figure 9.22c, $S[0] = [0, 1]$, $S[1] = [2]$ and $s = 2$.

The new algorithm based on split, i.e. Algorithm 9.7, is described as follows:

Algorithm 9.7 NewSRTP

```
Input:
     • L: an n-dimensional role's lower bound vector;
     • T: an mxn current role matrix; and
     • Qp: an mxn potential roles matrix.
Output:
     • Success: A sequence of matrices M^c0, M^c1, ..., M^cs-1,
                and a scale s.
     • Failure: A report for an unworkable group.
```

(Continued)

Algorithm 9.7 (Continued)

```
begin
   Step 1: Q := T + Qp;
   Step 2: Check matrix Q, if it is impossible to find a
           solution return failure;
   Step 3: initialize agent number (na) for splitting
           with m;
   Step 4: ObtainSplittingScheme(L, S, s, na, n);
   Step 5: Splitting (Q, s, S); //Get {Qk, Lk| 0≤k≤s-1}.
   Step 6: for (0≤i≤ s - 1)
   Step 6.1: result:=GRA (Li, Qi, T, m, n);
   Step 6.2: if (result = True)
                               obtain Ci from T;
             else
                               na = na-1 and goto Step 3;
             endif
   endfor;
   Step 7: return C0, C1, ..., Cs-1, and s.
end
```

Comparing Algorithm 9.7 with Algorithm 9.4 in Section 9.4.2, it is clear that there is no combination number affecting the complexity of the algorithm. The complexity of Splitting() is less than $O(m^3)$ because $s \leq m$ and $S[k]$.length $\leq m$ ($0 \leq k \leq s$). Therefore, the complexity of this algorithm is similar to that of the Kuhn-Munkres algorithm.

9.4.4   Performance Experiments

To verify the proposed algorithms, all the above algorithms are implemented in Java by adapting the implementation in (Zhu and Zhou 2008a) and tested using the random examples mentioned in (Zhu and Zhou 2008a, 2009). The experiments are run on a laptop with a CPU (T1800) of 2.10 GHz and with the development environment Microsoft Windows Vista (Home Edition) and Eclipse 3.2.2.

Experiments and complexity analysis show that all the role transfer problems except SRTP can be quickly solved. We only present the experiments for Algorithm 9.7 in this section.

In the experiments, there is an interesting phenomenon in the strong temporal transfer. The largest scale number obtained is exactly the same as n, which is the number of roles. This presents a fact that all the qualified agents playing a role at the same time is a definite solution for the SRTP if there are enough such agents.

A randomly organized group is formed by:

- Randomly setting the elements of L as 2 to 4 and guaranteeing that $\sum_{j=0}^{n-1} L[j] \geq m$.
- For the current matrix, randomly assigning role j to an agent i and guaranteeing that all the agents are assigned.
- For the potential matrix, randomly assigning role j to an agent i (if role j has not yet been assigned with a current agent). Each role j has at most $L[j]$ potential agents.

In the experiments, $m = 10, 20, \ldots, 50$, and $n = m/2$. Each experiment repeats for 300 randomly generated groups. The time (in milliseconds) is between 0 (<1) and 173 ms (Zhu and Zhou 2012) and it means Algorithm 9.7 is much more practical than Algorithm 9.4. For further interests, readers may make experiments to check the optimality of Algorithm 9.7 compared with Algorithm 9.4. Please note that because Algorithm 9.4 can only deal with small-scale problems, i.e. $m < 10$, it is very hard to compare them when the size of a group, i.e. m, is larger than 10.

9.5 Role Transfer Tool

We know that a group with roles and agents can be shown in a graph similar to Figures 9.1–9.3. Even though it is not difficult to draw such a graph with the current drawing software, it is difficult to modify the graph for different situations and to tell if agents are critical or if the group is workable. Our tool provides an easy way to form a group, express the assignment of roles to agents, and show the changes within a group. With the support of specially designed algorithms (Zhu and Zhou 2008a, 2009, 2012), our tool can tell if a group is workable; if there is a successful role transfer in a group; if an agent is critical in a group; and if there is a scheduling scheme when there are not enough agents to play all the roles.

The tool aims to assist decision makers in scheduling personnel for different positions in difficult situations. Such a tool must provide a list of functions for the management of a group. Using the E-CARGO model (Zhu and Zhou 2006a), we designed the following functions for groups ($\mathcal{G}$), agents ($\mathcal{A}$), and roles ($\mathcal{R}$):

Group:

- Create: create an empty group.
- Open: open an existing group.
- Save: save a group to a file.
- Print: print the current group.
- Check state: express whether the group is workable or not.

- Transfer: show the workable state of the group if there is a successful role transfer or a message if there is not.
- Temporal transfer: provide a temporal role transfer scheme when the agent number is less than the required number in a group.
- Critical agents: check if there are critical agents in the group.
- Zoom in/out: scale the size of the icons of agents and roles to accommodate the whole view of a group.

AGENT:

- Add: add an agent to the current group.
- Delete: delete an agent from the current group.
- Change: change the property/profile of an agent.

ROLE:

- Add: add a role to the current group.
- Delete: delete a role from the current group.
- Change: change the property of a role.

Role/Agent Assignment:

- Current role/agent: create a current assignment link between an agent and a role.
- Potential role/agent: create a potential assignment link between an agent and a role.
- Delete an assignment: delete the selected assignment link between an agent and a role.
  Note that classes ($\mathcal{C}$), objects ($\mathcal{O}$) and environments ($\mathcal{E}$) are the tool's internal components that are not visible to users.

The tool's internal architecture completely reflects the E-CARGO model (Zhu et al. 2008). The kernel mechanisms for RBC in Chapter 4 (Zhu and Zhou 2006a) can be directly applied in building this tool. The major challenge is to link the kernel mechanisms to the Graphical User Interface (GUI) components (Figure 9.28). In the GUI component, we create lists of GUI objects such as agents, roles, connections, and groups. These GUI objects are all used to store the information for the GUI and each has a link with its corresponding internal component.

To implement role transfer, a set of computational algorithms is required. Role transfer is a search for enough agents to fill every specific role under some conditions. Therefore, the algorithms applied in this tool are mainly based on exhaustive searches. To implement these algorithms, matrices are applied to express a group. The early solutions in (Zhu and Zhou 2008a) only solved a special case of role transfer problems. Instead, this tool applies newly developed algorithms that can be used to solve all the role transfer problems mentioned in this chapter.

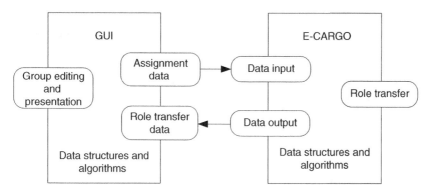

Figure 9.28 The architecture of the tool.

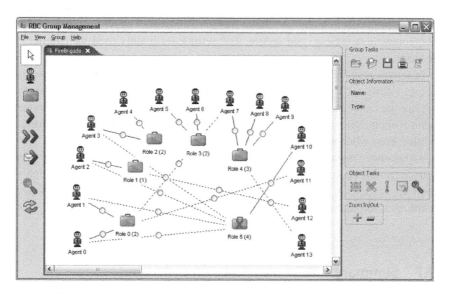

Figure 9.29 The example in Figure 9.12.

Figure 9.29 is the running screenshot example of the visualization tool for role transfer that corresponds to Figure 9.12.

9.6 Related Work

Although role transfer is evidently an important general problem in management (Black 1988), organizational behavior and performance (Ashforth 2001; Bhardwaj and Chandrakasan 2002; Cyert and MacCrimmon 1968; Neals and Griffin 2006; Nicholson 1984), system design (Odell et al. 2003), system construction

(Bhardwaj and Chandrakasan 2002), scheduling, training, and commanding (Turoff et al. 2004), there is no comprehensive research on role transfer theory and algorithms (Cormen et al. 2001; Diestel 2005). Some related research is in agent systems and wireless communications using the term *role assignment* (Chaimowicz et al. 2002; Dasgupta et al. 2008; Dastani et al. 2003; Odell et al. 2003; Stone and Veloso 1999; Vail and Veloso 2003), and others mainly investigate people's or organization's behaviors when role transfer happens and they use the term *role transition* (Ashforth 2001). Some other relevant research comes from psychologists, where the major concerns are the behavior of people and organizational performance when role transfers occur (Black 1988; Nicholson 1984).

Research on the delegation of rights (tasks, authorization, permissions, responsibilities, or even roles) (Crispo 1998; Ding et al. 1997; Na and Cheon 2000) also deals with the problem of transferring rights (permissions, responsibilities, or roles) to neighboring agents or subordinate users. It mainly provides policies, rules, or protocols to guarantee that a transfer (sometimes copy) process is possible, complete, trustable, and secure. The results are mainly used for computer security.

In failure resilience (Massoulie et al. 2003), the concerned problem is how to design agents such that they can dynamically join and leave an agent group to replace the faulty agents as needed. Related research in this direction concentrates on providing algorithms and protocols to guarantee that the system remains functioning even when some agents do not behave properly.

All the above research shows that there is indeed a strong need to investigate role transfer problems and their solutions. The results presented in this chapter have their applications in many different fields, such as information systems, management, production, and manufacturing industry.

9.7 Summary

Role transfer is important in an organization. Careful and effective assignments of current and potential roles help organizations prepare for future emergencies ahead of time. For situations where there are insufficient agents, temporal role transfers are a necessity. Based on the E-CARGO model, we define the properties of role transfer and design algorithms to facilitate successful role transfers. With these algorithms, we can precisely recognize whether a group is strong, check if a group is workable, and find a scheme of role transfer.

This chapter clarifies role transfer problems and proposes fundamental solutions for them. Role transfer problems may be simple for a small group, but they become more and more complicated as the number of roles and agents increases. Furthermore, an emergency requires definite role transfer schemes. The initial algorithms are based on an exhaustive search and the solution is only acceptable when the number of agents and roles is small. For cases with large numbers, a more efficient algorithm is proposed based on the GRA algorithm discussed in Chapter 5. The presented visualization tool can help managers deal with task assignment and management, especially in an emergency.

For future work, it is valuable to use visualized role transfer tools to teach students task assignment and management skills. Furthermore, these tools can be used to facilitate management tasks such as emergency management, group organization, and task distribution.

The second valuable direction for research is to investigate the efficiencies of the proposed algorithms in different organizational structures. Such analysis may help discover how the complexity of a role transfer problem is affected by different parameters in group organizations, e.g. the role-agent relationships (the role assignment matrices), the number of current/potential roles/agents, the number of roles that lose current agents and the number of lost agents.

The third meaningful direction for research is optimization. The presented M-M problem solution is only able to find a successful role transfer, but that solution might not be the optimal one. Similarly, the WRTP problem algorithm finds one scheme by copying the original agents. However, there may be a solution that requires copying only a part of all the agents. The solution to SRTP, i.e. algorithm ProcessSRTP, is optimal for the scale but may not be optimal for agents. For example, some agents may be heavily loaded while others may be lightly loaded. Investigations into further optimized solutions to the above problems will undoubtedly yield meaningful results.

References

Ashforth, B.E. (2001). *Role Transitions in Organizational Life: An Identity-based Perspective*. Mahwah, NJ: Lawrence Erlbaum Associates, Inc.

Baker, G. (2008). Java implementation of the classic Hungarian algorithm for the assignment problem. http://sites.google.com/site/garybaker/hungarian-algorithm/assignment (accessed 10 August 2020).

Bhardwaj, M. and Chandrakasan, A.P. (2002). Bounding the lifetime of sensor networks via optimal role assignments. *Proceedings of Twenty-First Annual Joint Conference of the IEEE Computer and Communications Societies (INFOCOM 2002)*, New York, USA (June 2002), vol. 3, pp. 1587–1596.

Black, J.S. (1988). Work role transitions: a study of American expatriate managers in Japan. *Journal of International Business Studies* 19 (2): 277–294.

Bostrom, R.P. (1980). Role conflict and ambiguity: critical variables in the MIS user-designer relationship. *Proceedings of the 17th Annual Computer Personnel Research Conference*, Miami, Florida, United States (June 1980), pp. 88–115.

Bourgeois, F. and Lassalle, J.C. (1971). An extension of the Munkres algorithm for the assignment problem to rectangular matrices. *Communications of the ACM* 14 (12): 802–804.

Chaimowicz, L., Campos, M.F.M., and Kumar, V. (2002). Dynamic role assignment for cooperative robots. *Proceedings of the IEEE International Conference on Robotics and Automation (ICRA '02)*, Washington, DC, USA (11–15 May 2002), pp. 293–298.

Cormen, T.H., Leiserson, C.E., Rivest, R.L., and Stein, C. (2001). *Introduction to Algorithms*, 2e. Cambridge, MA: The MIT Press.

Crispo, B. (1998). Delegation of responsibilities. Lecture Notes in Computer Science, *Proc. of 6th International Workshop on Security Protocols*, Cambridge, UK (April 1998), vol. 1550, pp. 624–625.

Cyert, R.M. and MacCrimmon, K.R. (1968). Organizations. In: *The Handbook of Social Psychology*, vol. 1 (eds. G. Lindzey and E. Aronson), 568–611. Addison-Wesley.

Dasgupta, D., Hernandez, G., Garrett, D., et al. (2008). A comparison of multiobjective evolutionary algorithms with informed initialization and Kuhn-Munkres algorithm for the sailor assignment problem. *Proceedings of the GECCO Conference Companion on Genetic and Evolutionary Computation*, Atlanta, GA, USA (12–16 July 2008), pp. 2129–2134.

Dastani, M., Dignum, V., and Dignum, F. (2003). Role-assignment in open agent societies. *Proceedings of the Second Int'l Joint Con. on Autonomous Agents and Multiagent Systems*, Melbourne, Australia (14–18 July 2003), pp. 489–496.

Diestel, R. (2005). *Graph Theory*, 3e. Heidelberg: Springer-Verlag.

Ding, Y., Horster, P., and Petersen, H. (1996). A new approach for delegation using hierarchical delegation tokens. In: *Communications and Multimedia Security II. IFIP Advances in Information and Communication Technology* (eds. P. Horster), 128–143, Springer.

Gilleland, M. (2007). Combination generator. http://www.merriampark.com/comb.htm (accessed 10 August 2020).

Kuhn, H.W. (1955). The Hungarian method for the assignment problem. *Naval Research Logistic Quarterly* 2: 83–97. (Reprinted in vol. 52, no. 1, 2005, pp. 7–21).

Massoulie, L., Kermarrec, A.M., and Ganesh, A.J. (2003), Network awareness and failure resilience in self-organizing overlay networks. *Proceedings of 22nd Int'l Symposium on Reliable Distributed Systems*, Florence, Italy (October 2003), pp. 47–55.

Munkres, J. (1957). Algorithms for the assignment and transportation problems. *Journal of the Society for Industrial and Applied Mathematics* 5 (1): 32–38.

Na, S.Y. and Cheon, S.H. (2000). Role delegation in role-based access control. *Proceedings of the Fifth ACM Workshop on Role-Based Access Control*, Berlin, Germany (26–28 July 2000), pp. 39–44.

Neals, M. and Griffin, M.A. (2006). A model of self-held work roles and role transitions. *Human Performance* 19 (1): 23–41.

Nicholson, N. (1984). A theory of work role transitions. *Administrative Science Quarterly* 29 (2): 172–191.

Odell, J.J., Van Dyke Parunak, H., Brueckner, S., and Sauter, J. (2003). Changing roles: dynamic role assignment. *Journal of Object Technology* 2 (5): 77–86.

Stone, P. and Veloso, M. (1999). Task decomposition, dynamic role assignment, and low-bandwidth communication for real-time strategic teamwork. *Artificial Intelligence* 110: 241–273.

Toroslu, I.H. and Üçoluk, G. (2007). Incremental assignment problem. *Information Sciences* 177 (6): 1523–1529.

Turoff, M., Chumer, M., Van de Walle, B., and Yao, X. (2004). The design of a dynamic emergency response management information system (DERMIS). *Journal of Information Technology Theory and Application (JITTA)* 5 (4): 1–35.

Vail, D. and Veloso, M. (2003). Multi-robot dynamic role assignment and coordination through shared potential fields. In: *Multi-Robot Systems* (eds. A. Schultz, L. Parkera and F. Schneider), 87–98. Kluwer.

Zhu, H. and Alkins, R. (2009a). Improvement to rated group role assignment algorithms. *Proceedings of the IEEE Int'l Conference on Systems, Man and Cybernetics*, San Antonio (October 2009), pp.4861–4866.

Zhu, H. and Alkins, R. (2009b). Group role assignment. *Proceedings of IEEE/ACM Int'l Symposium on Collaborative Technologies and Systems*, Baltimore, MA (May 2009), pp. 431–439.

Zhu, H. and Zhou, M.C. (2006a). Role-based collaboration and its kernel mechanisms. *IEEE Transactions on Systems, Man and Cybernetics, Part C* 36 (4): 578–589.

Zhu, H. and Zhou, M.C. (2006b). The role transferability in emergency management systems. *Proceedings of the 3rd Int'l Conf. on Information Systems for Crisis Response and Management (ISCRAM'06)*, Newark, NJ, USA (May 2006), pp. 487–496.

Zhu, H. and Zhou, M.C. (2008a). Role transfer problems and algorithms. *IEEE Transactions on Systems, Man and Cybernetics, Part A* 36 (6): 1442–1450.

Zhu, H. and Zhou, M. (2008b). Roles in information systems: a survey. *IEEE Transactions on Systems, Man and Cybernetics, Part C: Applications and Reviews* 38 (3): 377–396.

Zhu, H. and Zhou, M.C. (2009). M-M role transfer problems and solutions. *IEEE Transactions on Systems, Man and Cybernetics, Part A* 39 (2): 448–459.

Zhu, H. and Zhou, M.C. (2012). Efficient role transfer based on Kuhn–Munkres algorithm. *IEEE Transactions on Systems, Man and Cybernetics, Part A: Systems and Humans* 42 (2): 491–496.

Zhu, H., Grenier, M., Alkins, R., et al. (2008). A visualized tool for role transfer. *Proceedings of the IEEE International Conference on Systems, Man and Cybernetics (ICSMC)*, Singapore (12–15 October 2008), pp. 2225–2230.

Zhu, H., Zhou, M.C., and Alkins, R. (2012). Group role assignment via a Kuhn-Munkres algorithm-based solution. *IEEE Transactions on Systems, Man and Cybernetics, Part A* 42 (3): 739–750.

Exercises

1   What is the basic assumption for a role transfer problem to exist?
2   What are the major matrices/vectors used to define the role transfer problems?
3   Define the role transfer problem formally.
4   What do we mean by 1-1, 1-m, m-1, and m-m role transfer problems?
5   What is a temporal role transfer problem?
6   What are the two different restrictions for the temporal role transfer problem?
7   Why can a role transfer problem be solved by taking it as a GRA problem?
8   Do you think that the visualized role transfer tool is well designed? Do you have any ideas to improve this tool?
9   Do you think that the role transfer problem is well solved? Justify your answer.
10   Can you specify more role transfer problems by introducing other constraints?

10

More to Investigate

10.1 Role Negotiation

Role-Based Collaboration (RBC) and its Environments – Classes, Agents, Roles, Groups, and Objects (E-CARGO) model can be utilized to solve a number of complex problems in the real world. Group Role Assignment (GRA) is a major step before role-playing in RBC. Some important tasks must be accomplished prior to GRA. Role negotiation is such a task.

Negotiation is a common phenomenon in collaboration. For example, a recruiter (Kotlyar and Ades 2002) needs to negotiate with an applicant about their duties, responsibilities, salaries, and benefits for a posted position. In a fund application, the applicant and the authority need to negotiate the budget and objectives. To build a new enterprise, a board of governors needs to negotiate the budget and positions (roles). The primary objective of this section is to establish a mechanism to support negotiations prior to starting the cooperation process (i.e. agent evaluation).

Definition 10.1 *Role negotiation* is the process to determine the environment e for collaboration in the context of E-CARGO.

To organize a team, we need to know how many roles are and what roles are required. It is comparable to the planning, initiation, and requirement analysis phases in systems engineering (Blanchard and Fabrycky 2010; Pressman and Maxim 2014). Different from conventional approaches, RBC has a clear role concept to guide the process of planning, initiation, and requirement analysis. That is, our goal in these phases is to produce a set of roles that are clearly established and specified.

The process of role negotiation is used to determine the necessary details of the environment (Definition 4.7, i.e. $e := < \overline{d_e}, \; \mathcal{B}, \; \mathcal{D}_e >$) before establishing a team or a well-organized group. Specifically, role negotiation attempts to resolve the following questions:

E-CARGO and Role-Based Collaboration: Modeling and Solving Problems in the Complex World, First Edition. Haibin Zhu.
© 2022 by The Institute of Electrical and Electronics Engineers, Inc.
Published 2022 by John Wiley & Sons, Inc.

1) How many roles are needed to meet the collaboration requirement, i.e. n?
2) What is the class of the object elements in $\mathcal{D}_e$?
3) How many agents are required for each role, i.e. $L[j]$ $(0 \leq j < n)$?
4) How many agents are allowed to join the group, i.e. $U[j]$ $(0 \leq j < n)$?

10.2 Role Specification

Clear and exact role definitions and specifications are very important in collaboration systems. However, there is no efficient mechanism to support role specification in the current literature of collaboration research. Role specification aims to implement and construct fundamental mechanisms to specify roles. Such mechanisms can be used by a role facilitator who is in charge of collaboration. With these mechanisms, collaborators' roles can be tuned or changed through the role facilitator dynamically based on their requirements. Through this, collaboration can be conducted more smoothly, efficiently, and productively.

Our work on E-CARGO provides a mechanism to facilitate role definition and specification based on object systems. This work helps define precise methods for expressing roles that can be dynamically changed in a collaboration system. Under the proposed framework, users in a collaboration system can tune their roles and even create new ones to facilitate more productive and efficient collaboration.

Definition 10.2 *Role specification* aims to determine the details of the roles (Definition 4.6, i.e. $r := < \overline{d_r}, \mathcal{M}_{in}, \mathcal{M}_{out}, V_p^r, V_r^r, \mathcal{A}_c, \mathcal{A}_p, \mathcal{A}_o >$) necessary to establish a well-organized group or team.

To accomplish the task of role specification, we need to answer the following questions:

1) Which messages compose the incoming message set, i.e. $\mathcal{M}_{in}$?
2) Which messages compose the outgoing message set, i.e. $\mathcal{M}_{out}$?
3) What rights or supplies are provided to the player of this role, i.e. V_p^r?
4) What demands must the agent meet to play this role, i.e. V_r^r?

10.3 Agent Evaluation

Agent evaluation is rarely found in the literature on multi-agent systems. Moore et al. (2000) mention such problems but demonstrate the idea in a different way from RBC. They discuss a list of problems of selecting the appropriate agents to execute a specific task.

Similar research is recognized in the field of information extraction. Several resume extraction tools exist and claim to accurately extract data. Yu et al.

(2005) present an effective approach to resume information extraction that supports automatic resume management and routing. They design a cascaded information extraction framework: in the first pass, a resume is segmented into consecutive blocks attached with labels indicating the information types, and then in the second pass, the detailed information is extracted. They attempt to extract all information for categorization and classification. Ciravegna and Lavelli (2004) propose a toolkit of information extraction to learn information extraction rules for resumes written in English. The information extracted by their toolkit includes a structure of *Name, Street, City, Province, Email, Telephone, Fax*, and *Post Code*. However, their method is not appropriate for agent evaluation. Sitter and Daelemans (2003) perform information extraction on a document using a double classification method by first identifying where the relevant data may be located and then conducting an in-depth analysis of the areas of interest to extract the relevant data. Wu et al. (2007) propose the problem of resume mining in social network communities. Their aim in the study consists of three aspects: characterization of community, discrimination among communities, and community evolution mining.

Some related work can be found in operational research and production management. Gershoff et al. (2007) discuss the "Agent Evaluation" issue. Although their meaning of "agents" is different from those discussed in this section, their discussions shed light on the agent evaluation discussed in this section. Specifically, they state that it is a good idea to use customers' opinions to evaluate an agent's performance. Neely et al. (1997) propose a structure to measure the performance of an organization and list the requirements for performance measurement methods, e.g. a performance measurement method should be simple to understand, precise, objective, relate to specific goals, clearly defined, have an explicit purpose, etc. These requirements are good references to evaluate the soundness of a method for agent evaluation.

In the business world, people are required to have the appropriate knowledge, skills, and habits to be qualified for a new position. Successfully finding a new job is dependent on the similarities between new roles and those previously performed by the person (Ciravegna and Lavelli 2004), i.e. qualifications are the basic requirements for possible role-related activities. Agent evaluation is a fundamental problem that requires advanced methodologies, such as information classification, data mining, pattern searching, and matching.

In the research of multi-criteria decision making (MCDM), Zanakis et al. (1998) review and compare selected methods in the following areas: Simple Additive Weighting, Multiplicative Exponent Weighting, Analytic Hierarchy Process, the ELECTRE method, and the TOPSIS method. Their work offers a large collection of methods that can be used to support agent evaluation. It also provides some benchmarks for our specifically designed evaluation methods. We believe that MCDM can be applied to agent evaluation after roles are well specified.

Definition 10.3 *Agent Evaluation (AE)* is a process used to obtain the Q matrix for group g by combining all the factors related to group g, including its environment, related objects, roles, and agents.

To conduct agent evaluation, we need to review the definitions of roles (Definition 4.6, i.e. $r := < \overline{d_r}, M_{in}, M_{out}, V_p^r, V_r^r, A_e, A_{\#}, A_o >$) and agents (Definition 4.8, $a := < \overline{d_a}, e_a, s_a, V_p^a, V_r^a, R_e, R_{\#}, R_o, D_g >$).

From the two definitions, we notice that the elements V_p^r and V_r^r of a role and those of an agent, i.e. V_p^a and V_r^a, are used as the foundations for agent evaluation.

We first present the ordered enumeration forms of these four sets by supposing that V_p^r and V_r^a have the same cardinalities, and V_r^r and so do V_p^a, i.e. $u = |V_p^r| = |V_r^a|$, $v = |V_r^r| = |V_p^a|$.

- $V_p^r = \{v_{rp0}, v_{rp1}, ..., v_{rp(u-1)}\}$;
- $V_r^r = \{v_{rr0}, v_{rr1}, ..., v_{rr(v-1)}\}$;
- $V_p^a = \{v_{ap0}, v_{ap1}, ..., v_{ap(v-1)}\}$; and
- $V_r^a = \{v_{ar0}, v_{ar1}, ..., v_{ar(u-1)}\}$;

where $v_{rp0}, v_{rp1,...,} v_{ar(u-1)}$ are primitive objects that express the supplies and the demands. Here, "primitive" means that an object cannot be decomposed further into any finer-grain objects.

Now, the initial form of agent evaluation can be accomplished using the following formula. We use q^a to represent the qualification to meet the agent-side requests, q^r that to meet the role-side requests, and q^w the composed qualification with weight w for the agent-side requests.

$$q^a(a,r) = \sum_{k=0}^{u-1} sim(v_{rp(k)}, v_{ar(k)});$$

$$q^r(a,r) = \sum_{k=0}^{v-1} sim(v_{rr(k)}, v_{ap(k)});$$

$$q^w(a,r) = w \times \sum_{k=0}^{u-1} sim(v_{rp(k)}, v_{ar(k)}) + (1-w) \times \sum_{k=0}^{v-1} sim(v_{rr(k)}, v_{ap(k)})$$

where $sim(v_{rp(k)}, v_{ar(k)}) \in [0, 1]$ and $sim(v_{rr(k)}, v_{ap(k)}) \in [0, 1]$ mean the similarities between $v_{rp(k)}$ and $v_{ar(k)}$, and between $v_{rr(k)}$ and $v_{ap(k)}$, respectively, and w means the weight of the role-side qualifications.

To make the evaluation more accurate, we may spread the weight to each of the elements in the supplies and demands sets, i.e. $V_p^r, V_r^a, V_r^r,$ and V_p^a.

Agent evaluation is a dynamic and complex task. This section describes a simple approach to static agent evaluation by setting up an agent and role template. Agent evaluation is not difficult when the set of demands for each role is clearly defined. In fact, the difficulty lies primarily in formulating the set of well-accepted criteria and measuring how well the agent's qualification matches these criteria. Note that in a dynamic system, this set of criteria can be adaptively adjusted to reflect the changes.

An individual's qualification is dependent on many factors (Zhu and Feng 2013; Zhu and Grenier 2009), such as experience, personality, emotion, attitude, and situation. It is quite challenging to obtain the exact qualification of a user for a role. Clear and pertinent role specifications, real-time user information acquisition, pertinent user qualification models, and practical metrics are the keys to providing a successful evaluation. To effectively accomplish agent evaluation, the following problems must be solved:

- How can agent qualifications, i.e. V_p^a, be acquired and collected dynamically? Agents are changing, and their state varies as time elapses.
- How can role demands, i.e. V_r^r, be acquired and collected dynamically? In a system, the roles should be modified according to system changes. If the importance of roles is changing, the role demands also change. Some roles may be removed or added. Some are downgraded to take fewer responsibilities and disapproved of some rights, while others may be upgraded to take more responsibilities and be assigned more rights.
- How can the properties of agents be measured? After obtaining agent information, there is still a problem of transforming the information into meaningful data. That is to say, we must provide a well-designed metric system to quantify agent information in order to evaluate the agents' qualifications for the various roles.

10.4 Collective Group Role Assignment

As research effort in RBC continues, the number of significant challenges increases. Role assignment has a major impact on the efficiency of collaboration and the degree of satisfaction among the members involved. It is one of the most important steps in the process of RBC and Adaptive Collaboration (AC) (Zhu 2003, 2008, 2010, 2012, 2015, 2016; Zhu and Zhou 2006, Zhu et al. 2012a, b).

Group role assignment is a complex problem that can be solved if the agents' qualifications are provided (Zhu et al. 2012b). However, in the real world, there are many ways to accomplish role assignment for a group of agents. One typical situation in a group is that most agents are not fully qualified to play a specific role, but the group work requires that every responsibility of a role be taken by agents. We may need two or more agents to play a specific role together in order to complete the group work. We call this kind of role assignment collective group role assignment.

When the author had lunch at a conference in 2005 with Dr. Lotfi A. Zadeh, *the father of fuzzy logic*, he mentioned one important fact in scientific research, *the problems related to people and society are much more complex than problems in pure mathematics and technology.* Collective group role assignment (Zhu 2011) is highly related to human society, and, therefore, it is undoubtedly a complex problem.

This section formally defines the problem of collective group role assignment. Note that collective group role assignment is, in fact, a combination of simple agent evaluation and role assignment. The defined problem needs more investigation.

10.4.1 One-Way Collective Role Assignment

In order to deal with the role assignment problems in organizations and systems, we emphasize the static properties of roles and agents when forming a group and ignore the interactive properties of roles and agents after the group starts to work. In the following discussions, object identifiers are assumed to be unique strings, the same as we did in using E-CARGO.

Definition 10.4   We say *agent a meets one of the demands of role r*, if $\exists\, x \in a.V_p^a \wedge x \in r.V_r^r$.

Definition 10.5   We say *agent set A_r meets the requirement of role r*, if $r.V_r^r \subseteq \bigcup_{a \in A_r} a.V_p^a$.

Definition 10.6   Role r is called *one-way collectively workable*, if, for assignment $<A_r, r>$, A_r meets the requirement of role r, i.e. $r.V_r^r \subseteq \bigcup_{a \in A_r} a.V_p^a$.

Definition 10.7   Group g is called *one-way collectively workable* if all its roles are one-way collectively workable, i.e. $\forall j(r_j.V_r^r \subseteq \bigcup_{i=0}^{m-1} a_i.V_p^a \times T[i, j]))(0 \leq j \leq n-1)$, where $a_i.V_p^a \times T[i, j] = a_i.V_p^a$ if $T[i, j] = 1$; Φ if $T[i, j] = 0$.

Definition 10.8 *One-way collective group role assignment* (or simply, one-way assignment) is a process that aims to find an assignment matrix T to make the group one-way collectively workable.

10.4.2 Two-Way Collective Role Assignment

In the real world, we may face situations where a person may select one from several positions while a position may be sought by many applicants. Therefore, the selection is bi-directional. The best situation arises when an applicant possesses all the demands of a position, and the position satisfies all of the benefits sought by the individual. In the service world, this requirement is the same. For example, a server providing a service may request a client to provide enough budget, time, and information; a client may search for a service that meets their requirement with what they can provide. Therefore, we need to model this natural requirement, i.e. when forming a group, there must be considerate concerns for the above two-way selection. The two-way selection is a process to form a workable group.

Definition 10.9 We say agent a is satisfied with a supply of role r, if $\exists\, y \in a$. $V_r^a \wedge \exists x \in r.V_p^r \ni (y = x)$.

Definition 10.10 We say agent a is satisfied with the supplies of role r, if $r.V_p^r \supseteq a.V_r^a$.

Definition 10.11 Role r is called *simple two-way collectively workable* if, for assignment $<A_r, r>$, A_r meets the requirement of role r and each agent of A_r is satisfied with a supply of role r.

Definition 10.12 Role r is called *complete two-way collectively workable* if, for assignment $<A_r, r>$, A_r meets the requirement of role r and each agent of A_r is satisfied with the supplies of role r.

From Section 10.4.1, we know that not all agents meet all the demands of a role, and not all agents will accept a role only if all the requested rights are satisfied. Therefore, a two-way collective role assignment may work in a non-perfect situation.

Definition 10.13 Group g is *simple two-way collectively workable* if all its roles are simple two-way collectively workable.

Definition 10.14 Group g is *complete two-way collectively workable* if all its roles are complete two-way collectively workable.

Definition 10.15 For group g, tuple $<A_t, t>$ of $g.\mathcal{J}$ is called *a simple two-way collective role assignment* if A_t meets the requirement of role t, and role t meets at least one of the requested supplies of each agent in A_t, where $A_t = \{a_0, a_1, ..., a_{m_t - 1}\}$, and $m_r = |A_t|$.

In fact, the two-way collective role assignment can be expanded to a general form as defined in Definition 10.16, i.e. an agent will be happy to play a role only when the role provides all the agent's requests.

Definition 10.16 For group g, tuple $<A_t, t>$ of $g.\mathcal{J}$ is called a *complete two-way collective role assignment* if each agent of A_t meets the requirement of role t, and role t provides all the requested rights of each agent in A_t, where $A_t = \{a_0, a_1, ..., a_{m_t - 1}\}$, and $m_r = |A_t|$.

Please note that the discussed problems in this section have been solved initially in (Zhu 2011). However, the solutions are too complex to be practical. More investigations are required.

10.5 Role Engine

A role engine can be understood in the same way as a Prolog inference machine (Merritt 1989) or Python inference engine (Cariaggi 2019). For example, to use a Prolog system, people only need to input the rules and facts. The Prolog inference machine will search for the result. Similarly, to implement RBC on the proposed role engine, people only need to specify the roles and create agents based on the role specifications. When agents are put into the role engine, the engine will drive agents to obtain their goals by collaborating with other agents.

A role engine should do the following:

- Managing roles (create, delete, and modify);
- Managing agents (create, delete, and modify);
- Assigning roles to agents;
- Dispatching messages to agents; and
- Checking the consistency of the system.

A role engine is a platform for agents to conduct RBC. On this platform, agents work for the system by playing roles. By running the role engine, all the agents are driven to contribute and work diligently for the system. A role engine should possess role dynamics, facilitate role transfer, and support role assignments, interaction, and presentation. Therefore, all the discussed models, technologies, methodologies, and algorithms are fundamental elements for constructing a realistic role engine.

All the topics discussed in the previous chapters are aiming to implement a real role engine. Here, we present three other problems we did not talk about in the previous chapters and sections.

10.5.1 Role Dynamics

Dynamics are the forces that regulate the relationships among people or things and how those relationships evolve. Everything exists in the world for a particular reason. Every activity in the world is instigated for a specific reason. These reasons form the driving forces for creating new things and for new actions to be taken. In a society, people in an organization with positive dynamics will work actively and collaboratively toward the common goal of the organization. On the contrary, people in an organization with negative dynamics cannot work effectively and may not have a clear understanding of the goals necessary to make the organization competitive.

It is the same in a computational system. If agents are taken as computational components, a computational system is a multi-agent system. Multi-agent systems should encourage diversity of behavior in a population of cooperating agents. This is one kind of requirement for agents. There are definite reasons for the addition of new agents to the system. With well-built dynamics, agents will automatically follow the regulations of the RBC system and collaborate with each other to accomplish the common goal of the system.

To build an RBC system, the following questions should be answered: "How are agents created?" "Why are agents created?" "Why are agents transferred among different roles?" "How are agents made proactive?" and "How could a system obtain the best performance?". All these questions are concerned with the fundamental mechanisms used to build and operate a system, i.e. dynamics (Zhu 2012, 2015; Zhu et al. 2012a).

A role registration (Zhu 2007) should be processed in the role engine. All roles are registered. Agent productions are provided with role specifications from this process and they produce agents on demand. Recall what we discussed in Chapter 3, roles attract agents. The larger the number of vacant roles, the greater the attraction to agents. Such vacant roles drive related organizations to produce agents. The production of agents should be relative to the attraction generated by roles.

10.5.2 Role Interaction

With a role engine, agent interactions and collaborations are facilitated by roles. The role engine controls the messages exchanged among roles. Based on the E-CARGO model (Zhu and Zhou 2006), interaction is implemented by issuing messages. Dispatching messages to agents is a highly intelligent task accomplished by a role engine and its roles. The following properties should be considered when designing such a system:

- Fairness: the engine should dispatch messages evenly to peer agents. It should avoid starvation or overloading, where starvation means that an agent has not received any messages for a limited period of time, while overloading means that an agent receives too many messages in a limited period of time.
- Consistency: the engine should check for the consistency of role relations. When a new role relation is added, the system should be kept consistent.
- Completeness: All the outgoing messages specified by roles should be addressed by other roles. The engine should dispatch timely a specific message to test this completeness.

10.5.3 Role Presentation

Roles presented to users should be easily understandable. The specification of roles should consider the presentation aspect of roles. The aesthetic, intuitions, and human factors should be considered when describing a role (Zhu et al. 2008).

Similar to other human user interface requirements, role presentation has the following requirements:

- Easy to understand;
- Easy to remember;
- Used to support personalized or customized user interface; and
- Presented in a multimedia style, such as text, image, audio, video, and animation presentations.

To meet the requirements as above, iconizing different roles is beneficial to role presentation. Some roles can be easily expressed with icons such as a police officer

 , a waiter , and a worker . However, it is difficult to express a

computer professor, a software developer, and a system analyst. It is also very difficult to express the generalized concept of a "role" using an icon. Furthermore, it is a difficult task to design icons that contain all the information outlined in a role specification. Therefore, tables, lists, and graphs are needed to present roles.

Note that this topic leaves space for interested readers to contribute to the literature of RBC in the future.

10.6   Social Simulation

The people of this world cannot live alone. *The life and adventures of Robinson Crusoe* (Defoe 2019), although imaginary, support the idea that people need social lives. An organized society is essential for social life. Without joining a community or a society, individuals could not experience real social life. Social simulation is the reproduction of a real-world society using computer-based systems (Haradji et al. 2012).

A society is composed of individuals who are involved in social interactions (Bicchieri, et al. 2018, Birkin et al. 2010). It can be a group sharing the same geographical or social territory and can be formed under the same political authority or dominant cultural expectations. The nature of social activities is collaboration. Without collaboration, society cannot exist. Therefore, a society should have the same components as collaboration discussed in Chapter 1 (See Section 1.3).

Social simulation is a methodology used to map real-world social problems into a computer-based framework to seek new insights on social issues. To explore a social problem, we need to first establish a social simulation system, which can be taken as a grand software engineering (SE) project. From SE's viewpoint, every simulation system is a replica of a specific aspect of the real world. Therefore, a social simulation system should appropriately reflect aspects of real-world social systems.

According to the lifecycle of an SE project (Pressman and Maxim 2014), i.e. communication, planning, modeling, implementation, and deployment, the major product of the communication phase is the requirement analysis report. That is, we must first clearly state the requirements of a social simulation system. However, even though social simulation research has gone on for many years, few studies have approached the subject systematically. This section seeks to initiate this investigation process.

A social system is composed of numerous people, interacting through the establishment of relationships. Social system behaviors are dynamic and adaptive. Conventional techniques for the creation of social systems derive mainly from the humanities because many aspects of a social system are difficult to formalize. A social system possesses all the properties of a complex system. The complexity of social systems mainly comes from the aspects of collaboration, i.e. sharing, connections, communications, interactions, coordination, and cooperation. Computational social systems (Wang 2017) offer promising ways to address the problems faced when modeling real-world social systems.

In order to reproduce a society in a computer-based system, a social simulation methodology should be able to represent the following items:

1) Objective entities. A society is composed of not only people but also objective entities, which are materials that support people's lives.
2) Individuals. People or human beings are essential components of a society or community.
3) Rights and duties. These items are required for people in a community to take action to serve or be served.
4) Organization. It provides an abstract entity to express the structure or architecture of a society or community.
5) Collaboration. Collaboration activities, such as communications, interactions, coordination, and cooperation have to be expressed by computational structures.
6) Sharing. This is the nature of collaboration. Therefore, sharing must be expressed and supported by a computational simulation methodology or system.
7) Dynamics. It can be individual or collective dynamics. Pertinent computational methods should be provided to simulate various dynamics.
8) Trust, leadership, and social capital. Trust must be established when a society or a community starts to work (Dasgupta 1998; Gambetta 1998; Luhmann 2017; Ramchurn et al. 2004). It should be expressed and simulated in an appropriate computational way. Leadership significantly affects the collective behavior of a society or community (Day et al. 2014; Goleman 1998). A specialized computational structure should be designed by a simulation methodology for leadership. Social capital demonstrates the result of the collective behaviors of a society or community (Adler and Kwon 2002; Lin 2011). An abstract structure will help simulate social phenomena related to this factor.
9) Optimization and goal. In social simulation, decision making can be optimized with the assistance of computation. It should be an irreplaceable property of a computational simulation methodology compared with other simulation ways. A society or community should have a common goal for its members. An abstract entity should be provided by a computational structure to represent this.

From the requirement analysis, it is easy to assert that RBC, E-CARGO, and related formalizations are excellent tools for undertaking the development of a social simulation system.

Social simulation is a cutting-edge topic in an interdisciplinary field where Information Technology and Social Science overlap called computational social systems. In RBC and E-CARGO, individuals are agents, organizations are groups, structures are environments. A (social) system can be viewed as a collaboration of a group of agents working on an environment composed of classes of objects by playing different roles. Interested readers are encouraged to refer to our work related to this topic (Zhu 2020).

10.7 Adaptive Collaboration

Human communities, despite protecting individual interests and personal property, pursue collective interests when conducting teamwork. That is, in collaboration, the pursuit of optimal team performance is a stronger requirement than that of individual interests. Indeed, adaptive collaboration follows the social axiom *"every diligent person hopes to become a great member of a great team"* (Zhu 2005). For example, to be a member of a National Basketball Association (NBA) team is the ideal scenario for most basketball players in the world, and likewise, to be a member of a National Hockey League (NHL) team is ideal for most hockey players in the world.

After members join a team, their common goal is to maintain the significance of the team. To do so, they must adapt to the changes of the team and maintain the highest team performance. In many cases, it is not the team members that do not want to adapt to the team, but the reality is that the team's management regulations do not allow such adaptation or the team management does not know whether this adaptation works or not.

It is essential for members to be adaptive within a team since optimal team performance can only be achieved when everybody within the team contributes at their highest potential. Ultimately, this will require individuals to adapt and change their existing behavior. This is why traditional research on adaptation focuses on the adaptability of individual agents (de Wilde et al. 2003; Picard and Gleizes 2003; van Splunter et al. 2003) and the adaptability of machines to individual users (Atterer et al. 2006; Brusilovsky and Millán 2007; Fischer 2001; Gena 2005; Hou et al. 2007a, b; Jameson 2003). However, true Adaptive Collaboration (AC) concentrates on the adaptability of the team as a whole.

Team adaptation involves many aspects, such as team structure and individual adaptability. It reflects the ability of the team to work, live, socialize, and compete. Team adaptability is dependent on its organization, structure, culture, and regulations. Many factors affect collaborative performance in a team, such as culture, interests, personalities, health, equipment, hardware resources, benefits, abilities, powers, motivations, and situations. These factors make it difficult for researchers of man–machine systems and collaborative technologies to optimize team performance in a timely fashion.

Regulating people and agents to achieve better team performance is undoubtedly an essential function of multi-agent systems and adaptive systems (Atterer et al. 2006; Berman et al. 2003). However, there is little research that considers team performance in traditional intelligent systems and multi-agent systems. To take advantage of the dynamic nature of people, agents, and environments, adaptive collaboration systems are rational. AC facilitates this requirement by dynamically adjusting teamwork according to the changing situations in collaboration. An evident phenomenon that fosters AC can be observed in basketball or soccer games, where timely and effective shifts in player roles are essential to winning games.

The life cycle of RBC consists of three major tasks: *role negotiation, assignment, and execution.* As an important aspect of RBC, role assignment affects the efficiency of collaboration and the degree of satisfaction among the members involved. RBC is a promising methodology to conduct AC. Interested readers can refer to our work on this topic (Sheng et al. 2016; Zhu 2012, 2015; Zhu et al. 2012a).

10.8 Other Challenges in RBC and E-CARGO

Even though RBC and E-CARGO have been investigated for about two decades, we have not yet implemented a real RBC system due to the system's complexity. To implement real RBC systems, we still face many challenges.

One important problem in RBC is the specification and definition of the required relationships among roles. These relationships are the theoretical foundation for an RBC system running on a role engine. While we have specified the typical relations in an RBC system (Chapter 4; Zhu 2008), this problem remains open and requires more investigations in its formalizations and implementations. The implementation of a role engine or the proof of minimum and sufficient logic will verify the specified role relationships. By "minimum," we mean that removing any logic component from the system will lead to failure in establishing the application; and by "sufficient," we mean that such a system can support all the activities of agents on roles, and specify all the relationships among roles.

Role-based software engineering (Zhu and Zhou 2006) and role-based programming (Steimann 2000, 2008) are innovative and promising methodologies that have the potential to improve the productivity of software development teams and the quality of software. Although Aspect-Oriented Programming (AOP) (Kendal 1999) and Subject-Oriented Programming (SOP) (Harrison and Ossher 1993) claim to provide similar approaches to software engineering and programming, the author believes that AOP and SOP are not widely accepted after two decades because of the lack of formal specifications for roles/role players, and the lack of a role engine design that supports high-level role-based design and agent development.

Role-based chatting is supported by a web-based tool (Zhu et al. 2008). It provides a trade-off between anonymity and trust. It helps shy people present their ideas and significantly improves the satisfaction of those participating in the collaboration. Named collaboration allows aggressive contributors to dominate the process, and thus potentially excluding useful ideas of shy people. More applications and empirical studies are required to present and verify such promises in role-based chatting.

Adaptive collaboration systems (ACSs) (Zhu 2015, Zhu et al. 2012a) are another important branch that needs to be investigated. In our previous work, we demonstrated that RBC and E-CARGO provide a solid foundation for establishing adaptive systems. The RBC process and dynamic parameters in the E-CARGO model show significant promise in supporting the development of ACSs.

Because RBC is a well-specified methodology and the E-CARGO model is well-defined, it is not difficult to use these tools to discover new challenges in the research of collaboration systems. We suggest the following streams of investigation:

1) By introducing new parameters and conditions into role assignment, one may discover different constraints that affect the quality of team organization. This stream may discover many challenges that belong to the categories of Linear Programming (LP) or Non-Linear Programming (NLP), such as GRA^+ and GRA^{++}(Chapters 6 and 7; Zhu 2016).

2) By introducing relationships among the components into the E-CARGO model, one may discover problems related to their coordination, interaction, and management (Zhu 2003, 2010). These challenges may overlap with those in logical systems or algebraic systems (Liu et al. 2014).

3) By detailing or adding parameters to the components of E-CARGO, one may find problems related to the facilitation of collaboration. The solutions to these problems will facilitate the implementation of collaboration systems, such as multi-agent systems. For example, if individual role-playing logic is added to an agent and group role-playing logic to the environment, one may discover many challenges in intelligent agent systems or multi-agent systems (Alonso et al. 2003, Al-Zaghameem and Alfraheed 2013; Cabri 2007, 2012; Caetano et al. 2009; Campbell and Wu 2011; Ferber et al. 2004; Li et al. 2012; Mao et al. 2014; Nair et al. 2003).

4) By applying RBC and E-CARGO to related areas, such as Computer-Supported Cooperative Work (CSCW) (Zhu and Zhou 2006), software engineering (Kühn et al. 2014), multi-agent systems (Al-Zaghameem and Alfraheed 2013; Cabri 2007, 2012; Caetano et al. 2009; Ferber et al. 2004; Li et al. 2012; Mao et al. 2014; Nair et al. 2003), scheduling (Conway et al. 2012), cloud computing (Armbrust et al. 2010; Teng et al. 2014), Internet of Things (Fortino et al. 2018; Savaglio et al. 2020), and web services (Gomez and Passerini 2007; Liu et al. 2011; Shen and Sun 2011; Wang et al. 2011; Weerawarana et al. 2008; Zhang et al. 2008), one may formalize the problems and propose innovative solutions to resolve them.

It is noted that the above innovative discoveries are not intuitively deduced in related research methodologies other than RBC, such as NLP, logic systems, algebraic systems, and multi-agent systems. It is RBC and E-CARGO that provide such opportunities for researchers to discover these challenges because RBC and E-CARGO can help illustrate and clarify new requirements in the sense of collaboration.

10.8.1 Optimizations

It is interesting to note that many RBC problems involve optimization. In fact, RBC research bridges the concepts of collaboration and optimization. Optimization methods can be applied to problem solving only when the problem is well-defined in terms of optimizations (Rardin 1997; Wolsey and Nemhauser 1999), such as Integer Programming (IP), LP, or NLP. RBC provides an approach to specifying many complex problems in terms of optimization. In fact, RBC expands the application areas of the optimization theory and models to include collaboration. The difference between RBC and optimization research is that RBC presents a new viewpoint for researchers in their investigations of collaboration systems.

On the other hand, not all RBC problems are optimizations. In RBC, some problems are so complex that we only need to provide a feasible solution. For example, the role transfer problem (Chapter 9; Zhu and Zhou 2008, 2009, 2012) aims only to find a feasible solution. In role transfer, even though it is hard to set up an optimized group that has no critical agents, we could use a computer-based solution, such as the role transfer tool, to check whether an existing group has any critical agents. The GRA$^+$ problems can be expressed with LP formulations, but a solution does not have to use LP (Zhu 2016).

10.8.2 Agent-Oriented Software Engineering (AOSE)

It is advocated by many researchers in Agent-Oriented Software Engineering (AOSE) that agents and groups are considered to be the fundamental concepts in software development (Cossentino et al. 2007; Ferber et al. 2004; Fortino et al. 2005; Padgham et al. 2009; Wooldridge et al. 2000). Compared with AOSE, RBC offers the advantages of well-defined components and formalizations. AOSE lacks formalizations in spite of previous efforts to formalize it (Pradel et al. 2012). This shortcoming impedes the research in AOSE because the lack of formalization makes it difficult to clearly formalize problems, verify, and validate the proposed solutions to these problems. Software development is of such complexity that it is difficult to undertake in the absence of well-defined specification tools. Unified

Modeling Language (UML) is grammatical but not fully formalized. RBC and the E-CARGO model present a promising way to overcome this obstacle.

10.8.3 Multi-Agent Systems

Multi-agent systems (MASs) have been a hot cutting-edge research topic for many years (Ferber et al. 2004; Ferrari and Zhu 2009, 2011, 2012; Nair et al. 2003; Wooldridge et al. 2000). Compared with the MASs' approach that encourages individualism, RBC argues that there are benefits in emphasizing collectivism in collaboration. In MASs, it is considered essential not only to design highly intelligent agents that collaborate, but also to prioritize self-benefit. RBC counters this by emphasizing the group benefit. Although there are some trials in the formalizations of role allocations in MASs (Campbell and Wu 2011), RBC promises more supplementary research compared to MASs, i.e. from Definition 4.11 in Chapter 4, if $\mathcal{H}$ is empty, then system Σ becomes a multi-agent system.

10.9 Not the End

RBC and E-CARGO have kicked off the investigation in collaboration theory and practice and have achieved inspiring results. Although E-CARGO has established a stepping stone, there is a long way to go in RBC research.

In the field of social psychology, although there is a huge literature on role theory, there is no conclusion on how roles affect the satisfaction, effectiveness, and results of the collaboration. We need to conduct more studies to answer these questions. The following are some questions that should be addressed in the field of social psychology:

- Do people enjoy working in an environment with clearly tagged roles?
- Do people enjoy working with process roles or interface roles?
- Do people enjoy working with people or with roles (people are hidden by roles)?
- Under what circumstances do people enjoy RBC?

From the viewpoint of CSCW system design, CSCW-roles cannot effectively support role content, role specification, role assignment, and role relationships. Innovative presentation tools are also required for collaboration facilitators. We need to consider the social and psychological requirements of human users and design roles that facilitate collaboration among them. The following questions should be resolved to facilitate effective collaboration among people:

- How do we present roles with a special design?
- How do we remind people of their roles?

- How do we evaluate the effectiveness of RBC?
- How do we support role negotiations, including role assignment and role specification?
- How do we build a role hierarchy?
- How do we schedule people with different roles?

From the viewpoint of agent systems, agent-roles cannot effectively support role relationships and role content. We need to find ways to organize agents to form intelligent agent systems that allow agents to become autonomous, autonomic, reactive, active, pro-active, social, and adaptable. To accomplish this, we need to answer the following questions:

- How can we build a role grid or role network that enables agents to live and grow (sociality)?
- How can we make roles the dynamics of agents (autonomy)?
- How can we design agents so that they learn to play new roles (learning capacity)?
- How do we use roles to encourage agents to actively work in an agent community (activity)?
- How can we make roles form easier environments for agents to adapt (adaptability)?

From the viewpoint of object modeling, current modeling-roles are poorly defined for properly supporting role content, role assignment, and role relationships. We need to develop roles into a mechanism similar to classes, objects, and functions. To make roles a first-class modeling mechanism, we need to answer the following questions:

- What should a role express in role specification?
- How should we make roles a well-defined mechanism that accommodates all the aspects related to roles?
- As for modeling languages, what language mechanisms should we provide to help specify roles?
- How do we extract roles from a problem domain, i.e. role mining?
- How do we express and process role assignment in software development?

From the viewpoint of trusted computing, we need to answer the following questions:

- How do we use roles to express the trustworthiness of agents?
- How do we express the trustworthiness of a group?
- How do we set up the initial trustworthiness of an agent and a group?
- How do we adjust the trustworthiness of an agent and a group?
- What criteria should we establish?

We may ask more questions if we delve into the details of the above questions. The author believes that E-CARGO is a promising model for investigating the above mentioned challenges. Interested readers are welcome to allocate time and effort in related investigations and make innovative contributions to the literature.

References

Adler, P.S. and Kwon, S.-W. (2002). Social capital: prospects for a new concept. *The Academy of Management Review* 27 (1): 17–40.

Alonso, E., Kudenko, D., and Kazakov, D. (eds.) (2003). *Adaptive Agents and Multi-Agent Systems*, LNAI, vol. 2636. Berlin, Germany: Springer.

Al-Zaghameem, A.O. and Alfraheed, M. (2013). An expressive role-based approach for improving distributed collaboration transparency. *International Journal of Computer Science Issues* 10 (4–2): 61–67.

Armbrust, M., Fox, A., Griffith, R. et al. (2010). A view of cloud computing. *Communications of the ACM* 53 (4): 50–58.

Atterer, R., Wnuk, M., and Schmidt, A. (2006). Knowing the user's every move: user activity tracking for website usability evaluation and implicit interaction. *Proceedings of the 15th International Conference on World Wide Web*, Cambridge, UK (23–26 May 2006), pp. 203–212.

Berman, F., Wolski, R., Casanova, H. et al. (2003). Adaptive computing on the grid using AppLeS. *IEEE Transactions on Parallel and Distributed Systems* 14 (4): 369–382.

Bicchieri, C., Muldoon, R., and Sontuoso, A. (2018). Social norms. In: *Stanford Encyclopedia of Philosophy (Winter 2018 Edition)* (ed. E.N. Zalta). Stanford University. plato.stanford.edu/archives/win2018/entries/social-norms/.

Birkin, M., Procter, R., Allan, R. et al. (2010). Elements of a computational infrastructure for social simulation. *Philosophical Transactions on the Royal Society (A)* 368: 3797–3812.

Blanchard, B.S. and Fabrycky, W.J. (2010). *Systems Engineering and Analysis*, 5e. Pearson.

Brusilovsky, P. and Millán, E. (2007). User models for adaptive hypermedia and adaptive educational systems. In: *Proceedings of the International Conference on the Adaptive Web: Methods and Strategies of Web Personalization*, LNCS 4321 (eds. P. Brusilovsky, A. Kobsa and W. Nejdl), 3–53. Berlin, Heidelberg, Germany: Springer.

Cabri, G. (2007). Environment-supported roles to develop complex systems. In: *Engineering Environment-Mediated Multi-Agent Systems*, vol. 5049 (eds. D. Weyns, S. A. Brueckner and Y. Demazeau), 284–295. Lecture Notes in Computer Science (LNCS).

Cabri, G. (2012). Agent roles for context-aware P2P systems. In: *Agents and Peer-to-Peer Computing*, vol. 6573 (eds. S. Joseph, Z. Despotovic, G. Moro and S. Bergamaschi), 104–114. LNCS.

Caetano, A., Silva, R., and Tribolet, J. (2009). A role-based enterprise architecture framework. *The 24th Annual ACM Symposium on Applied Computing*, Honolulu, Hawaii, USA (8–12 March 2009), pp. 253–258.

Campbell, A. and Wu, A.S. (2011). Multi-agent role allocation: issues, approaches, and multiple perspectives. *Autonomous Agents and Multi-Agent Systems* 22: 317–355.

Cariaggi, F. (2019). Inference engine python API documentation. https://software. intel.com/en-us/forums/intel-distribution-of-openvino-toolkit/topic/807755.

Ciravegna, F. and Lavelli, A. (2004). Learning Pinocchio: adaptive information extraction for real world applications. *Journal of Natural Language Engineering* 10 (2): 145–165.

Conway, R.W., Maxwell, W.L., and Miller, L.W. (2012). *Theory of scheduling*. Mineola, NY: Courier Dover Publications.

Cossentino, M., Fortino, G., Garro, A. et al. (2007). PASSIM: a simulation-based process for the development of multi-agent systems. *International Journal of Agent-Oriented Software Engineering* 2 (2): 132–170.

Dasgupta, P. (1998). Trust as a commodity. In: *Trust: Making and Breaking Cooperative Relations* (ed. D. Gambetta), 49–72. Blackwell.

Day, D.V., Fleenor, J.W., Atwater, L.E. et al. (2014). Advances in leader and leadership development: a review of 25 years of research and theory. *The Leadership Quarterly* 25: 63–82.

Defoe, D. (2019). *The Life and Adventures of Robinson Crusoe*. San Diego, CA: Canterbury Classics.

Ferber, J., Gutknecht, O., and Michel, F. (2004). From agents to organizations: an organizational view of multi-agent systems. In: *Agent-Oriented Software Engineering (AOSE) IV*, LNCS, vol. 2935 (eds. P. Giorgini, J. Müller and J. Odell), 214–230. Berlin, Heidelberg, Germany: Springer.

Ferrari, L. and Zhu, H. (2009). Making agent roles perceivable through proxy bytecode manipulation. *Proceedings of the ACM/IEEE Int'l Symposium on CTS*, Baltimore, USA (18–22 May 2009), pp. 408–416.

Ferrari, L. and Zhu, H. (2011). Enabling dynamic roles for agents. *Proceedings of ACM/IEEE Int'l Conference on Collaborative Technologies and Systems (CTS)*, Philadelphia, PL, USA (23–27 May 2011), pp. 500–507.

Ferrari, L. and Zhu, H. (2012). Autonomous role discovery for collaborating agents. *Journal of Software P&E* 42 (6): 707–731.

Fischer, G. (2001). User modeling in human–computer interaction. *User Modeling and User-Adapted Interaction* 11: 65–86.

Fortino, G., Garro, A., and Russo, W. (2005). An integrated approach for the development and validation of multi-agent systems. *International Journal of Computer Systems Science & Engineering* 20 (4): 259–271.

Fortino, G., Russo, W., Savaglio, C. et al. (2018). Agent-oriented cooperative smart objects: from IoT system design to implementation. *IEEE Transactions on Systems, Man and Cybernetics: Systems* 48 (11): 1939–1956.

Gambetta, D. (1998). Can we trust trust? In: *Trust: Making and Breaking Cooperative Relations* (ed. D. Gambetta), 213–237. Basil, UK: University of Oxford.

Gena, C. (2005). Methods and techniques for the evaluation of user-adaptive systems. *The Knowledge Engineering Review* 20 (1): 1–37.

Gershoff, A.D., Mukherjee, A., and Mukhopadhyay, A. (2007). Few ways to love, but many ways to hate: attribute ambiguity and the positivity effect in agent evaluation. *Journal of Consumer Research* 33 (4): 499–505.

Goleman, D. (1998). What makes a leader? *Harvard Business Review* 76: 93–102.

Gomez, E.A. and Passerini, K. (2007). Service-based crisis management: local and global community roles and communication options. *International Journal of Intelligent Control and Systems* 12 (2): 198–207.

Haradji, Y., Poizat, G., and Sempé, F. (2012). Human activity and social simulation. In: *Advances in Applied Human Modeling and Simulation* (ed. V.G. Duffy), 416–425. Boca, Raton, FL: CRC Express.

Harrison, W. and Ossher, H. (1993). Subject-oriented programming - a critique of pure objects. *Proceedings of Conference on Object Oriented Programming Systems Languages and Applications*, Washington, DC, USA (26 September 1993–1 October 1993), pp. 411–428.

Hou, M., Gauthier, M.S., and Banbury, S. (2007a). Development of a generic design framework for intelligent adaptive systems. *Proceedings of the 12th International Conference on Human-Computer Interaction*, Part III, Beijing, China (22–27 July 2007), pp. 313–320.

Hou, M., Kobierski, R.D., and Brown, M. (2007b). Intelligent adaptive interfaces for the control of multiple UAVs. *Journal of Cognitive Engineering and Decision Making* 1 (3): 327–362.

Jameson, A. (2003). Adaptive interface and agents. In: *The Human Computer Interaction Handbook* (eds. J.A. Jacko and A. Sears), 305–330. Mahwah, NJ: Lawrence Erlbaum.

Kendal, E.A. (1999). Role model design and implementation with aspect-oriented programming. *ACM SIGPLAN Notices* 34 (10): 353–369.

Kotlyar, I. and Ades, K. (2002). Don't overlook recruiting tools. *HR Magazine* 47: 97–102.

Kühn, T., Leuthäuser, M., Götz, S. et al. (2014). A metamodel family for role-based modeling and programming languages. In: *Software Language Engineering*, vol. 8706 (eds. B. Combemale, D.J. Pearce, O. Barais and J.J. Vinju), 141–160. LNCS.

Li, X., Chen, Y., and Dong, Z. (2012). Qualitative description and quantitative optimization of tactical reconnaissance agents system organization. *International Journal of Computational Intelligence Systems* 5 (4): 723–734.

Lin, N. (2011). *Social Capital: A Theory of Social Structure and Action*. Cambridge, England: Cambridge University Press.

Liu, L., Zhu, H., and Huang, Z. (2011). Analysis of minimal privacy disclosure in web services collaboration with role mechanisms. *Expert Systems with Applications* 38 (4): 4540–4549.

Liu, D., Teng, S., Zhu, H., and Tang, Y. (2014). Minimal role playing logic in role-based collaboration. *The IEEE Int'l Conference on SMC*, San Diego, USA (5–8 October 2014), pp. 1439–1444.

Luhmann, N. (2017). *Trust and Power*. Konstanz, Germany: Wiley.

Mao, X., Dong, M., and Zhu, H. (2014). A two-layer approach to developing self-adaptive multi-agent systems in open environment. *International Journal of Agent Technologies and Systems (IJATS)* 6 (1): 65–85.

Merritt, D. (1989). Using prolog's inference engine. In: *Building Expert Systems in Prolog* (ed. D. Merritt), 15–31. New York, NY: Springer Compass International. Springer.

Moore, J., Inder, R., Chung, P., et al. (2000). Who does what? Matching agents to tasks in adaptive workflow. *International Conference on Enterprise Information Systems*, Stafford, UK (4–7 July 2000). http://citeseer.ist.psu.edu/old/moore00who.html.

Nair, R., Tambe, M., and Marsella, S. (2003). Role allocation and reallocation in multiagent teams: towards a practical analysis. *Proceedings of the Second Int'l Joint Conference on Autonomous Agents and Multiagent Systems*, ACM, Melbourne, Australia (14–18 July 2003). http://teamcore.usc.edu/papers/2003/nair-aamas03.pdf.

Neely, A., Richards, H., Mills, J. et al. (1997). Designing performance measures: a structured approach. *International Journal of Operations & Production Management* 17 (11): 1131–1152.

Padgham, L., Winikoff, M., DeLoach, S., and Cossentino, M. (2009). A unified graphical notation for AOSE. In: *Agent-Oriented Software Eng. IX*, vol. 5386 (eds. M. Luck and J.J. Gomez-Sanz), 116–130. LNCS.

Picard, G. and Gleizes, M.–.P. (2003). An agent architecture to design self-organizing collectives: principles and application. In: *Adaptive Agents and MultiAgent Systems: Adaptation and Multi-Agent Learning*, LNCS 2636 (eds. E. Alonso, D. Kudenko and D. Kazakov), 110–124. Berlin, Heidelberg, Germany: Springer.

Pradel, M., Henriksson, J., and Aßmann, U. (2012). A good role model for ontologies. In: *Enterprise Information Systems and Advancing Business Solutions* (ed. M. Tavana), 225–235. IGI Global.

Pressman, R. and Maxim, B. (2014). *Software Engineering: A Practitioner's Approach*, 8e. Columbus, OH: McGraw-Hill Education.

Ramchurn, S.D., Huynh, D., and Jennings, N.R. (2004). Trust in multi-agent systems. *The Knowledge Engineering Review* 19 (1): 1–25.

Rardin, R.L. (1997). *Optimization in Operations Research*. Upper Saddle River, NJ: Prentice Hall.

Savaglio, C., Ganzha, M., Paprzycki, M. et al. (2020). Agent-based internet of things: State-of-the-art and research challenges. *Future Generation Computer Systems* 102: 1038–1053.

Shen, H. and Sun, C. (2011). Achieving data consistency by contextualization in web-based collaborative applications. *ACM Transactions on Internet Technology* 10 (4): 13–37.

Sheng, Y., Zhu, H., Zhou, X., and Hu, W. (2016). Effective approaches to adaptive collaboration via dynamic role assignment. *IEEE Transactions on Systems, Man, and Cybernetics: Systems* 46 (1): 76–92.

Sitter, A. and Daelemans, W. (2003). Information extraction via double classification. *Proceedings of the International Workshop and Tutorial on Adaptive Text Extraction and Mining*, Cavtat–Dubrovnik, Croatia (22 September 2003), pp. 62–73. http://www.dcs.shef.ac.uk/~fabio/ATEM03/ATEM03-Proceedings.pdf#page=68.

van Splunter, S., Wijngaards, N.J.E., and Brazier, F.M.T. (2003). Structuring agents for adaptation. In: *Adaptive Agents and Multi-Agent Systems*, LNAI 2636 (eds. E. Alonso, D. Kudenko and D. Kazakov), 174–186. Berlin: Springer-Verlag.

Steimann, F. (2000). On the representation of roles in object-oriented and conceptual modelling. *Data & Knowledge Engineering* 35 (1): 83–106.

Steimann, F. (2008). Role + counter-role = relationship + collaboration. *Workshop on Rel. and Asso. in Object-Oriented Languages, OOPSLA*, Nashville, Tennessee, USA (19–23 October 2008). http://deposit.fernuni-hagen.de/2201/1/Role_Counter_Role_Relationship_Collaboration.pdf.

Teng, S., Zheng, C., Zhu, H. et al. (2014). A cooperative intrusion detection model for cloud computing networks. *International Journal of Security and Its Applications* 8 (3), 2014: 107–118.

Wang, F. (2017). Computational social systems in a new period: a fast transition into the third axial age. *IEEE Transactions on Computational Social Systems* 4 (3): 52–53.

Wang, S., Li, J., and Ma, S. (2011). Dynamic and secure business data exchange model for SaaS-based collaboration supporting platform of industrial chain. *International Journal of Advanced Pervasive and Ubiquitous Computing* 3 (3), 14 pages, DOI: https://doi.org/10.4018/japuc.2011070105.

Weerawarana, S., Curbera, F., Leymann, F. et al. (2008). *Web Services Platform Architecture*. Upper Saddle River, NJ: Prentice Hall.

de Wilde, P., Chli, M., Correia, L. et al. (2003). Adapting populations of agents. In: *Adaptive Agents and Multi-Agent Systems*, LNAI 2636 (eds. E. Alonso, D. Kudenko and D. Kazakov), 110–124. Berlin: Springer-Verlag.

Wolsey, L.A. and Nemhauser, G.L. (1999). *Integer and Combinatorial Optimization*. New York: Wiley-Interscience.

Wooldridge, M., Jennings, N.R., and Kinny, D. (2000). The Gaia methodology for agent-oriented analysis and design. *Journal of Autonomous Agents and Multi-Agent Systems* 3 (3): 285–312.

Wu, B., Pei, X., Tan, J.B., and Wang, Y. (2007). Resume mining of communities in social network. *7th IEEE International Conference on Data Mining Workshop*, Omaha, NE, USA (28–31 October 2007), pp. 435–440.

Yu, K., Guan, G., and Zhou, M. (2005). Resume information extraction with cascaded hybrid model. *Proceedings of the 43rd Annual Meeting of the Association for Computational Linguistics*, Ann Arbor, USA (June 2005), pp. 499–506.

Zanakis, S.H., Solomon, A., Wishart, N., and Dublish, S. (1998). Multi-attribute decision making: a simulation comparison of select methods. *European Journal of Operational Research* 7: 507–529.

Zhang, T., Ying, S., Cao, S., and Zhang, J. (2008). A modelling approach to service-oriented architecture. *Enterprise Information Systems* 2 (3): 239–257.

Zhu, H. (2003). Some issues in role-based collaboration. *Proceedings of IEEE Canada Conference on Electrical and Computer Engineering (CCECE'03)*, Montreal, Canada (4–7 May 2003), vol. 2, pp.687–690.

Zhu, H. (2005). Encourage participants' contributions by roles. *Proceedings of The IEEE Int'l Conference on Systems, Man and Cybernetics*, Big Island, Hawaii (9–12 October 2005), pp. 1574–1579.

Zhu, H. (2007). Role as dynamics of agents in multi-agent systems. *System and Informatics Science Notes* 1 (2): 165–171.

Zhu, H. (2008). Fundamental issues in the design of a role engine. *Proceedings of ACM/ IEEE Int'l Symposium on CTS*, Irvine, CA, USA (19–23 May 2008), pp. 399–407.

Zhu, H. (2010). Role-based autonomic systems. *International Journal of Software Science and Computational Intelligence* 2 (3): 32–51.

Zhu, H. (2011). Collective role assignment and its complexity. *The IEEE Int'l Conference on Systems, Man and Cybernetics (ICSMC)*, Anchorage, AK, USA (October 2011), pp. 1897–1902.

Zhu, H. (2012). Agent training plan for sustainable groups. *Proceedings of ACM/IEEE Int'l Conference on CTS*, Denver, CO, USA (21–25 May 2012), pp. 322–329.

Zhu, H. (2015). Adaptive collaboration systems. *IEEE Systems, Man, and Cybernetics Magazine* 1 (4): 8–15.

Zhu, H. (2016). Avoiding conflicts by group role assignment. *IEEE Transactions on Systems, Man, and Cybernetics: Systems* 46 (4): 535–547.

Zhu, H. (2020). Computational social simulation with E-CARGO: comparison between collectivism and individualism. *IEEE Transactions on Computational Social Systems* 7 (6): 1345–1357.

Zhu, H. and Feng, L. (2013). Agent evaluation in distributed adaptive systems. *The IEEE Int'l Conference on SMC*, Manchester, UK (13–16 October 2013), pp.752–757.

Zhu, H. and Grenier, M. (2009). Agent evaluation for role assignment. *Proceedings of the IEEE 8th International Conference on Cognitive Informatics (ICCI'09)*, Hong Kong, China (July 2013), pp. 405–411.

Zhu, H. and Zhou, M. (2006). Role-based collaboration and its kernel mechanisms. *IEEE Transactions on SMC, Part C* 36 (4): 578–589.

Zhu, H. and Zhou, M. (2008). Role transfer problems and their algorithms. *IEEE Transactions on SMC(A)* 38 (6): 1442–1450.

Zhu, H. and Zhou, M. (2009). M–M role-transfer problems and their solutions. *IEEE Transactions on SMC(A)* 39 (2): 448–459.

Zhu, H. and Zhou, M.C. (2012). Efficient role transfer based on Kuhn–Munkres algorithm. *IEEE Transactions on Systems, Man and Cybernetics, Part A: Systems and Humans* 42 (2): 491–496.

Zhu, H., Alkins, R., and Grenier, M. (2008). Role-based chatting. *Journal of Software* 3 (6): 69–78.

Zhu, H., Hou, M., and Zhou, M.C. (2012a). Adaptive collaboration based on the E-CARGO model. *International Journal of Agent Technologies and Systems* 4 (1): 59–76.

Zhu, H., Zhou, M.C., and Alkins, R. (2012b). Group role assignment via a Kuhn-Munkres algorithm-based solution. *IEEE Transactions on Systems, Man and Cybernetics, Part A* 42 (3): 739–750.

Index

E-CARGO and Role-Based Collaboration: Modeling and Solving Problems in the Complex World,
First Edition. Haibin Zhu.
© 2022 by The Institute of Electrical and Electronics Engineers, Inc.
Published 2022 by John Wiley & Sons, Inc.

IEEE Press Series on Systems Science and Engineering

Editor:

MengChu Zhou, *New Jersey Institute of Technology and Tongji University*

Co-Editors:

Han-Xiong Li, *City University of Hong-Kong*
MargotWeijnen, *Delft University of Technology*

The focus of this series is to introduce the advances in theory and applications of systems science and engineering to industrial practitioners, researchers, and students. This series seeks to foster system-of-systems multidisciplinary theory and tools to satisfy the needs of the industrial and academic areas to model, analyze, design, optimize and operate increasingly complex man-made systems ranging from control systems, computer systems, discrete event systems, information systems, networked systems, production systems, robotic systems, service systems, and transportation systems to Internet, sensor networks, smart grid, social network, sustainable infrastructure, and systems biology.

Printed and bound by CPI Group (UK) Ltd, Croydon, CR0 4YY

27/10/2024

14580670-0002